Universitext

AF307371

Zhi-Ming Ma   Michael Röckner

Introduction to the Theory of (Non-Symmetric)
Dirichlet Forms

Springer-Verlag
Berlin Heidelberg New York
London Paris Tokyo
Hong Kong Barcelona
Budapest

Zhi-Ming Ma
Institute of Applied Mathematics
Academia Sinica, P. O. Box 2734
100080 Beijing
People's Republic of China

Michael Röckner
Institut für Angewandte Mathematik
Universität Bonn
Wegelerstr. 6
W-5300 Bonn 1, FRG

Mathematics Subject Classification (1991)
Primary: 31C25
Secondary: 60J45, 60J40

Library of Congress Cataloging-in-Publication Data
Ma, Zhi-Ming, 1948-   Introduction to the theory of (non-symmetric) Dirichlet forms / Zhi-Ming
Ma, Michael Röckner.   p.   cm. – (Universitext)
Includes bibliographical references and index.

ISBN-13: 978-3-540-55848-4   e-ISBN-13: 978-3-642-77739-4
DOI: 10.1007/978-3-642-77739-4

1. Dirichlet forms.   2. Markov processes.   I. Ma, Zhi-Ming.   II. Title.   QA274.2.R63   1992
519.2–dc20   92-28120   CIP

This work is subject to copyright. All rights are reserved, whether the whole or part of the material
is concerned, specifically the rights of translation, reprinting, reuse of illustrations, recitation,
broadcasting, reproduction on microfilms or in any other way, and storage in data banks.
Duplication of this publication or parts thereof is permitted only under the provisions of the
German Copyright Law of September 9, 1965, in its current version, and permission for use must
always be obtained from Springer-Verlag. Violations are liable for prosecution under the German
Copyright Law.

© Springer-Verlag Berlin Heidelberg 1992

Typesetting: Camera ready by author
41/3140 - 5 4 3 2 1 0 - Printed on acid-free paper

Z.-M. M. dedicates this book to his wife.

M. R. dedicates this book to his father
on the occasion of his 65th birthday.

This work was supported by the Chinese National Natural Science Foundation, the Sonderforschungsbereich 256, Bonn, and the Max-Planck-Gesellschaft.

Contents

Contents

Chapter 0

Introduction

The purpose of this book is to give a streamlined introduction to the theory of
(not necessarily symmetric) Dirichlet forms on general state spaces. It includes
both the analytic and the probabilistic part of the theory up to and including
the construction of an associated Markov process. It is based on recent joint
work of S. Albeverio and the two authors and on a one-year-course on Dirichlet
forms taught by the second named author at the University of Bonn in 1990/91.
It addresses both researchers and graduate students who require a quick but
complete introduction to the theory. Prerequisites are a basic course in probability theory (including elementary martingale theory up to the optional sampling
theorem) and a sound knowledge of measure theory (as, for example, to be
found in Part I of H. Bauer [B 78]). Furthermore, an elementary course on linear operators on Banach and Hilbert spaces (but without spectral theory) and
a course on Markov processes would be helpful though most of the material
needed is included here.

The success of the modern theory of Dirichlet forms is based on the rich in-
terplay between its analytic and probabilistic components. The analytic part of
the theory goes back to the pioneering papers of A. Beurling and J. Deny [BeDe
58, 59], [De 70], whereas the more recent probabilistic part was really initiated
by the fundamental work of M. Fukushima [F 71, 73, 76, 80] and M.L. Silverstein [Si 74] combining symmetric Markov processes and Dirichlet forms. This
program was then extended to the non-symmetric case by S. Carillo-Menendez
[Ca-Me 75] and Y. LeJan [Le 77, 78, 82]. The theory developed by the above-
mentioned authors was carried out only for locally compact state spaces. More
recently, the emphasis has shifted to attempts towards extending the theory
to also cover infinite dimensional state spaces (cf., for example, the papers by
S. Albeverio/R. Høegh-Krohn [AH-K 75, 77a,b], P. Paclet [Pa 78], S. Kusuoka
[Ku 82], E.B. Dynkin [Dy 82], N. Bouleau/F. Hirsch [BH 86a,b, 91], D. Feyel/A.
de La Pradelle [FdL 79], T. Hida/J. Potthoff/L. Streit [HPS 88], P. Fitzsim-
mons/R. Getoor [FiG 88], P. Fitzsimmons [Fi 89], S. Albeverio/M. Röckner

[AR 89a,b, 90a,b, 91], M. Röckner [R 90a,b], M. Röckner/T.S. Zhang [RZ 90], B. Schmuland [S 90], S. Albeverio/Z.M. Ma [AM 91a-f] and I. Shigekawa/T. Taniguchi [ShiTa 91]. Here, as in the "classical" case, the combination of analytic and probabilistic methods provides a powerful instrument which has resulted in substantial progress being made. All in all, it can be said that also in the "non-classical" theory the two components have developed into equal counterparts. This book provides a framework which covers the classical case together with all the above extensions. The contents of Chapters I-IV supplemented by the Appendix are suitable for a one-year-course for students in their ($\geq$) 4th year of studies and would prepare them for the study of recent publications in this area. Many exercises are included and an entire chapter consists of examples whose overlap with example-sections in the standard literature (cf. e.g. [F 80]) has been minimized.

Since the content of each chapter is summarized at its beginning we only give a brief outline here: In Chapter I we provide the necessary background from functional analysis and present the analytic framework of the theory of Dirichlet forms on arbitrary measure spaces. Chapter II is devoted to examples. Chapter III treats the analytic potential theory of Dirichlet forms and can be considered as the link between the analytic and the probabilistic part of the theory. In Chapter IV we prove a necessary and sufficient condition for the existence of a corresponding "nice" Markov process and prove that this correspondence is one-to-one. In addition, we identify those examples from Chapter II which thus have an associated process. In Chapter V we study Dirichlet forms for which the corresponding Markov processes have special additional path properties. In Chapter VI we develop a general technique how to reduce our more general framework of Dirichlet forms to the "classical" theory on locally compact spaces. Finally, supplements to both the analytic and probabilistic parts of the book are given in the Appendix. In a special section at the end of each chapter we give the sources in the literature which we have used and add some further comments. It is, however, beyond the scope of this book to give a precise historic overview mentioning all mathematicians who have contributed to the development and present state of the theory. We would like to apologize at this point to all those who should have been included in the Bibliography at the end of this book, but have not been.

Now we would like to describe those aspects we have particularly aimed to emphasize in this book:

1. We have separated the part of the theory which can be done on state spaces which are merely measure spaces (cf. Chapter I) from that part which requires a topology on the state space (cf. Chapters III-VI).

2. We keep the topological assumptions on the state space E (when they are needed) as general as possible. A priori, we only suppose E to be Hausdorff and, for convenience that its Borel σ-algebra is generated by the continuous functions on E. In particular, we cover the examples with infinite dimensional state spaces which have attracted a lot of attention in recent years.

3. The most important question about a Dirichlet form is whether it has a probabilistic counterpart, i.e., a "nice" Markov process associated with it. This question has been answered positively in fundamental works by M. Fukushima (cf. [F 71, 80], see also [Ca-Me 75], [FdL 78], and [Le 82] for the non-symmetric case) provided the state space is a locally compact separable metric space and the Dirichlet form is "regular". But it has been known for some time that both the topological and the "regularity" assumptions are too strong. In this book we give a complete proof of an analytic characterization of all (not necessarily symmetric) Dirichlet forms on arbitrary state spaces which have an associated "nice" Markov process. These Dirichlet forms are called "quasi-regular". The process construction is presented in detail and done via "Ray resolvents/compactification" (cf. Chapter IV).

4. In order to keep the book to a reasonable size we shall not reprove all "classical" results for regular Dirichlet forms in our more general case. We only extend those results needed for the above characterization together with those needed to develop a method that we call "local compactification" or "regularization" and which enables us to reduce our situation to the classical framework (see Chapter VI). Thus we can take advantage of the already existing excellent exposition [F 80] and all its results can be made available by virtue of this general procedure. For instance, all the potential theory in [F 80] (see also [O 88]) not covered by our Chapter III or the results on additive functionals and stochastic calculus in [F 80] (or [O 88]) can be carried over to quasi-regular Dirichlet forms on general state spaces in a straightforward manner. The same applies, of course, to recently published results still obtained in the "classical" framework. We emphasize that in this way also the construction of the process can be reduced to the classical construction (see [F 80,§6]). But a complete self-contained exposition thereof would not have been shorter than the one given in Chapter IV. We, however, do not tackle the problem of extending the boundary theory of Dirichlet forms by our "regularization" method though this part of the theory has been developed substantially in the "classical" framework (cf. [Si 76], [Le 77, 78]).

5. Apart from the prerequisites mentioned above we have tried to keep this book reasonably self-contained up to and including Chapter IV, Sect. 4. Subsequently, more advanced probability theory is required of which the proofs are not included, but only references given.

In addition to those mentioned in 1.– 5. above the following new results are contained in this book: an elementary proof of A. Beurling's and J. Deny's result that the "Markov property" and the "unit contraction property" are equivalent for closed symmetric forms which also works in the non-symmetric case (cf. I.4.7); a proof of the equivalence of the sector condition and the analyticity of the corresponding contraction semigroup (cf. I.2.21); a proof of the one-to-one correspondence between "quasi-regular" Dirichlet forms and pairs of "m-sectorial" right-processes (cf. Chapter IV, Sect. 6); an analytic characterization of those Dirichlet forms on general state spaces which are associated with diffusions and Hunt processes (cf. Chapter V); a result on the existence of "localizing functions" in Dirichlet spaces (cf. V.1.7); the result that by "local compactification" every right-process on a general state space properly associated with a quasi-regular Dirichlet form can be considered as a Hunt process on a locally compact separable metric space (cf. VI.1.6).

In order to ease the reading we now include some remarks concerning the style adapted in this book. We have included many exercises in the text which are used subsequently. For those we consider a little harder we have included references. Basic definitions that we assume the reader to be familiar with are nevertheless recalled in the text. For reasons of economy we are quite brief on some parts of the potential theory involved. For instance, we do not include a systematic study of semigroups or resolvents of kernels, but only prove those facts we need. For the reader who is not interested in the details of the analytic part of the theory Diagrams 1,2,3 in Chapter I provide a means to get to the probabilistic part faster. Occasionally, quite detailed arguments of similar type appear several times, but in different proofs. This is done deliberately, in order to make partial reading possible to some extent. Arabic numbers without brackets (e.g. 2.8) in the text refer to a definition, theorem, remark etc. within the same chapter, those with brackets (e.g. (2.8)) refer to formulae within the same chapter. If a different chapter is referred to, the respective Roman number is added (e.g. III.2.8 or III.(2.8)). The reader interested in the "local compactification" can skip Chapter V and read directly Chapter VI. Only the proof of Theorem VI.1.6 depends on Chapter V, but it is not difficult to deduce it directly from Chapter IV.

Finally, we turn to our notational conventions though they are quite standard. $\mathbb{C}$, $\mathbb{R}$, $\mathbb{Q}$, $\mathbb{N}$ denote the complex, real, rational, and natural numbers respectively. A *function* on some set is always meant to be (extended) real-valued. We use the notion *positive* and *increasing* in the sense of "greater than or equal", otherwise we say *strictly positive* and *strictly increasing*. The notation "$L^p_{(loc)}$" is meant in the usual way but we always mean spaces of (classes of) functions, otherwise e.g. in the case of $\mathbb{R}^d$-valued *mappings* on a measure space $(E, \mathcal{B}, m)$ we write $L^p(E \to \mathbb{R}^d; m)$. In order to avoid overloading the notation, we do not distinguish between m-equivalence classes of functions on E and a particular representative or *m-version* (which is always assumed to be $\mathcal{B}$-measurable) if there is no confusion possible. But if we write $\mathcal{B}_b \cap L^2(E; m)$ we mean the set of all bounded m-square integrable functions on E (cf. [F 80]). The beginner is advised to be careful in this respect (e.g. in the proof of II.4.2).

During the preparation of this book we have benefited to a "non-measurable" extent from the advice, support and expertise of a large number of people whom we would like to thank at this point. First of all we thank our friend and teacher S. Albeverio for his steady encouragement and help during the whole period of preparation of this book. In addition, we are also grateful to P. Fitzsimmons, M. Fukushima, R. Getoor, W. Hansen, Y. LeJan, T. Lyons, P.A. Meyer, T. Oshima and B. Schmuland who considerably influenced us through their work and fruitful discussions. Furthermore, we would like to thank the following colleagues for working through (parts of) the manuscript and for many helpful comments:

Y. Chen, Z. Dong, B.K. Driver, D. Feyel, N. Gantert, P. Imkeller, N. Jacob, A. de La Pradelle, V. Metz, L. Overbeck, I. Mc Gillivray, C. Preston, A. Schied, M. Schweitzer, I. Shigekawa, P. Stollmann, A. Stremme; furthermore those students who participated in lectures and seminars held by the two authors at the Academia Sinica in Beijing and the Universities of Bielefeld and Bonn or at the spring resp. winter schools in Pasecky (CSFR) and Friedrichroda (Germany). We also thank I. Pützer and U. Goldschmitt for the excellent typing of a "constantly changing" manuscript. Finally, for their understanding and support for our work, we want to express our sincere gratitude to our respective families: Xinhua Yang, her mother, and Liang Ma, and Elisabeth and Melanie Röckner. At times, it was something of a burden for them, also!

Chapter I

Functional Analytic Background

In this chapter we present some background from functional analysis. The proofs of a few specific facts, not necessary right away for understanding the text, are however, postponed to the Appendix (cf. A. Sections 1,2). We start in Section 1 with the relationship between strongly continuous contraction resolvents $(G_\alpha)_{\alpha>0}$, strongly continuous contraction semigroups $(T_t)_{t>0}$ and their generators $(L, D(L))$ on an arbitrary Banach space. In particular, we prove the Hille-Yosida theorem. These results are summarized in the diagram on p. 14. In Section 2 we study coercive closed forms $(\mathcal{E}, D(\mathcal{E}))$ on a Hilbert space $\mathcal{H}$ and their relation first with strongly continuous contraction resolvents and their generators on $\mathcal{H}$ and subsequently their relation with the corresponding semigroups (cf. the diagram on p. 27). Section 3 is devoted to the crucial notion of closability which is important for applications and the examples to be treated in this book. In Section 4 we specialize to the case where $\mathcal{H}$ is an L^2-space over an arbitrary measure space and study the relations between respective "contraction properties" of the four corresponding objects $(\mathcal{E}, D(\mathcal{E}))$, $(G_\alpha)_{\alpha>0}$, $(T_t)_{t>0}$ and $(L, D(L))$ from Section 2.

1 Resolvents, semigroups, generators

In this section we recall some basic facts on semigroups, resolvents and generators. We fix a Banach space $(B, \| \ \|)$ over $\mathbb{K} := \mathbb{R}$ or $\mathbb{C}$ with (topological) dual B' . We call a pair $(L, D(L))$ a *linear operator* on B if $D(L)$ is a linear subspace of B and $L : D(L) \to B$ a $\mathbb{K}$-linear map. We shall sometimes write L instead of $(L, D(L))$. If L is one-to-one, L^{-1} is defined by $D(L^{-1}) = L(D(L))$ and $L^{-1}u = v$, where $v \in B$ such that $Lv = u$. Given two linear operators L_1, L_2 on B and α, $\beta \in \mathbb{K}$, the linear operator $\alpha L_1 + \beta L_2$ on B is defined by $D(\alpha L_1 + \beta L_2) = D(L_1) \cap D(L_2)$ and $(\alpha L_1 + \beta L_2)u = \alpha L_1 u + \beta L_2 u$. $L_2 L_1$ is defined as $D(L_2 L_1) = \{u \in D(L_1) | L_1 u \in D(L_2)\}$ and $L_2 L_1 u = L_2(L_1 u)$. We write $L_1 = L_2$ if $D(L_1) = D(L_2)$ and $L_1 u = L_2 u$ for all $u \in D(L_1) = D(L_2)$. We denote the identity operator on B by Id_B and abbreviate $\alpha\,\mathrm{Id}_B$ by α if $\alpha \in \mathbb{K}$. Recall that a linear operator L on B is *continuous* (or *bounded*) if

$$\|L\| := \sup\{\|Lu\| \,|\, u \in B,\ \|u\| \leq 1\} < \infty .$$

a)   Resolvents

Definition 1.1. Let L be a linear operator on B. The *resolvent set* $\rho(L)$ of L is defined to be the set of all $\alpha \in \mathbb{K}$ such that $(\alpha - L) : D(L) \to B$ is one-to-one and for its inverse $(\alpha - L)^{-1}$ we have

(a) $D((\alpha - L)^{-1}) = B$

(b) $(\alpha - L)^{-1}$ is continuous on B.

$\sigma(L) := \mathbb{K}\backslash\rho(L)$ is called the *spectrum* of L and $\{(\alpha - L)^{-1} | \alpha \in \rho(L)\}$ the *resolvent* of L .

Lemma 1.2. *Let L be a linear operator on B. If $\rho(L) \neq \emptyset$ then L is closed (i.e., $D(L)$ is complete w.r.t. the graph norm $\|Lu\| + \|u\|$, $u \in D(L)$).*

Proof. Let $\alpha \in \rho(L)$ and $u_n \in D(L)$, $n \in \mathbb{N}$, such that $u_n \xrightarrow[n\to\infty]{} u$ and $Lu_n \xrightarrow[n\to\infty]{} v$ in B for some $u, v \in B$. Then $\lim_{n\to\infty}(\alpha - L)u_n = \alpha u - v$ and hence $u = \lim_{n\to\infty} u_n = \lim_{n\to\infty}(\alpha - L)^{-1}(\alpha - L)u_n = (\alpha - L)^{-1}(\alpha u - v)$. Therefore, $u \in (\alpha - L)^{-1}(B) = D(L)$ and $v = \alpha u - (\alpha - L)u = Lu$. $\qquad\square$

Proposition 1.3. *Let L be a linear operator on B and set for $\alpha \in \rho(L)$,*

$$G_\alpha := (\alpha - L)^{-1} .$$

(i) *Let $\alpha, \beta \in \rho(L)$ then*

$$(1.1) \qquad\qquad G_\alpha - G_\beta = (\beta - \alpha)G_\alpha G_\beta = (\beta - \alpha)G_\beta G_\alpha .$$

(ii) *Suppose L is densely defined (i.e., $D(L)$ is dense in B), $]0, \infty[\subset \rho(L)$ and each αG_α is a contraction on B, i.e., $\|\alpha G_\alpha\| \leq 1$ for all $\alpha \in]0, \infty[$. Then*

$$(1.2) \qquad\qquad \lim_{\alpha\to\infty} \alpha G_\alpha u = u \quad \text{for all } u \in B .$$

Proof. (i): Note that $G_\beta u \in D(L)$ for all $u \in B$. Hence $(\beta - \alpha)G_\alpha G_\beta = G_\alpha(\beta - L + L - \alpha)G_\beta = G_\alpha(\beta - L)G_\beta - G_\alpha(\alpha - L)G_\beta = G_\alpha - G_\beta$. Interchanging the roles of α and β we complete the proof.

(ii): Let $u \in D(L)$ and $\alpha \in]0, \infty[$, then $\alpha G_\alpha u - u = G_\alpha(\alpha u - (\alpha - L)u) = G_\alpha Lu$. Hence $\|\alpha G_\alpha u - u\| \leq \alpha^{-1}\|Lu\|$ and (1.2) follows for all $u \in D(L)$. Since $\|\alpha G_\alpha u\| \leq \|u\|$ for all $u \in B$ and $D(L)$ is dense in B, a "3ϵ-argument" now implies (1.2) for all $u \in B$. $\qquad\square$

Remark. (1.1) is called the *first resolvent equation*.

Definition 1.4. A family $(G_\alpha)_{\alpha>0}$ of linear operators on B with $D(G_\alpha) = B$ for all $\alpha \in]0, \infty[$ is called a *strongly continuous contraction resolvent* (on B) if

(i) $\lim_{\alpha\to\infty} \alpha G_\alpha u = u$ for all $u \in B$ *(strong continuity).*

(ii) αG_α is a contraction on B for all $\alpha > 0$.

(iii) $G_\alpha - G_\beta = (\beta - \alpha)G_\alpha G_\beta$ for all $\alpha, \beta > 0$.

Proposition 1.5. *Let $(G_\alpha)_{\alpha>0}$ be a strongly continuous contraction resolvent on B. Then there exists exactly one linear operator $(L, D(L))$ on B such that $]0, \infty[\subset \rho(L)$ and $G_\alpha = (\alpha - L)^{-1}$ for all $\alpha > 0$. This operator is closed and densely defined.*

Proof. Observe that $G_\alpha(B)$ is independent of α by 1.4 (iii) and that 1.4 (iii) together with 1.4 (i) implies that each G_α is one-to-one. So, any L as in the assertion would have to satisfy

$$(1.3) \qquad D(L) = G_\alpha(B) \text{ and } L = \alpha - G_\alpha^{-1} \text{ for all } \alpha > 0 .$$

This shows uniqueness, but also existence if we can show that $\alpha - G_\alpha^{-1} = \beta - G_\beta^{-1}$ on $G_\alpha(B)(= G_\beta(B))$ for all $\alpha, \beta > 0$. So, let $u \in G_\alpha(B)$. Then $u = G_\alpha v$ for some $v \in B$ and

$$G_\beta \left((\alpha - G_\alpha^{-1})u - (\beta - G_\beta^{-1})u\right) = \alpha G_\beta G_\alpha v - G_\beta v - \beta G_\beta G_\alpha v + G_\alpha v$$

which is equal to zero by 1.4 (iii). Since G_β is one-to-one we conclude that $\alpha - G_\alpha^{-1} = \beta - G_\beta^{-1}$. Clearly, $D(L) = G_\alpha(B)$ is dense in B by 1.4 (i). The closedness follows by 1.2. $\qquad\qquad\Box$

Remark. (i) Note that we have not used 1.4 (ii) in the above proof.
(ii) L in 1.5 is called the *generator* of $(G_\alpha)_{\alpha>0}$.

b)   Semigroups and generators

Definition 1.6. A family $(T_t)_{t>0}$ of linear operators on B with $D(T_t) = B$ for all $t > 0$ is called a *strongly continuous contraction semigroup* (on B) if

(i) $\lim_{t\to 0} T_t u = u$ for all $u \in B$ *(strong continuity).*

(ii) T_t is a contraction on B for all $t > 0$.

(iii) $T_t T_s = T_{t+s}$ for all $t, s > 0$ *(semigroup property).*

Exercise 1.7. Consider the situation of 1.6 and set $T_0 := \mathrm{Id}_B$. Prove that then 1.6 (i) is equivalent with

(i)' $t \mapsto T_t u$ is continuous on $[0, \infty[$ for all $u \in B$.

Definition 1.8. Given a strongly continuous contraction semigroup $(T_t)_{t>0}$ on B, the linear operator $(L, D(L))$ on B defined by

$$D(L) \; := \; \{u \in B \mid \lim_{t \downarrow 0} \frac{1}{t}(T_t u - u) \text{ exists } \}$$

$$Lu \; := \; \lim_{t \downarrow 0} \frac{1}{t}(T_t u - u), \; u \in D(L),$$

is called the *(infinitesimal) generator* of $(T_t)_{t>0}$.

Exercise 1.9. Prove that if $u \in D(L)$ then for every $t \geq 0$, $T_t u \in D(L)$ and $\frac{d}{dt} T_t u = L T_t u = T_t L u$.

Proposition 1.10. *Let $(T_t)_{t>0}$ be a strongly continuous contraction semigroup on B with generator L. Then L is closed, densely defined, $]0, \infty[\subset \rho(L)$ and if $G_\alpha := (\alpha - L)^{-1}$, $\alpha > 0$, then*

$$(1.4) \qquad G_\alpha u = \int_0^\infty e^{-\alpha s} T_s u \, ds , \quad u \in B , \quad \alpha > 0 .$$

In particular, $(G_\alpha)_{\alpha>0}$ is a strongly continuous contraction resolvent.

Proof. To prove that L is closed let $u_n \in D(L)$, $n \in \mathbb{N}$, such that $u_n \xrightarrow[n \to \infty]{} u$ and $L u_n \xrightarrow[n \to \infty]{} v$ in B for some $u, v \in B$. Then for every $t > 0$ by 1.7, 1.9 and the fundamental theorem of calculus

$$\frac{1}{t}(T_t u - u) \; = \; \lim_{n \to \infty} \frac{1}{t}(T_t u_n - u_n)$$

$$= \; \lim_{n \to \infty} \frac{1}{t} \int_0^t T_s L u_n \, ds$$

$$= \; \frac{1}{t} \int_0^t T_s v \, ds .$$

Applying the fundamental theorem of calculus again we obtain that $u \in D(L)$ and $Lu = v$.

For $\alpha > 0$ define $\tilde{G}_\alpha u := \int_0^\infty e^{-\alpha s} T_s u \, ds$, $u \in B$. Then $\alpha \tilde{G}_\alpha$ is a contraction on B and for all $t > 0$, $u \in B$,

$$\frac{1}{t}(T_t \tilde{G}_\alpha u - \tilde{G}_\alpha u) \; = \; \frac{1}{t} e^{\alpha t} \int_t^\infty e^{-\alpha s} T_s u \, ds - \frac{1}{t} \int_0^\infty e^{-\alpha s} T_s u \, ds$$

$$= \; \frac{1}{t}(e^{\alpha t} - 1) \int_t^\infty e^{-\alpha s} T_s u \, ds - \frac{1}{t} \int_0^t e^{-\alpha s} T_s u \, ds .$$

Hence $\tilde{G}_\alpha u \in D(L)$ and $L\tilde{G}_\alpha u = \alpha \tilde{G}_\alpha u - u$, i.e., $(\alpha - L)\tilde{G}_\alpha u = u$. For $u \in D(L)$ we have in addition that

$$L \int_0^\infty e^{-\alpha s} T_s u \, ds = \int_0^\infty e^{-\alpha s} L T_s u \, ds = \int_0^\infty e^{-\alpha s} T_s L u \, ds .$$

The first equality follows by approximation with Riemann sums from the facts that $s \mapsto e^{-\alpha s} T_s u$ and $s \mapsto e^{-\alpha s} L T_s u = e^{-\alpha s} T_s L u$ are integrable over $]0, \infty[$, and that L is closed. Thus, if $u \in D(L)$, then $L\tilde{G}_\alpha u = \tilde{G}_\alpha L u$, hence $\tilde{G}_\alpha(\alpha - L)u = (\alpha - L)\tilde{G}_\alpha u = u$. Consequently,

$$\tilde{G}_\alpha = (\alpha - L)^{-1} \, ,$$

in particular, $]0, \infty[\subset \rho(L)$ and (1.4) is shown.

Finally, we note that for $u \in B$

$$\alpha \tilde{G}_\alpha u = \int_0^\infty e^{-s} T_{s/\alpha} u \, ds \, ,$$

hence $\lim_{\alpha \to \infty} \alpha \tilde{G}_\alpha u = u$. Since $\tilde{G}_\alpha(B) = D(L)$, it follows that L is densely defined. The last part of the assertion is now clear by 1.3(i). $\square$

Remark 1.11. (1.4) says that $(G_\alpha)_{\alpha>0}$ is the Laplace transform of $(T_t)_{t>0}$. Solving informally the ordinary differential equation in 1.9 yields $T_t = e^{tL}$, $t > 0$, and (1.4) becomes plausible immediately. Furthermore, one thus gets an idea how to prove the converse of 1.10.

The following theorem is due to E. Hille [H 48], [H 52] and K. Yosida [Y 48].

Theorem 1.12. *A necessary and sufficient condition that a densely defined linear operator L on B is the generator of a strongly continuous contraction semigroup is that*

(i) $]0, \infty[\subset \rho(L)$

(ii) $\|\alpha(\alpha - L)^{-1}\| \leq 1$ *for all $\alpha > 0$.*

In this case the strongly continuous semigroup is uniquely determined by L and L is closed.

Proof. The necessity part is contained in 1.10, so we only have to prove sufficiency and uniqueness.

Sufficiency: For $\alpha > 0$ define $L^{(\alpha)} := \alpha(\alpha G_\alpha - 1)$ with $G_\alpha := (\alpha - L)^{-1}$. Observe that by 1.3 (ii) for $u \in D(L)$, $L^{(\alpha)}u = \alpha G_\alpha(\alpha - G_\alpha^{-1})u = \alpha G_\alpha L u \xrightarrow[\alpha \to \infty]{} Lu$. Since each $L^{(\alpha)}$ is a bounded linear operator on B with $D(L^{(\alpha)}) = B$, we can define for $u \in B$, $t \geq 0$,

$$T_t^{(\alpha)}u := e^{tL^{(\alpha)}}u = e^{-t\alpha} \sum_{n=0}^\infty \frac{(t\alpha)^n}{n!}(\alpha G_\alpha)^n u \, .$$

Clearly,

$$\|T_t^{(\alpha)}\| \leq e^{-t\alpha} \sum_{n=0}^\infty \frac{(t\alpha)^n}{n!}\|\alpha G_\alpha\|^n \leq 1 \, ,$$

hence they are contractions. Furthermore, since for α, β, $t > 0$ and $u \in B$

$$e^{tL^{(\alpha)}}u - e^{tL^{(\beta)}}u = \int_0^t \frac{d}{ds}\left(e^{sL^{(\alpha)}}e^{(t-s)L^{(\beta)}}u\right)ds\ ,$$

we obtain for $u \in D(L)$ (since $e^{tL^{(\alpha)}}e^{sL^{(\beta)}} = e^{sL^{(\beta)}}e^{tL^{(\alpha)}}$ for all α, β, t, $s > 0$),

$$\begin{aligned}
\|T_t^{(\alpha)}u - T_t^{(\beta)}u\| &\leq \int_0^t \|e^{sL^{(\alpha)}}e^{(t-s)L^{(\beta)}}\|\ \|L^{(\alpha)}u - L^{(\beta)}u\|ds \\
&\leq t\|L^{(\alpha)}u - L^{(\beta)}u\|\ .
\end{aligned}$$

Hence for all $u \in D(L)$, $T_t u := \lim\limits_{\alpha \to \infty} T_t^{(\alpha)}u$ exists locally uniformly in $t \in [0, \infty[$. Since $D(L)$ is dense in B we can extend each T_t uniquely to a bounded linear operator with domain B. Denoting this extension again by T_t we easily see that $(T_t)_{t>0}$ is a strongly continuous contraction semigroup. It remains to show that its generator, say $\tilde{L}$, is equal to L. Let $u \in D(L)$, t, $\alpha > 0$, then

$$e^{tL^{(\alpha)}}u - u = \int_0^t e^{sL^{(\alpha)}}L^{(\alpha)}u\ ds\ .$$

Hence, since $L^{(\alpha)}u \xrightarrow[\alpha \to \infty]{} Lu$, we have that

$$T_t u - u = \int_0^t T_s Lu\ ds\ .$$

Consequently, $u \in D(\tilde{L})$ and $\tilde{L}u = Lu$, i.e., $D(L) \subset D(\tilde{L})$ and $\tilde{L}_{|D(L)} = L$. In particular, $D(L)$ is mapped bijectively onto B by $\alpha - \tilde{L}(= \alpha - L)$, $\alpha > 0$. By the necessity part, also $D(\tilde{L})$ is mapped bijectively onto B by $\alpha - \tilde{L}$, $\alpha > 0$. Hence $D(L) = D(\tilde{L})$.

Uniqueness: Assume $(\tilde{T}_t)_{t>0}$ is another strongly continuous contraction semigroup with generator L. Since for all $u \in B$ by 1.10

$$\int_0^\infty e^{-\alpha s}T_s u\ ds = (\alpha - L)^{-1}u = \int_0^\infty e^{-\alpha s}\tilde{T}_s u\ ds\ ,\quad \alpha > 0\ ,$$

it follows that for all $u \in B$, $l \in B'$

$$\int_0^\infty e^{-\alpha s}l(T_s u)\ ds = \int_0^\infty e^{-\alpha s}l(\tilde{T}_s u)\ ds\ ,\quad \alpha > 0\ ,$$

i.e., the Laplace transforms of the continuous functions $s \mapsto l(T_s u)$ and $s \mapsto l(\tilde{T}_s u)$, $s > 0$, coincide, which are therefore equal. By the Hahn-Banach theorem (cf. e.g. [Ch 69b]) this implies that $T_s u = \tilde{T}_s u$, $s > 0$, for all $u \in B$. $\qquad\square$

Remark 1.13. Note that in the situation of 1.12 we have that if B is a real Hilbert space with inner product $(\ ,\ )$, then

$$(T_t u, u) \leq (u, u) \text{ for all } t > 0\ ;$$

hence $(-Lu, u) = \lim\limits_{t \downarrow 0}\frac{1}{t}(u - T_t u, u) \geq 0$ for all $u \in D(L)$. In fact, it can be shown that if B is a real Hilbert space, 1.12 (i), (ii) can be replaced by (i)′ $(Lu, u) \leq 0$ for all $u \in D(L)$ *(negative definite)*

(ii)′ $(\alpha - L)(D(L)) = B$ for some $\alpha > 0$

provided L is closed (cf. [ReS 75, Theorem X.48]). Furthermore, then also

$$(1.5) \quad (G_\alpha f, f) = (G_\alpha f, -L\, G_\alpha f) + \alpha(G_\alpha f, G_\alpha f) \geq 0 \text{ for all } f \in B, \ \alpha > 0 \ .$$

The following diagram summarizes what we have achieved in this section, i.e., the unique correspondence between strongly continuous contraction semigroups, strongly continuous contraction resolvents and certain (in general unbounded) linear operators on B.

Diagram 1

$B =$ Banach space over $\mathbb{K} = \mathbb{R}, \mathbb{C}$

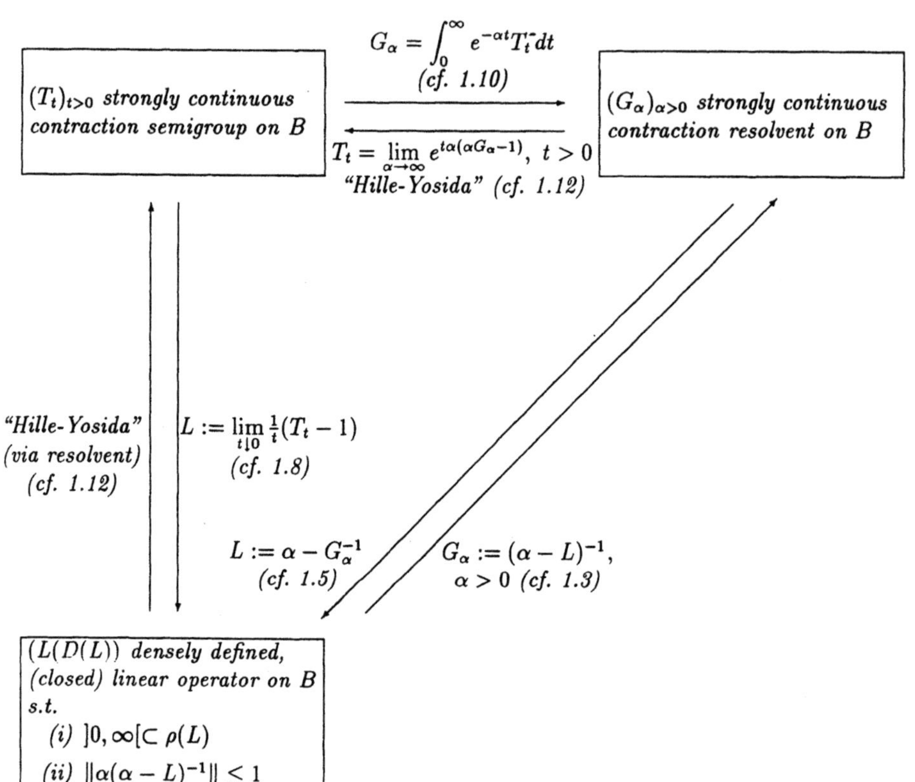

To conclude this section we briefly discuss the "symmetric case": assume B is a real Hilbert space (i.e., its norm comes from an inner product $(\ ,\ )$) and L is a *self-adjoint, negative definite* operator (i.e., L coincides with its adjoint $\hat{L}$ on B and $(Lu, u) \leq 0$ for all $u \in D(L)$). Then (i) and (ii) in the above diagram always hold and the above correspondences remain true if one adds that both, all T_t, $t > 0$, and all G_α, $\alpha > 0$, are self-adjoint. In this case the informal expression $T_t = e^{tL}$, $t > 0$, becomes rigorous since e^{tL} is defined by the spectral theory for self-adjoint operators on a Hilbert space.

2   Coercive bilinear forms

In this section we want to discuss the connection between coercive bilinear forms and the objects in the preceding section in the case where B is replaced by some real Hilbert space $\mathcal{H}$ with inner product $(\ ,\ )$ and norm $\|\ \| := (\ ,\ )^{1/2}$ which we fix in this section.

Let D be a linear subspace of $\mathcal{H}$ and $\mathcal{E} : D \times D \to \mathbb{R}$ a bilinear map. We define its *symmetric part* and *antisymmetric part* $(\tilde{\mathcal{E}}, D), (\check{\mathcal{E}}, D)$ respectively by

$$(2.1) \quad \tilde{\mathcal{E}}(u, v) := \frac{1}{2}(\mathcal{E}(u, v) + \mathcal{E}(v, u)) ; \quad \check{\mathcal{E}}(u, v) := \frac{1}{2}(\mathcal{E}(u, v) - \mathcal{E}(v, u)) ,$$

$u, v \in D$. Clearly, $\mathcal{E} = \tilde{\mathcal{E}} + \check{\mathcal{E}}$. For $\alpha \geq 0$ we set

$$(2.2) \qquad\qquad \mathcal{E}_\alpha(u, v) := \mathcal{E}(u, v) + \alpha(u, v) ; \quad u, v \in D .$$

Assume $(\mathcal{E}, D)$ is *positive definite* (i.e., $\mathcal{E}(u, u) \geq 0$ for all $u \in D$). Then $(\mathcal{E}, D)$ is said to satisfy the *weak sector condition* if

(2.3) there exists a constant $K > 0$ (called *continuity constant*) such that
$$|\mathcal{E}_1(u, v)| \leq K\ \mathcal{E}_1(u, u)^{1/2}\mathcal{E}_1(v, v)^{1/2} \text{ for all } u, v \in D .$$

Clearly, (2.3) just says that $\mathcal{E}_1$ is continuous w.r.t to the norm $\tilde{\mathcal{E}}_1^{1/2}$ on D.

Exercise 2.1. Let $(\mathcal{E}, D)$ be as above, $(\mathcal{E}, D)$ positive definite.

(i) Show that the norms $\tilde{\mathcal{E}}_\alpha^{1/2}$, $\alpha > 0$, on D are all equivalent.

(ii) Prove that the following assertions are equivalent:

    (a) $(\mathcal{E}, D)$ satisfies the weak sector condition.

    (b) For every $\alpha > 0$ there exists $K_\alpha \in {]0, \infty[}$ such that

$$|\mathcal{E}_\alpha(u, v)| \leq K_\alpha\ \mathcal{E}_\alpha(u, u)^{1/2}\mathcal{E}_\alpha(v, v)^{1/2} \text{ for all } u, v \in D .$$

    (c) For every $\alpha > 0$ there exists $K'_\alpha \in {]0, \infty[}$ such that

$$|\mathcal{E}(u, v)| \leq K'_\alpha\ \mathcal{E}_\alpha(u, u)^{1/2}\mathcal{E}_\alpha(v, v)^{1/2} \text{ for all } u, v \in D .$$

(iii) Show that K_α in (ii)(b) can be chosen independently of $\alpha > 0$ if and only if $(\mathcal{E}, D)$ satisfies the *(strong) sector condition*, i.e.,

$$(2.4) \qquad \text{there exists } K \in {]0, \infty[} \text{ such that}$$
$$|\mathcal{E}(u,v)| \leq K\, \mathcal{E}(u,u)^{1/2}\mathcal{E}(v,v)^{1/2} \quad \text{for all } u, v \in D .$$

(iv) Show that for all $\alpha \geq 0$ the following assertions are equivalent.

    (a) There exists $K_\alpha \in {]0, \infty[}$ such that

$$|\mathcal{E}(u,v)| \leq K_\alpha \mathcal{E}_\alpha(u,u)^{1/2}\mathcal{E}_\alpha(v,v)^{1/2} \quad \text{for all } u, v \in D .$$

    (b) There exists $K'_\alpha \in {]0, \infty[}$ such that

$$|\check{\mathcal{E}}(u,v)| \leq K'_\alpha \mathcal{E}_\alpha(u,u)^{1/2}\mathcal{E}_\alpha(v,v)^{1/2} \quad \text{for all } u, v \in D .$$

Correspondingly, a positive definite linear operator $(L, D(L))$ on $\mathcal{H}$ is said to satisfy the *(strong) sector condition* if

$$(2.5) \qquad \text{there exists a constant } K > 0 \text{ such that}$$
$$|(Lu, v)| \leq K(Lu, u)^{1/2}(Lv, v)^{1/2} \quad \text{for all } u, v \in D(L) .$$

Note that if L is *strictly positive definite* (i.e., there exists $c \in {]0, \infty[}$ such that $(Lu, u) \geq c(u, u)$ for all $u \in D(L)$) and bounded, then L satisfies (2.5).

Remark 2.2. The reason for calling (2.4) or (2.5) sector condition is given in 2.17, 2.18 below. If $\mathcal{E} = \tilde{\mathcal{E}}$ then (2.4) holds with $K = 1$. This is just Cauchy-Schwarz's inequality.

Definition 2.3. A pair $(\mathcal{E}, D(\mathcal{E}))$ is called a *symmetric closed form* (on $\mathcal{H}$) if $D(\mathcal{E})$ is a dense linear subspace of $\mathcal{H}$ and $\mathcal{E} : D(\mathcal{E}) \times D(\mathcal{E}) \to \mathbb{R}$ is a positive definite bilinear form which is *symmetric* (i.e., $\mathcal{E} = \tilde{\mathcal{E}}$) and *closed* on $\mathcal{H}$ (i.e., $D(\mathcal{E})$ is complete w.r.t. the norm $\mathcal{E}_1^{1/2}$).

Definition 2.4. A pair $(\mathcal{E}, D(\mathcal{E}))$ is called a *coercive closed form* (on $\mathcal{H}$) if $D(\mathcal{E})$ is a (dense) linear subspace of $\mathcal{H}$ and $\mathcal{E} : D(\mathcal{E}) \times D(\mathcal{E}) \to \mathbb{R}$ is a bilinear form such that the following two conditions hold.

    (i) Its symmetric part $(\tilde{\mathcal{E}}, D(\mathcal{E}))$ is a symmetric closed form on $\mathcal{H}$.

    (ii) $(\mathcal{E}, D(\mathcal{E}))$ satisfies the weak sector condition (2.3).

Remark 2.5. (i) A continuous bilinear form $(\mathcal{E}, D(\mathcal{E}))$ on some Hilbert space $(\mathcal{H}_1, \langle\, ,\, \rangle_{\mathcal{H}_1})$ is called *coercive* if there exists a constant $c > 0$ such that $\mathcal{E}(u,u) \geq c\langle u, u \rangle_{\mathcal{H}_1}$ for all $u \in D(\mathcal{E})$. Clearly, $(\mathcal{E}, D(\mathcal{E}))$ is then a coercive closed form in the sense of 2.4 which even satisfies (2.4). Conversely, a coercive closed form as in 2.4 satisfying (2.4) is coercive on $(\mathcal{H}_1, \langle\, ,\, \rangle_{\mathcal{H}_1})$ with $\mathcal{H}_1 = D(\mathcal{E})$, $\langle\, ,\, \rangle_{\mathcal{H}_1} = \tilde{\mathcal{E}}$. (ii) Because of the weak sector condition coercive closed forms in the sense of 2.4 are also sometimes called *quasi-symmetric*. But we avoid this terminology since (2.3) is less restrictive than it appears.

Below we shall use the following result due to G. Stampacchia (cf. [St 64]) in an essential way. From now on (unless otherwise stated) we consider $D(\mathcal{E})$ to be equipped with one of the equivalent norms $\tilde{\mathcal{E}}_\alpha^{1/2}$, $\alpha > 0$.

Theorem 2.6. *Let $(\mathcal{E}, D(\mathcal{E}))$ be a coercive closed form on $\mathcal{H}$ and let C be a non-empty closed convex subset of $D(\mathcal{E})$. Let J be a continuous linear functional on $D(\mathcal{E})$ and $\alpha > 0$. Then there exists a unique $v \in C$ such that*

$$(2.6) \qquad \mathcal{E}_\alpha(v, w - v) \geq J(w - v) \quad \text{for all } w \in C .$$

Proof. Uniqueness: Let $v_1, v_2 \in C$ satisfy (2.6). Then

$$\mathcal{E}_\alpha(v_1, v_2 - v_1) \geq J(v_2 - v_1) \qquad \text{and} \qquad \mathcal{E}_\alpha(v_2, v_1 - v_2) \geq J(v_1 - v_2) .$$

Hence $0 \leq \mathcal{E}_\alpha(v_2 - v_1, v_2 - v_1) = \mathcal{E}_\alpha(v_2, v_2 - v_1) - \mathcal{E}_\alpha(v_1, v_2 - v_1) \leq 0$, i.e., $v_2 = v_1$.

Existence: Let us first consider the case where $\check{\mathcal{E}} \equiv 0$. Define for $v \in D(\mathcal{E})$

$$I(v) := \mathcal{E}_\alpha(v, v) - 2J(v) , \quad d := \inf_{v \in C} I(v) .$$

Then, since $I(v) \geq \mathcal{E}_\alpha(v, v) - 2\|J\|\mathcal{E}_\alpha(v, v)^{1/2} = \left(\mathcal{E}_\alpha(v, v)^{1/2} - \|J\|\right)^2 - \|J\|^2 \geq -\|J\|^2$, $v \in D(\mathcal{E})$, we have $d > -\infty$. Hence we can find $v_n \in C$ such that $d \leq I(v_n) < d + \frac{1}{n}, n \in \mathbb{N}$. Then

$$\begin{aligned}
\mathcal{E}_\alpha(v_n - v_m, v_n - v_m) &= 2\mathcal{E}_\alpha(v_n, v_n) + 2\mathcal{E}_\alpha(v_m, v_m) - 4\mathcal{E}_\alpha\left(\frac{v_n + v_m}{2}, \frac{v_n + v_m}{2}\right) \\
&= 2\,I(v_n) + 2\,I(v_m) - 4\,I\left(\frac{v_n + v_m}{2}\right) \\
&\leq 2(d + \frac{1}{n}) + 2(d + \frac{1}{m}) - 4d \xrightarrow[n,m\to\infty]{} 0 .
\end{aligned}$$

Hence, we can find $v \in C$ such that $v_n \xrightarrow[n\to\infty]{} v$ in $D(\mathcal{E})$ and $I(v_n) \xrightarrow[n\to\infty]{} I(v) = d$. Now, let $w \in C$ and $\epsilon \in]0, 1[$, then $v + \epsilon(w - v) = \epsilon w + (1 - \epsilon)v \in C$; hence

$$0 \leq I(v + \epsilon(w - v)) - I(v) = 2\epsilon\,\mathcal{E}_\alpha(v, w - v) - 2\epsilon\,J(w - v) + \epsilon^2 \mathcal{E}_\alpha(w - v, w - v) .$$

Dividing by ϵ and then letting ϵ tend to zero we obtain (2.6) for $\mathcal{E}$ if $\check{\mathcal{E}} \equiv 0$.
To prove the general case define for $\beta \geq 0$, $B^{(\beta)} := \tilde{\mathcal{E}} + \beta\check{\mathcal{E}}$, $D(B^{(\beta)}) = D(\mathcal{E})$. Then $(B^{(\beta)}, D(\mathcal{E}))$ is also a coercive closed form on $\mathcal{H}$ and J is continuous on $D(\mathcal{E})$ w.r.t. $\tilde{\mathcal{E}}_\alpha^{1/2} = (\tilde{B}_\alpha^{(\beta)})^{1/2}$. Since we know that the assertion holds for $\beta = 0$, it suffices to show that if the assertion holds for $(B^{(\beta_0)}, D(\mathcal{E}))$ then it holds for $(B^{(\beta)}, D(\mathcal{E}))$ with $\beta_0 \leq \beta < \beta_0 + K_\alpha^{-1}$ where K_α is as in 2.1(ii)(b).
Let $u \in D(\mathcal{E})$. Applying the theorem for $(B^{(\beta_0)}, D(\mathcal{E}))$ with the continuous (w.r.t. $(\tilde{B}_\alpha^{(\beta_0)})^{1/2} = \tilde{\mathcal{E}}_\alpha^{1/2}$) linear functional $w \mapsto J(w) - (\beta - \beta_0)\check{\mathcal{E}}(u, w), w \in D(\mathcal{E})$, we conclude that there exists $Tu \in C$ such that for all $w \in C$

$$(2.7) \qquad B_\alpha^{(\beta_0)}(Tu, w - Tu) \geq J(w - Tu) - (\beta - \beta_0)\check{\mathcal{E}}(u, w - Tu) .$$

Let $u_1, u_2 \in D(\mathcal{E})$. Applying (2.7) to $u = u_1$, $w = Tu_2$ and $u = u_2$, $w = Tu_1$ and adding the two inequalities one obtains

$$\mathcal{E}_\alpha(Tu_1 - Tu_2, Tu_1 - Tu_2) \le (\beta - \beta_0)\, \check{\mathcal{E}}(u_2 - u_1, Tu_1 - Tu_2) .$$

By the weak sector condition the right hand side is dominated by

$$(\beta - \beta_0)K_\alpha\, \mathcal{E}_\alpha(u_2 - u_1, u_2 - u_1)^{1/2}\mathcal{E}_\alpha(Tu_1 - Tu_2, Tu_1 - Tu_2)^{1/2} .$$

Hence

$$\mathcal{E}_\alpha(Tu_1 - Tu_2, Tu_1 - Tu_2) \le (\beta - \beta_0)^2 K_\alpha^2\, \mathcal{E}_\alpha(u_1 - u_2, u_1 - u_2) ,$$

and if $(\beta - \beta_0)K_\alpha < 1$ this means that the map $u \mapsto Tu$ is a contraction. Then by the Banach fixed point theorem we can find $v \in D(\mathcal{E})$ such that $v = T(v) \in C$. By (2.7) we have that for all $w \in C$

$$B_\alpha^{(\beta_0)}(v, w - v) \ge J(w - v) - (\beta - \beta_0)\check{\mathcal{E}}(v, w - v) ,$$

that is

$$B_\alpha^{(\beta)}(v, w - v) \ge J(w - v) .$$

$\square$

Exercise 2.7. Prove that if in 2.6 C is a closed linear subspace of $D(\mathcal{E})$ then (2.6) is equivalent with

$$\mathcal{E}_\alpha(v, w) = J(w) \quad \text{for all} \ \ w \in C .$$

The following two theorems give the connection between closed forms and strongly continuous contraction resolvents resp. semigroups on $\mathcal{H}$ and their generators (studied in the preceding section).

Theorem 2.8. *Let $(\mathcal{E}, D(\mathcal{E}))$ be a coercive closed form on $\mathcal{H}$ with continuity constant K . Then there exist unique strongly continuous contraction resolvents $(G_\alpha)_{\alpha>0}$, $(\hat{G}_\alpha)_{\alpha>0}$ on $\mathcal{H}$ such that*

$$\text{(2.8)} \qquad \begin{aligned} &G_\alpha(\mathcal{H}),\ \hat{G}_\alpha(\mathcal{H}) \subset D(\mathcal{E}) \quad \text{and} \quad \mathcal{E}_\alpha(G_\alpha f, u) = (f, u) = \mathcal{E}_\alpha(u, \hat{G}_\alpha f) \\ &\text{for all } f \in \mathcal{H},\ u \in D(\mathcal{E}),\ \alpha > 0 . \end{aligned}$$

In particular,

$$\text{(2.9)} \qquad (G_\alpha f, g) = (f, \hat{G}_\alpha g) \quad \text{for all} \ \ f, g \in \mathcal{H} ,$$

i.e., $\hat{G}_\alpha$ is the adjoint (cf. Appendix A, Sect. 1 below) of G_α for all $\alpha > 0$. And if $(T_t)_{t>0}$, $(\hat{T}_t)_{t>0}$ denote the strongly continuous contraction semigroups corresponding to $(G_\alpha)_{\alpha>0}$, $(\hat{G}_\alpha)_{\alpha>0}$ respectively, then

$$\text{(2.10)} \qquad (T_t f, g) = (f, \hat{T}_t g) \quad \text{for all} \ \ f, g \in \mathcal{H} ,\ t > 0 .$$

Proof. Fix $f \in \mathcal{H}$. Applying 2.7 with $J(u) := (f, u)$, $u \in D(\mathcal{E})$ and $C := D(\mathcal{E})$, we obtain that for any $\alpha > 0$ there exists a unique $G_\alpha f \in D(\mathcal{E})$ such that $\mathcal{E}_\alpha(G_\alpha f, u) = (f, u)$ for all $u \in D(\mathcal{E})$. Clearly, $f \mapsto G_\alpha f$ is linear on $\mathcal{H}$ and if $f \in \mathcal{H}$ such that $G_\alpha f = 0$ then $0 = \mathcal{E}_\alpha(G_\alpha f, u) = (f, u)$ for all $u \in D(\mathcal{E})$, hence $f = 0$, since $D(\mathcal{E})$ is dense in $\mathcal{H}$. The same arguments give rise to injective linear operators $\hat{G}_\alpha$, $\alpha > 0$, such that $\mathcal{E}_\alpha(u, \hat{G}_\alpha f) = (f, u)$ for all $u \in D(\mathcal{E})$. Then clearly, (2.9) holds since

$$(G_\alpha f, g) = \mathcal{E}_\alpha(G_\alpha f, \hat{G}_\alpha g) = (f, \hat{G}_\alpha g) \quad \text{for all} \quad f, g \in \mathcal{H}.$$

If $\alpha, \beta > 0$ then for all $f \in \mathcal{H}$, $u \in D(\mathcal{E})$

$$\begin{aligned}
\mathcal{E}_\alpha(G_\beta f - (\alpha - \beta)G_\alpha G_\beta f, u) &= \mathcal{E}_\beta(G_\beta f, u) + (\alpha - \beta)(G_\beta f, u) \\
&\quad -(\alpha - \beta)(G_\beta f, u) \\
&= (f, u) = \mathcal{E}_\alpha(G_\alpha f, u),
\end{aligned}$$

and the first resolvent equation for $(G_\alpha)_{\alpha > 0}$ follows. In particular, $G_\alpha(\mathcal{H})$ is independent of α and as in the proof of 1.5 one sees that for $\alpha > 0$ the linear operator

$$L := \alpha - G_\alpha^{-1}, \quad D(L) := G_\alpha(\mathcal{H})$$

is defined independently of α. $G_\alpha(\mathcal{H})$ is dense in $\mathcal{H}$ since for $g \in \mathcal{H}$ with $(G_\alpha f, g) = 0$ for all $f \in \mathcal{H}$, (2.9) implies that $\hat{G}_\alpha g = 0$, hence $g = 0$. Furthermore, since (by 2.4 (i))

$$\|f\| \, \|G_\alpha f\| \geq (f, G_\alpha f) = \mathcal{E}_\alpha(G_\alpha f, G_\alpha f) \geq \alpha(G_\alpha f, G_\alpha f),$$

it follows that $\|\alpha G_\alpha\| \leq 1$ for all $\alpha > 0$. By definition $]0, \infty[\subset \rho(L)$, hence all assumptions on L in 1.3 (ii) are fulfilled, so $(G_\alpha)_{\alpha > 0}$ is also strongly continuous. The proof that $(\hat{G}_\alpha)_{\alpha > 0}$ is a strongly continuous contraction resolvent on $\mathcal{H}$ is analogous. The existence of $(T_t)_{t > 0}$, $(\hat{T}_t)_{t > 0}$ now follows from the Hille-Yosida Theorem (cf. 1.12). (2.10) follows from (2.9) by (1.4) and the fact that the Laplace transform is one-to-one on continuous functions (cf. the uniqueness part of the proof for 1.12). $\qquad\square$

Remark 2.9. (i) Note that since $\|G_\alpha\| \leq \alpha^{-1}$, it follows from (2.8) that each G_α is a continuous linear operator from $\mathcal{H}$ to $D(\mathcal{E})$.

(ii) We could have defined $\hat{G}_\alpha$ as the adjoint of G_α, $\alpha > 0$, and then checked that $(\hat{G}_\alpha)_{\alpha > 0}$ has the desired properties (cf. Appendix A, Sect. 1). $(G_\alpha)_{\alpha > 0}$, $(T_t)_{t > 0}$ are called *resolvent* resp. *semigroup associated with* $(\mathcal{E}, D(\mathcal{E}))$ and $(\hat{G}_\alpha)_{\alpha > 0}$, $(\hat{T}_t)_{t > 0}$ are called *coresolvent* resp. *cosemigroup associated with* $(\mathcal{E}, D(\mathcal{E}))$. Below we shall prove a complete one-to-one correspondence between $(G_\alpha)_{\alpha > 0}$ and $(\mathcal{E}, D(\mathcal{E}))$ (and hence between $(\hat{G}_\alpha)_{\alpha > 0}$ and $(\mathcal{E}, D(\mathcal{E}))$). Therefore, we shall call the corresponding generators $L, \hat{L}$ of $(G_\alpha)_{\alpha > 0}$, $(\hat{G}_\alpha)_{\alpha > 0}$ respectively, the *generator* and *cogenerator* of $(\mathcal{E}, D(\mathcal{E}))$.

Now we turn to the question whether a resolvent $(G_\alpha)_{\alpha>0}$ (with adjoint $(\hat{G}_\alpha)_{\alpha>0}$) as above, uniquely determines the form $(\mathcal{E}, D(\mathcal{E}))$ via (2.8), and under which assumptions there always exists a coercive closed form $(\mathcal{E}, D(\mathcal{E}))$ on $\mathcal{H}$ satisfying (2.8) if $(G_\alpha)_{\alpha>0}$ is a given strongly continuous contraction resolvent on $\mathcal{H}$. We begin with the following

Corollary 2.10. *Let* $(\mathcal{E}, D(\mathcal{E}))$, $(G_\alpha)_{\alpha>0}$ *be as in 2.8 and let* $(L, D(L))$ *be the generator of* $(G_\alpha)_{\alpha>0}$. *Then*

(2.11) $D(L) \subset D(\mathcal{E})$ *and* $\mathcal{E}(u,v) = (-Lu, v)$ *for all* $u \in D(L)$, $v \in D(\mathcal{E})$.

In particular, $1 - L$ *satisfies the sector condition (2.5).*
Corresponding statements hold for the generator $\hat{L}$ *of* $(\hat{G}_\alpha)_{\alpha>0}$.

Proof. Let $\alpha > 0$, $u \in D(L)$, then $u \in G_\alpha(\mathcal{H}) \subset D(\mathcal{E})$ and for all $v \in D(\mathcal{E})$

$$\mathcal{E}(u,v) = \mathcal{E}_\alpha(G_\alpha G_\alpha^{-1} u, v) - \alpha(u,v) = (G_\alpha^{-1} u, v) - \alpha(u,v) = (-Lu, v) .$$

$$\square$$

Now we want to show that $(\mathcal{E}, D(\mathcal{E}))$ can be reconstructed from its associated resolvent $(G_\alpha)_{\alpha>0}$.
Define for a given strongly continuous contraction resolvent on $\mathcal{H}$ for $\beta > 0$

(2.12) $$\mathcal{E}^{(\beta)}(u,v) := \beta(u - \beta G_\beta u, v) ; \quad u, v \in \mathcal{H} .$$

Observe that for $\beta > 0$, $u \in \mathcal{H}$, since

$$\mathcal{E}^{(\beta)}(u, \beta G_\beta u) = \mathcal{E}^{(\beta)}(u,u) - \beta(u - \beta G_\beta u, u - \beta G_\beta u)$$

and

$$\mathcal{E}^{(\beta)}(\beta \hat{G}_\beta u, u) = \mathcal{E}^{(\beta)}(u,u) - \beta(u - \beta \hat{G}_\beta u, u - \beta \hat{G}_\beta u),$$

we have

(2.13) $\mathcal{E}^{(\beta)}(u, \beta G_\beta u) \leq \mathcal{E}^{(\beta)}(u,u)$ and $\mathcal{E}^{(\beta)}(\beta \hat{G}_\beta u, u) \leq \mathcal{E}^{(\beta)}(u,u)$.

Lemma 2.11. *Let* $(\mathcal{E}, D(\mathcal{E}))$; $(G_\alpha)_{\alpha>0}$, $(\hat{G}_\alpha)_{\alpha>0}$ *be as in 2.8. Then for* $\beta > 0$

(i) $\mathcal{E}^{(\beta)}(u,v) = \mathcal{E}(\beta G_\beta u, v)$ *for all* $u \in \mathcal{H}$, $v \in D(\mathcal{E})$.

(ii) $\mathcal{E}(\beta G_\beta u, \beta G_\beta u) \leq \mathcal{E}^{(\beta)}(u,u)$ *and* $\mathcal{E}(\beta \hat{G}_\beta u, \beta \hat{G}_\beta u) \leq \mathcal{E}^{(\beta)}(u,u)$ *for all* $u \in \mathcal{H}$.

(iii) $|\mathcal{E}_1^{(\beta)}(u,v)| \leq (K + 1)\mathcal{E}_1(u,u)^{1/2}\mathcal{E}_1^{(\beta)}(v,v)^{1/2}$ *for all* $u \in D(\mathcal{E})$, $v \in \mathcal{H}$.

(iv) $\mathcal{E}_1(\beta G_\beta u, \beta G_\beta u) \leq (K + 1)^2 \mathcal{E}_1(u,u)$ *for all* $u \in D(\mathcal{E})$.

Proof. (i): $\mathcal{E}^{(\beta)}(u,v) = \beta\mathcal{E}_\beta(G_\beta u, v) - \beta(\beta G_\beta u, v) = \mathcal{E}(\beta G_\beta u, v)$.
(ii): $\mathcal{E}(\beta G_\beta u, \beta G_\beta u) = \mathcal{E}^{(\beta)}(u, \beta G_\beta u) \le \mathcal{E}^{(\beta)}(u, u)$ by (i) and (2.13). The proof
for $\hat{G}_\beta$ is similar.
(iii):

$$
\begin{aligned}
|\mathcal{E}^{(\beta)}(u,v)| &= |\mathcal{E}(u, \beta\hat{G}_\beta v)| \\
&\le K\, \mathcal{E}_1(u,u)^{1/2}\mathcal{E}_1(\beta\hat{G}_\beta v, \beta\hat{G}_\beta v)^{1/2} \\
&\le K\, \mathcal{E}_1(u,u)^{1/2}\mathcal{E}_1^{(\beta)}(v,v)^{1/2}
\end{aligned}
$$

by (i), (ii), since $\|\beta G_\beta\| \le 1$. Now (iii) follows easily.
(iv): $\mathcal{E}_1(\beta G_\beta u, \beta G_\beta u) \le \mathcal{E}_1^{(\beta)}(u,u) \le (K+1)^2\mathcal{E}_1(u,u)$ by (ii) and (iii) . $\square$

The following lemma is a consequence of the well-known Banach-Saks resp.
Banach-Alaoglu theorem. Since it will be used many times throughout this
book we include the proofs of both underlying theorems in the Appendix (cf.
Appendix A, Sect. 2).

Lemma 2.12. *Let $(\mathcal{E}, D(\mathcal{E}))$ be a coercive closed form and $u_n \in D(\mathcal{E})$, $n \in \mathbb{N}$,
such that*

$$
\sup_{n\in\mathbb{N}} \mathcal{E}(u_n, u_n) < \infty \ .
$$

*If $u \in \mathcal{H}$ such that $u_n \to u$ in $\mathcal{H}$ as $n \to \infty$, then $u \in D(\mathcal{E})$ and $u_n \to u$
weakly in the Hilbert space $(D(\mathcal{E}), \tilde{\mathcal{E}}_1)$ and there exists a subsequence $(u_{n_k})_{k\in\mathbb{N}}$
of $(u_n)_{n\in\mathbb{N}}$ such that its Cesaro mean $w_n := \frac{1}{n}\sum_{k=1}^n u_{n_k} \to u$ in $D(\mathcal{E})$ as $n \to \infty$.
Moreover,*

$$
\mathcal{E}(u,u) \le \liminf_{n\to\infty} \mathcal{E}(u_n, u_n) \ .
$$

Proof. Since

$$
\sup_{n\in\mathbb{N}} \mathcal{E}_1(u_n, u_n) < \infty \ ,
$$

by the Banach-Alaoglu theorem (cf. A.2.1) there exists $v \in D(\mathcal{E})$ such that
$u_{n_k} \to v$ weakly in $(D(\mathcal{E}), \tilde{\mathcal{E}}_1)$ as $n \to \infty$, for some subsequence $(n_k)_{k\in\mathbb{N}}$ of
$(n)_{n\in\mathbb{N}}$. By the Banach-Saks theorem (cf. A.2.2) the Cesaro mean $(w_n)_{n\in\mathbb{N}}$ of a
subsequence of $(u_{n_k})_{k\in\mathbb{N}}$ converges to v in $D(\mathcal{E})$, hence in $\mathcal{H}$. Since $w_n \to u$ in
$\mathcal{H}$ as $n \to \infty$, $u = v$. Since this reasoning holds for every subsequence, $u_n \to u$
weakly in $(D(\mathcal{E}), \tilde{\mathcal{E}}_1)$ as $n \to \infty$. Furthermore,

$$
\mathcal{E}(u,u) = \lim_{n\to\infty} \tilde{\mathcal{E}}(u, u_n) \le \liminf_{n\to\infty}(\tilde{\mathcal{E}}(u,u)^{1/2}\tilde{\mathcal{E}}(u_n, u_n)^{1/2})
$$

and consequently,

$$
\mathcal{E}(u,u)^{1/2} \le \liminf_{n\to\infty} \mathcal{E}(u_n, u_n)^{1/2}
$$

and the last part of the assertion follows. $\square$

Theorem 2.13. *Let $(\mathcal{E}, D(\mathcal{E}))$, $(G_\alpha)_{\alpha>0}$ be as in 2.8 and let $(L, D(L))$ be the
generator of $(G_\alpha)_{\alpha>0}$.*

(i) Let $u \in \mathcal{H}$. Then $u \in D(\mathcal{E})$ if and only if $\sup_{\beta>0} \mathcal{E}^{(\beta)}(u,u) < \infty$.

(ii) $D(L)$ is dense in $D(\mathcal{E})$ and moreover, for all $u \in D(\mathcal{E})$

$$(2.14) \qquad \lim_{\beta \to \infty} \mathcal{E}_1(\beta G_\beta u - u, \beta G_\beta u - u) = 0 .$$

In particular, the closed form $(\mathcal{E}, D(\mathcal{E}))$ is uniquely determined by $(G_\alpha)_{\alpha>0}$ via (2.8).

(iii) $\lim_{\beta \to \infty} \mathcal{E}^{(\beta)}(u,v) = \mathcal{E}(u,v)$ for all $u,v \in D(\mathcal{E})$.

Proof. (i): Let $u \in D(\mathcal{E})$ then by 2.11 (iii)

$$\mathcal{E}^{(\beta)}(u,u) \leq (K+1)^2 \mathcal{E}_1(u,u) \quad \text{for all } \beta > 0 .$$

Conversely, if $u \in \mathcal{H}$ such that $\sup_{\beta>0} \mathcal{E}^{(\beta)}(u,u) < \infty$, then by 2.11 (ii) and since $\|\beta G_\beta\| \leq 1$,

$$\sup_{\beta>0} \mathcal{E}_1(\beta G_\beta u, \beta G_\beta u) < \infty .$$

By the strong continuity of $(G_\beta)_{\beta>0}$, $\beta G_\beta u \xrightarrow[\beta \to \infty]{} u$ in $\mathcal{H}$, hence $u \in D(\mathcal{E})$ by 2.12 and (i) is proved.

(ii): Let $u \in D(\mathcal{E})$. Then by 2.11 (iv), $\sup_{\beta>0} \mathcal{E}_1(\beta G_\beta u, \beta G_\beta u) < \infty$, hence (by 2.12) there exists a sequence $(\beta_n G_{\beta_n} u)_{n \in \mathbb{N}}$ with $\beta_n \uparrow \infty$ whose Cesaro mean converges to u in $D(\mathcal{E})$. But $\beta_n G_{\beta_n} u \in G_{\beta_n}(\mathcal{H}) = D(L)$ for all $n \in \mathbb{N}$. Hence $D(L)$ is dense in $D(\mathcal{E})$. To prove (2.14), by 2.11 (iv) (and a "3ϵ-argument") it suffices now to consider $u = G_1 f$, $f \in \mathcal{H}$. But then $\mathcal{E}_1(\beta G_\beta u - u, \beta G_\beta u - u) = (\beta G_\beta f - f, \beta G_\beta u - u) \to 0$ as $\beta \to \infty$.

(iii) follows immediately from (ii) by 2.11 (i). $\qquad\qquad\square$

To prove the converse of 2.8 (cf. 2.15 below) we first need the following

Lemma 2.14. *Let $(G_\alpha)_{\alpha>0}$ be a strongly continuous contraction resolvent on $\mathcal{H}$ with generator L . Then the following are equivalent:*

(i) $(1-L)$ satisfies the sector condition (2.5).

(ii) G_α satisfies (2.5) for one (resp. all) $\alpha > 0$.

Proof. As in 2.1(ii) it follows that in (i) we may replace 1 by any $\alpha > 0$. Then the assertion is obvious since $G_\alpha = (\alpha - L)^{-1}$. $\qquad\qquad\square$

Theorem 2.15. *Let $(G_\alpha)_{\alpha>0}$ be a strongly continuous contraction resolvent on $\mathcal{H}$ such that each G_α satisfies the sector condition (2.5) and let L be its generator. Define*

$$\mathcal{E}(u,v) := (-Lu, v) \; ; \; u,v \in D(L) .$$

Let $D(\mathcal{E})$ be the completion of $D(L)$ w.r.t. $\tilde{\mathcal{E}}_1^{1/2}$ and denote the unique bilinear extension of $\mathcal{E}$ to $D(\mathcal{E})$ which is continuous w.r.t. $\tilde{\mathcal{E}}_1^{1/2}$ again by $\mathcal{E}$. Then $(\mathcal{E}, D(\mathcal{E}))$ is a coercive closed form on $\mathcal{H}$ such that

$$(2.15) \qquad \mathcal{E}(u,v) = (-Lu, v) \text{ for all } u \in D(L) , v \in D(\mathcal{E}) .$$

Furthermore, $(G_\alpha)_{\alpha>0}$ and $(\mathcal{E}, D(\mathcal{E}))$ are related by (2.8).

Proof. If $(u_n)_{n\in\mathbb{N}}$ is a Cauchy sequence in $D(L)$ w.r.t. $\tilde{\mathcal{E}}_1^{1/2}$ such that $u_n \xrightarrow[n\to\infty]{} 0$ in $\mathcal{H}$, then by 2.14

$$
\begin{aligned}
0 \le \mathcal{E}_1(u_n, u_n) &= \mathcal{E}_1(u_n - u_m, u_n - u_m) + \mathcal{E}_1(u_n - u_m, u_m) + ((1 - L)u_m, u_n) \\
&\le \mathcal{E}_1(u_n - u_m, u_n - u_m) + ((1 - L)u_m, u_n) \\
&\quad + K\, \mathcal{E}_1(u_n - u_m, u_n - u_m)^{1/2} \mathcal{E}_1(u_m, u_m)^{1/2}
\end{aligned}
$$

which can be made arbitrarily small for m, n large. Hence the map mapping every $\tilde{\mathcal{E}}_1^{1/2}$-Cauchy sequence in $D(L)$ to the corresponding Cauchy sequence i.e., limit in $\mathcal{H}$, gives rise to an inclusion of the completion $D(\mathcal{E})$ of $D(L)$ w.r.t. $\tilde{\mathcal{E}}_1^{1/2}$ into $\mathcal{H}$. The unique bilinear extension of $\mathcal{E}$ satisfying

$$
|\mathcal{E}_1(u, v)| \le K\, \mathcal{E}_1(u, u)^{1/2}\mathcal{E}_1(v, v)^{1/2} \quad \text{for all } u, v \in D(\mathcal{E})
$$

with domain $D(\mathcal{E})$ is then a closed form on $\mathcal{H}$. We then have that for every $\alpha > 0$, $G_\alpha(\mathcal{H}) = D(L) \subset D(\mathcal{E})$ and for all $u \in \mathcal{H}$, $v \in D(\mathcal{E})$

$$
\mathcal{E}_\alpha(G_\alpha u, v) = (-L\, G_\alpha u, v) + \alpha(G_\alpha u, v) = (u, v)\,.
$$

(2.15) follows now by (2.11). $\qquad\square$

Finally, one can obtain L directly from $(\mathcal{E}, D(\mathcal{E}))$.

Proposition 2.16. *Let $(\mathcal{E}, D(\mathcal{E}))$ be a coercive closed form on $\mathcal{H}$. Define*

(2.16) $D(L) := \{u \in D(\mathcal{E}) \mid v \mapsto \mathcal{E}(u, v)$ is continuous w.r.t. $(\,,\,)^{1/2}$ on $D(\mathcal{E})\}$.

For $u \in D(L)$, let Lu denote the unique element in $\mathcal{H}$ such that $(-Lu, v) = \mathcal{E}(u, v)$ for all $v \in D(\mathcal{E})$. Then L is the generator of the strongly continuous contraction resolvent $(G_\alpha)_{\alpha>0}$ which is related to $(\mathcal{E}, D(\mathcal{E}))$ by (2.8). In particular, $1 - L$ satisfies the sector condition (2.5) and we have a one-to-one correspondence between such L and coercive closed forms $(\mathcal{E}, D(\mathcal{E}))$ specified by (2.11).

Proof. Let $(G_\alpha)_{\alpha>0}$ with generator $\tilde{L}$ be the strongly continuous contraction resolvent on $\mathcal{H}$ associated with $(\mathcal{E}, D(\mathcal{E}))$ by 2.8. Then by (2.11) it follows that $D(\tilde{L}) \subset D(L)$ and since $D(\mathcal{E})$ is dense in $\mathcal{H}$, that $\tilde{L} = L$ on $D(\tilde{L})$. But for $\alpha > 0$, $(\alpha - L)(D(L)) \supset (\alpha - L)(D(\tilde{L})) = (\alpha - \tilde{L})(D(\tilde{L})) = \mathcal{H}$ and $\alpha - L$ is one-to-one since $(\alpha - L)u = 0$ for some $u \in D(L)$ implies $((\alpha - L)u, u) = \mathcal{E}_\alpha(u, u) = 0$ and hence $u = 0$. Consequently, $D(L) = D(\tilde{L})$. $\qquad\square$

So far we have not included the semigroup $(T_t)_{t>0}$ in our considerations of this section. In order to describe the additional property of $(T_t)_{t>0}$ which corresponds to the sector condition of $1 - L$ (where L is the generator) resp. to the sector condition of the resolvent operators G_α, we need to complexify.

Let $(\mathcal{H}_{\mathbb{C}}, (\,,\,))$ be the *complexification* of $(\mathcal{H}, (\,,\,))$, i.e., $\mathcal{H}_{\mathbb{C}} = \mathcal{H} \times \mathcal{H}$ with addition given by $[f_1, g_1] + [f_2, g_2] = [f_1 + f_2, g_1 + g_2]$, scalar multiplication given by $(a + ib)[f, g] = [af - bg, ag + bf]$, and inner product given by $([f_1, g_1], [f_2, g_2]) := (f_1, f_2) + i(f_1, g_2) - i(g_1, f_2) + (g_1, g_2)$. Again $\|\,\| := (\,,\,)^{1/2}$

denotes the corresponding norm on $\mathcal{H}_{\mathbb{C}}$. Any linear operator $(L, D(L))$ on $\mathcal{H}$ can be "extended" to an operator $(L^{\mathbb{C}}, D(L^{\mathbb{C}}))$ on $\mathcal{H}_{\mathbb{C}}$ by

$$L^{\mathbb{C}}([u, v]) = [Lu, Lv] \text{ for } [u, v] \in D(L^{\mathbb{C}}) := \{[u, v] \in \mathcal{H}_{\mathbb{C}} \mid u, v \in D(L)\} \ .$$

Define for $K \in]0, \infty[$

$$(2.17) \qquad\qquad S(K) := \{z \in \mathbb{C} \mid |\mathrm{Im}\, z| \leq K \mathrm{Re}\, z\} \ .$$

Proposition 2.17. *Let $(L, D(L))$ be a positive definite linear operator on $\mathcal{H}$. Then L satisfies the sector condition (2.5) if and only if*

$$(2.18) \ \{(L^{\mathbb{C}}([u, v]), [u, v]) \mid [u, v] \in D(L^{\mathbb{C}})\} \subset S(K) \text{ for some } K \in]0, \infty[\ .$$

Proof. We first note that

$$(L^{\mathbb{C}}([u, v]), [u, v]) \in S(K)$$

if and only if

$$(2.19) \qquad\qquad |(Lu, v) - (Lv, u)| \leq K((Lu, u) + (Lv, v)) \ .$$

But if L satisfies (2.5) with K then for all $u, v \in D(L)$

$$|(Lu, v) - (Lv, u)| \leq 2K(Lu, u)^{1/2}(Lv, v)^{1/2} \leq K((Lu, u) + (Lv, v)) \ .$$

Conversely, if $u, v \in D(L)$ satisfy (2.19), then

$$\begin{aligned}
(Lu, v) &= \frac{1}{2}((Lu, v) - (Lv, u)) + \frac{1}{2}((Lu, v) + (Lv, u)) \\
&\leq \frac{K}{2}((Lu, u) + (Lv, v)) + (Lu, u)^{1/2}(Lv, v)^{1/2}
\end{aligned}$$

(where we applied Cauchy-Schwarz's inequality to $[u, v] \mapsto \frac{1}{2}((Lu, v) + (Lv, u))$). This implies that

$$(Lu, v) \leq \frac{\sup(K, 1)}{2} \left((Lu, u)^{1/2} + (Lv, v)^{1/2}\right)^2 \ .$$

Replacing u, v by $\pm u[(Lu, u) + \varepsilon]^{-1/2}$ resp. $v[(Lv, v) + \varepsilon]^{-1/2}$, $\varepsilon > 0$, we easily obtain (2.5) with constant $2 \sup(K, 1)$.. $\qquad\qquad \square$

Remark 2.18. The set on the left hand side of (2.18) is called the *numerical range* of $L^{\mathbb{C}}$. So, (2.18) means that the numerical range of $L^{\mathbb{C}}$ is contained in the *sector $S(K)$*. Hence 2.17 justifies the terminology sector condition for (2.5).

Definition 2.19. Let $K \in]0, \infty[$. A family $(T_z)_{z \in S(K)}$ of bounded linear operators on $\mathcal{H}_{\mathbb{C}}$ is called a *holomorphic contraction semigroup* on $S(K)$ if

(i) $T_0 = id_{\mathcal{H}_\mathbb{C}}$

(ii) $T_{z_1} T_{z_2} = T_{z_1+z_2}$ for all $z_1, z_2 \in S(K)$.

(iii) $T_z f \to f$ in $\mathcal{H}_\mathbb{C}$ as $z \to 0$ in $S(\tilde{K})$ for all $f \in \mathcal{H}_\mathbb{C}$ and all $\tilde{K} < K$.

(iv) $\|T_z\| \leq 1$ for all $z \in S(K)$.

(v) $z \mapsto (T_z f, g)$ is analytic on the interior of $S(K)$ for all $f, g \in \mathcal{H}_\mathbb{C}$.

We need the following modification of the Hille-Yosida theorem.

Theorem 2.20. *For a linear operator $(L, D(L))$ on $\mathcal{H}_\mathbb{C}$ and $K \in]0, \infty[$ the following are equivalent:*

(i) *L is the generator of a holomorphic contraction semigroup $(T_z)_{z \in S(K^{-1})}$ on $S(K^{-1})$, i.e., for all $\tilde{K} > K$, $Lu = \lim\limits_{\substack{z \to 0 \\ z \in S(\tilde{K}^{-1})}} \frac{1}{z}(T_z u - u)$ for $u \in D(L)$*

where

$$D(L) := \{v \in \mathcal{H}_\mathbb{C} \mid \lim_{\substack{z \to 0 \\ z \in S(\tilde{K}^{-1})}} \frac{1}{z}(T_z v - v) \ exists.\} \ .$$

(ii) *L is m-dissipative of type $S(K^{-1})$, i.e., $1 \in \rho(L)$ and for all $\lambda \in \mathbb{C} \setminus (-S(K))$ and all $u \in D(L)$*

$$\|(\lambda - L)u\| \geq \mathrm{dis}(\lambda, -S(K))\|u\|$$

where $\mathrm{dis}(\lambda, -S(K))$ denotes the distance from λ to $-S(K)$ in $\mathbb{C}$.

In this case for all $t > 0$, $m \in \mathbb{N}$, and $f \in \mathcal{H}$, one has $T_t f \in D(L^m)$ and $\|L^m T_t f\| \leq C\|f\|/t^m$ for some $C \in]0, \infty[$ independent of f and t.

We do not give a proof of 2.20 here but refer instead to [Go 85, Sect. 5], [C 75, Theorem 1.3 and Proposition 1.4], and [ReS 75, Corollary 2 of Theorem X.52]. Now we are prepared to identify the property of $(T_t)_{t>0}$ corresponding to the weak sector condition.

Corollary 2.21. *Let $(L, D(L))$ be the generator of a strongly continuous contraction semigroup on $\mathcal{H}$. Then the following are equivalent:*

(i) *$1 - L$ satisfies the sector condition (2.5).*

(ii) *$(e^{-t} T_t^\mathbb{C})_{t>0}$ is the restriction of a holomorphic contraction semigroup on the sector $S(K^{-1})$ for some $K \in]0, \infty[$.*

Proof. $(i) \Rightarrow (ii)$: Suppose (i) with constant K. Let $L' := (L - 1)^\mathbb{C}$. Clearly, $1 \in \rho(L')$ and for all $u \in D(L') \setminus \{0\}$ and $\lambda \in \mathbb{C} \setminus (-S(K))$

$$\begin{aligned}
\|(\lambda - L')u\| &\geq \|u\|^{-1}|((\lambda - L')u, u)| \\
&= |\lambda + \|u\|^{-2}(-L'u, u)|\|u\| \\
&\geq \mathrm{dis}(\lambda, -S(K))\|u\|
\end{aligned}$$

by 2.17. Hence L' is m-dissipative of type $S(K^{-1})$ and thus generates a holomorphic contraction semigroup $(T'_z)_{z \in S(K^{-1})}$ on $S(K^{-1})$ by 2.20. Clearly, $(L-1)^{\mathbb{C}}$ is the generator of both $(T'_t)_{t>0}$ and $(e^{-t}T_t^{\mathbb{C}})_{t>0}$ and (ii) is proved.

$(ii) \Rightarrow (i)$: Let $(T'_z)_{z \in S(K^{-1})}$ be the holomorphic contraction semigroup on $S(K^{-1})$ extending $(e^{-t}T_t^{\mathbb{C}})_{t>0}$. Then $L' := (L-1)^{\mathbb{C}}$ is the generator of $(T'_z)_{z \in S(K^{-1})}$ (cf. [Go 85, Remark 5.7]), thus L' is m-dissipative of type $S(K^{-1})$ by 2.20. Hence for each $\alpha \in S(K^{-1})$ and each $u \in D(L') = D(L^{\mathbb{C}})$

$$\|(\frac{1}{\alpha t} - L')u\| \geq \frac{1}{\alpha t}\|u\| \quad \text{for all } t > 0 \; ,$$

i.e., $\|(1 - \alpha t L')u\| \geq \|u\|$ for all $t > 0$. Consequently,

$$\begin{aligned}
0 \; \leq \; & \frac{d}{dt}_{\big|_{t=0}} \sqrt{\|u\|^2 + |\alpha|^2 t^2 \|L'u\|^2 + 2t\mathrm{Re}(-\alpha L'u, u)} \\
= \; & \|u\|^{-1} \mathrm{Re}(-\alpha L'u, u) \; .
\end{aligned}$$

This implies that

$$\mathrm{Re}\alpha \, \mathrm{Re}(-L'u, u) \geq -\mathrm{Im}\alpha \, \mathrm{Im}(-L'u, u) \; .$$

Choosing $\alpha \in S(K^{-1}) \backslash \{0\}$ such that $\mathrm{Im}\alpha = \pm K^{-1}\mathrm{Re}\alpha$, we obtain that

$$|\mathrm{Im}(-L'u, u)| \leq K\mathrm{Re}(-L'u, u)$$

i.e., $(-L'u, u) \in S(K)$. Now (i) follows by 2.17. $\qquad\qquad\qquad\square$

Let us summarize what we have achieved so far in the following diagram (extending Diagram 1 on p. 14).

Diagram 2

$\mathcal{H}$ = Hilbert space over $\mathbb{R}$ with inner product $(\ ,\ )$ and norm $\|\ \| := (\ ,\ )^{1/2}$.

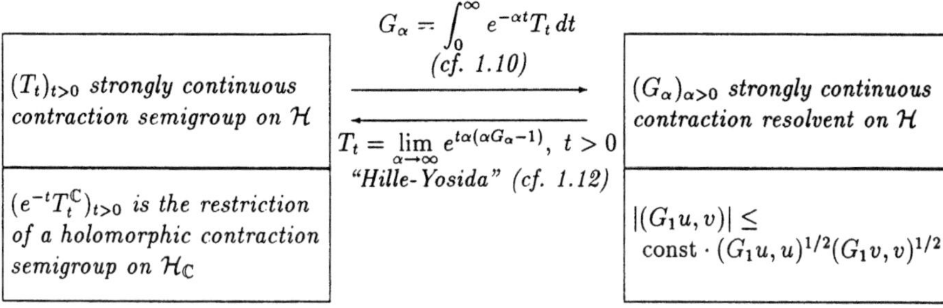

$$G_\alpha = \int_0^\infty e^{-\alpha t} T_t\, dt$$
(cf. 1.10)

$$T_t = \lim_{\alpha \to \infty} e^{t\alpha(\alpha G_\alpha - 1)},\ t > 0$$
"Hille-Yosida" (cf. 1.12)

$(T_t)_{t>0}$ *strongly continuous contraction semigroup on* $\mathcal{H}$

$(e^{-t}T_t^{\mathbb{C}})_{t>0}$ *is the restriction of a holomorphic contraction semigroup on* $\mathcal{H}_{\mathbb{C}}$

$(G_\alpha)_{\alpha>0}$ *strongly continuous contraction resolvent on* $\mathcal{H}$

$$|(G_1 u, v)| \leq \text{const} \cdot (G_1 u, u)^{1/2}(G_1 v, v)^{1/2}$$

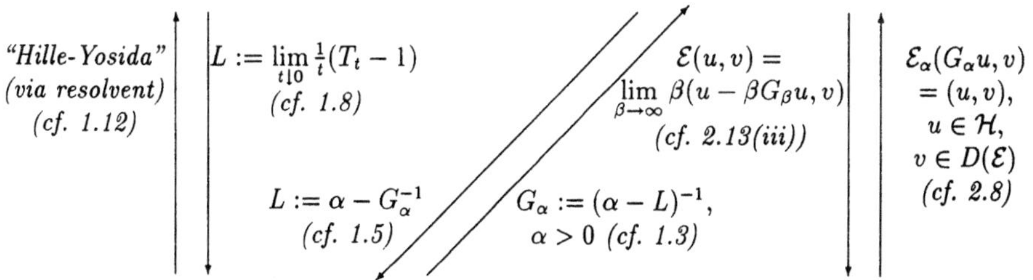

"Hille-Yosida" (via resolvent) (cf. 1.12)

$$L := \lim_{t \downarrow 0} \tfrac{1}{t}(T_t - 1)$$
(cf. 1.8)

$$L := \alpha - G_\alpha^{-1}$$
(cf. 1.5)

$$\mathcal{E}(u,v) = \lim_{\beta \to \infty} \beta(u - \beta G_\beta u, v)$$
(cf. 2.13(iii))

$$G_\alpha := (\alpha - L)^{-1},\ \alpha > 0\ (cf.\ 1.3)$$

$$\mathcal{E}_\alpha(G_\alpha u, v) = (u,v),\ u \in \mathcal{H},\ v \in D(\mathcal{E})$$
(cf. 2.8)

$(L(D(L)))$ *densely defined, (closed) linear operator on* $\mathcal{H}$ *s.t.*

  (i) $]0, \infty[\subset \rho(L)$

  (ii) $\|\alpha(\alpha - L)^{-1}\| \leq 1$

  (iii) $|((1 - L)u, v)| \leq \text{const} \cdot ((1 - L)u, u)^{1/2} \cdot ((1 - L)v, v)^{1/2}$

$$D(L) := \{u \in D(\mathcal{E}) \mid \exists Lu \in \mathcal{H}\ s.t.\ \mathcal{E}(u,v) = (-Lu, v)\ \forall v \in D(\mathcal{E})\}\ (cf.\ 2.16)$$

$$\mathcal{E}(u,v) := (-Lu, v),\ u, v \in D(L)$$
& completition (cf. 2.15)

$(\mathcal{E}, \mathcal{D}(\mathcal{E}))$ *coercive closed form on* $\mathcal{H}$ *(cf. 2.4)*

Now let us discuss the special symmetric case. All facts appearing below which follow from spectral theory on $\mathcal{H}$ will not be used in the sequel, but are included for completeness. So, if $\mathcal{E}(u,v) = \mathcal{E}(v,u)$ for all $u,v \in D(\mathcal{E})$, then the corresponding property of the generator L is symmetry and hence self-adjointness, since $(\alpha - L)(D(L)) = \mathcal{H}$ (cf. [ReS 72, Theorem VIII.3]). In fact, in this case properties (i)-(iii) in the above diagram can be replaced by the assumptions that L is negative definite and self-adjoint on $\mathcal{H}$ (cf. 1.13 and [ReS 72, Theorem VIII.3]). The corresponding property for $(G_\alpha)_{\alpha>0}$ is again that each G_α is symmetric. The most important observation, however, is that all sector conditions automatically hold by Cauchy-Schwarz's inequality.

Exercise 2.22. Prove that in the symmetric case $\mathcal{E}^{(\beta)}(u,u) \uparrow \mathcal{E}(u,u)$ as $\beta \to \infty$ for all $u \in D(\mathcal{E})$.

We note that L (or equivalently $\mathcal{E}$, $(G_\alpha)_{\alpha>0}$) is symmetric if and only if the associated semigroup $(T_t)_{t>0}$ is symmetric. In this case we have a corresponding approximation in terms of $(T_t)_{t>0}$. Define for $t > 0$

$$^{(t)}\mathcal{E}(u,v) = \frac{1}{t}(u - T_t u, v) \; ; \quad u,v \in \mathcal{H} \;.$$

Using the spectral representation of the generator L

$$-L = \int_0^\infty \lambda \, dE_\lambda \;,$$

one has that

$$T_t = e^{tL} = \int_0^\infty e^{-t\lambda} dE_\lambda \;, \quad t > 0 \;,$$

and

$$\sqrt{-L} = \int_0^\infty \sqrt{\lambda} \, dE_\lambda \;.$$

Then one easily shows that $D(\sqrt{-L}) = D(\mathcal{E})$ and that

$$\mathcal{E}(u,v) = (\sqrt{-L}u, \sqrt{-L}v) \; ; \quad u,v \in D(\sqrt{-L}) \;.$$

Furthermore, $u \in D(\mathcal{E})$ if and only if $\sup_{t>0} {}^{(t)}\mathcal{E}(u,u) < \infty$ and ${}^{(t)}\mathcal{E}(u,u) \uparrow \mathcal{E}(u,u)$ as $t \to 0$ in this case. We refer to [F 80, §1.3] for details.

3   Closability

This section is devoted to the notion of *closability* which in applications turns out to be crucial, since in examples the given positive bilinear form is almost never closed.

Definition 3.1. Let $\mathcal{E}$ with domain D be a positive definite bilinear form on $\mathcal{H}$. $(\mathcal{E}, D)$ is called *closable* (on $\mathcal{H}$) if for all $u_n \in D$, $n \in \mathbb{N}$, such that $(u_n)_{n\in\mathbb{N}}$ is $\mathcal{E}$-Cauchy (i.e., $\mathcal{E}(u_n - u_m, u_n - u_m) \xrightarrow[n,m\to\infty]{} 0$) and $u_n \xrightarrow[n\to\infty]{} 0$ in $\mathcal{H}$, it follows that $\mathcal{E}(u_n, u_n) \xrightarrow[n\to\infty]{} 0$.

Remark 3.2. (i) Note that $(\mathcal{E}, D)$ as in 3.1 is closable if and only if its symmetric part $(\tilde{\mathcal{E}}, D)$ is. If $\mathcal{E}(u, v) = (Tu, Tv); u, v \in D$, for some linear operator T on $\mathcal{H}$ with domain D, then $(\mathcal{E}, D)$ is closable if and only if T is a closable operator on $\mathcal{H}$.

(ii) Let $(\mathcal{E}, D)$ be as in 3.1. Consider the pre-Hilbert space D with inner product $\tilde{\mathcal{E}}_1$ and denote its abstract completion (w.r.t. $\tilde{\mathcal{E}}_1^{1/2}$) by $\overline{D}$ equipped with (the unique extension) of $\tilde{\mathcal{E}}_1^{1/2}$. There is a unique continuous map $i : \overline{D} \to \mathcal{H}$ extending the inclusion map $D \subset \mathcal{H}$ (cf. the proof of 2.15). $(\mathcal{E}, D)$ is closable if and only if $i : \overline{D} \to \mathcal{H}$ is injective.

(iii) If $(\mathcal{E}, D)$ is closable, then its symmetric part $\tilde{\mathcal{E}}$ extends uniquely to $\overline{D}$ (cf. (ii)). Suppose $(\mathcal{E}, D)$ satisfies the weak sector condition (2.3), then also $\mathcal{E}$ extends uniquely to $\overline{D}$. (Note that we already used this fact in the proof of 2.15). We denote this extension again by $\mathcal{E}$. Clearly if D is dense in $\mathcal{H}$, $(\mathcal{E}, \overline{D})$ is the smallest coercive closed form (in the sense of 2.4) extending $(\mathcal{E}, D)$, i.e., for any coercive closed form $(\mathcal{E}^{(1)}, D(\mathcal{E}^{(1)}))$ such that $D \subset D(\mathcal{E}^{(1)})$ and $\mathcal{E}^{(1)}(u, v) = \mathcal{E}(u, v)$ for all $u, v \in D$ we have that $\overline{D} \subset D(\mathcal{E}^{(1)})$ and $\mathcal{E}^{(1)}(u, v) = \mathcal{E}(u, v)$ for all $u, v \in \overline{D}$. $(\mathcal{E}, \overline{D})$ is called the *closure* of $(\mathcal{E}, D)$.

Proposition 3.3. *Let S be a negative definite linear operator on $\mathcal{H}$ such that $1 - S$ satisfies the sector condition (2.5). Define*

$$\mathcal{E}(u, v) := (-Su, v) \; ; \; u, v \in D(S) \; .$$

Then $(\mathcal{E}, D(S))$ is closable on $\mathcal{H}$. In particular, this holds for any negative definite linear operator on $\mathcal{H}$ which is symmetric (i.e., $(Su, v) = (u, Sv)$ for all $u, v \in D(S)$).

The last part of the assertion is clear, while the first part is an immediate consequence of the following

Lemma 3.4. *Let $(\mathcal{E}, D)$ be a positive definite bilinear form on $\mathcal{H}$ satisfying the weak sector condition (2.3) and such that for all $u_n \in D$, $n \in \mathbb{N}$, with $u_n \xrightarrow[n \to \infty]{} 0$ in $\mathcal{H}$, it follows that $\mathcal{E}(v, u_n) \xrightarrow[n \to \infty]{} 0$ for all $v \in D$. Then $(\mathcal{E}, D)$ is closable.*

Proof. (cf. the proof of 2.15). Let $u_n \in D$, $n \in \mathbb{N}$, such that $(u_n)_{n \in \mathbb{N}}$ is $\mathcal{E}$-Cauchy and $u_n \xrightarrow[n \to \infty]{} 0$ in $\mathcal{H}$. Then

$$
\begin{aligned}
\mathcal{E}_1(u_n, u_n) &= \mathcal{E}_1(u_n - u_m, u_n - u_m) + \mathcal{E}_1(u_n - u_m, u_m) + \mathcal{E}_1(u_m, u_n) \\
&\leq \mathcal{E}_1(u_n - u_m, u_n - u_m) + K \, \mathcal{E}_1(u_n - u_m, u_n - u_m)^{1/2} \mathcal{E}_1(u_m, u_m)^{1/2} \\
&\quad + \mathcal{E}_1(u_m, u_n)
\end{aligned}
$$

which can be made arbitrarily small for n, m large. $\qquad\square$

Proposition 3.5. *Let $(\mathcal{E}, D)$ be a positive definite bilinear form on $\mathcal{H}$. Suppose there exists a symmetric closable form $(\mathcal{E}^{(1)}, D)$ on $\mathcal{H}$ such that for some constant $c_1 > 0$*

$$(3.1) \qquad c_1^{-1}\mathcal{E}_1^{(1)}(u,u) \leq \mathcal{E}_1(u,u) \leq c_1\mathcal{E}_1^{(1)}(u,u) \quad \text{for all } u \in D .$$

Then $(\mathcal{E}, D)$ is closable on $\mathcal{H}$. If moreover, D is dense in $\mathcal{H}$ and for some constant $c_2 > 0$

$$(3.2) \qquad |\mathcal{E}_1(u,v)| \leq c_2\mathcal{E}_1^{(1)}(u,u)^{1/2}\mathcal{E}_1^{(1)}(v,v)^{1/2} \quad \text{for all } u,v \in D ,$$

then its closure $(\mathcal{E}, \overline{D})$ is a coercive closed form on $\mathcal{H}$ in the sense of 2.4.

Proof. Let $u_n \in D$, $n \in \mathbb{N}$, such that $(u_n)_{n\in\mathbb{N}}$ is $\mathcal{E}$-Cauchy and $u_n \xrightarrow[n\to\infty]{} 0$ in $\mathcal{H}$. Then by (3.1), $(u_n)_{n\in\mathbb{N}}$ is $\mathcal{E}^{(1)}$-Cauchy, hence by assumption $\mathcal{E}^{(1)}(u_n, u_n) \xrightarrow[n\to\infty]{} 0$ and again by (3.1), $\mathcal{E}(u_n, u_n) \to 0$. Consequently, $(\mathcal{E}, D)$ is closable and if (3.2) holds we have that for all $u, v \in D$

$$|\mathcal{E}_1(u,v)| \leq c_2 \cdot c_1\mathcal{E}_1(u,u)^{1/2}\mathcal{E}_1(v,v)^{1/2} .$$

By continuity this inequality extends to all $u, v \in \overline{D}$, thus $(\mathcal{E}, \overline{D})$ is a coercive closed form in this case, if D is dense in $\mathcal{H}$. $\square$

Remark 3.6. We shall see in the examples in Chapter II that in some cases it is enough to have the lower bound in (3.1) to obtain closability of $(\mathcal{E}, D)$.

Also the following will be useful below.

Proposition 3.7. *Let $(\mathcal{E}^{(k)}, D^{(k)})$, $k \in \mathbb{N}$, be closable symmetric bilinear forms on $\mathcal{H}$.*

(i) Define $D := \left\{ u \in \bigcap\limits_{k\in\mathbb{N}} D^{(k)} \mid \sum\limits_{k=1}^{\infty} \mathcal{E}^{(k)}(u,u) < \infty \right\}$ and

$$\mathcal{E}(u,v) := \sum_{k=1}^{\infty} \mathcal{E}^{(k)}(u,v) \quad ; u, v \in D$$

Then $(\mathcal{E}, D)$ is closable on $\mathcal{H}$.

(ii) Suppose $D^{(k)} \supset D^{(k+1)}$ and $\mathcal{E}^{(k)}(u,u) \leq \mathcal{E}^{(k+1)}(u,u)$ for all $u \in D^{(k+1)}$, $k \in \mathbb{N}$. Define $D := \left\{ u \in \bigcap\limits_{k\in\mathbb{N}} D^{(k)} \mid \sup\limits_{k\in\mathbb{N}} \mathcal{E}^{(k)}(u,u) < \infty \right\}$ and for $u, v \in D$

$$\mathcal{E}(u,v) := \frac{1}{4}\left[\mathcal{E}(u+v, u+v) - \mathcal{E}(u-v, u-v)\right]$$

where $\mathcal{E}(u,u) := \sup\limits_{k\in\mathbb{N}} \mathcal{E}^{(k)}(u,u)$. Then $(\mathcal{E}, D)$ is closable on $\mathcal{H}$.

Proof. Let $u_n \in D$, $n \in \mathbb{N}$, such that $(u_n)_{n\in\mathbb{N}}$ is $\mathcal{E}$-Cauchy and $u_n \xrightarrow[n\to\infty]{} 0$ in $\mathcal{H}$. Then by assumption in both (i) and (ii), $\lim\limits_{n\to\infty} \mathcal{E}^{(k)}(u_n, u_n) = 0$ for all $k \in \mathbb{N}$. Hence in case (i) by Fatou's Lemma for all $n \in \mathbb{N}$

$$\begin{aligned}
\mathcal{E}(u_n, u_n) &= \sum_{k=1}^{\infty} \lim_{m \to \infty} \mathcal{E}^{(k)}(u_n - u_m, u_n - u_m) \\
&\leq \liminf_{m \to \infty} \mathcal{E}(u_n - u_m, u_n - u_m) \ .
\end{aligned}$$

Similarly in case (ii) for all $n \in \mathbb{N}$

$$\begin{aligned}
\mathcal{E}(u_n, u_n) &= \sup_{k \in \mathbb{N}} \lim_{m \to \infty} \mathcal{E}^{(k)}(u_n - u_m, u_n - u_m) \\
&\leq \liminf_{m \to \infty} \mathcal{E}(u_n - u_m, u_n - u_m) \ .
\end{aligned}$$

In both cases this $\liminf$ can be made arbitrarily small for large n. $\qquad\square$

Remark 3.8. Note that in 3.7 it might be that $D = \{0\}$.

Exercise 3.9. Prove that 3.7 remains true if "closable" is replaced by "closed".

4   Contraction properties

In this section we replace $\mathcal{H}$ by the concrete Hilbert space $L^2(E; m) := L^2(E; \mathcal{B}; m)$ with usual inner product $(,)$ where $(E; \mathcal{B}; m)$ is a measure space. As usual we set for $u, v : E \to \mathbb{R}$

$$u \vee v := \sup(u, v), \ u \wedge v := \inf(u, v), \ u^+ := u \vee 0, \ u^- := -(u \wedge 0) \ .$$

We write $f \leq g$ or $f < g$ for $f, g \in L^2(E; m)$ (or any m-classes f, g of functions on E) if the inequality holds m-a.e. for corresponding representatives.

Definition 4.1. (i) Let G be a bounded linear operator on $L^2(E; m)$ with $D(G) = L^2(E; m)$. G is called *sub-Markovian* if for all $f \in L^2(E; m)$, $0 \leq f \leq 1$ implies $0 \leq Gf \leq 1$. A strongly continuous contraction resolvent $(G_\alpha)_{\alpha>0}$ resp. semigroup $(T_t)_{t>0}$ is called *sub-Markovian* if all αG_α, $\alpha > 0$, resp. T_t, $t > 0$, are sub-Markovian.
(ii) A closed densely defined linear operator L on $L^2(E; m)$ is called *Dirichlet operator* if $(Lu, (u - 1)^+) \leq 0$ for all $u \in D(L)$.

Exercise 4.2. (i) Let G be sub-Markovian. Prove that $Gf \geq 0$ for all $f \in L^2(E; m)$ with $f \geq 0$.
(ii) Let L be a Dirichlet operator. Prove that L is negative definite.

Proposition 4.3. *Let $(G_\alpha)_{\alpha>0}$ be a strongly continuous contraction resolvent on $L^2(E; m)$ with corresponding generator L and semigroup $(T_t)_{t>0}$. Then the following are equivalent.*

  (i) $(G_\alpha)_{\alpha>0}$ is sub-Markovian.

  (ii) $(T_t)_{t>0}$ is sub-Markovian.

  (iii) L is a Dirichlet operator.

Proof. $(i) \Rightarrow (ii)$: Since for all $u \in D(L)$, $t > 0$,

$$T_t u = \lim_{\alpha \to \infty} \exp(t\alpha(\alpha G_\alpha - 1))u$$

(cf. the proof of 1.12), we have that $0 \le T_t u \le 1$ if $u \in D(L)$ with $0 \le u \le 1$. Since for $f \in L^2(E;m)$, $\beta G_\beta f \xrightarrow[\beta \to \infty]{} f$ in $L^2(E;m)$ and $\beta G_\beta f \in D(L)$ and since βG_β is sub-Markovian, (ii) now easily follows.

$(ii) \Rightarrow (iii)$: Let G be any sub-Markovian contraction operator on $L^2(E;m)$, then for all $f \in L^2(E;m)$, since $f = (f-1)^+ + f \wedge 1$ and $G(f \wedge 1) \le G(|f| \wedge 1) \le 1$,

$$
\begin{aligned}
(Gf, (f-1)^+) &= (G(f-1)^+, (f-1)^+) + (G(f \wedge 1), (f-1)^+) \\
&\le ((f-1), (f-1)^+) + \int (f-1)^+ dm \\
&= (f, (f-1)^+) .
\end{aligned}
$$

Hence, for all $u \in D(L)$

$$(Lu, (u-1)^+) = \lim_{t \downarrow 0} \frac{1}{t}(T_t u - u, (u-1)^+) \le 0 .$$

$(iii) \Rightarrow (i)$: Let $f \in L^2(E;m)$ and $v := \alpha G_\alpha f$. If $f \le 1$, then

$$
\begin{aligned}
\alpha(v, (v-1)^+) &= (\alpha v - Lv, (v-1)^+) + (Lv, (v-1)^+) \\
&\le \alpha(f, (v-1)^+) \le \alpha \int (v-1)^+ dm .
\end{aligned}
$$

Hence

$$\int ((v-1)^+)^2 dm \le 0$$

i.e., $v \le 1$. If $f \ge 0$, then $-nf \le 1$ hence $-nv \le 1$ for all $n \in \mathbb{N}$. Consequently, $v \ge 0$. $\qquad \square$

Theorem 4.4. *Suppose $(\mathcal{E}, D(\mathcal{E}))$ is a coercive closed form on $L^2(E;m)$ with continuity constant K and corresponding resolvent $(G_\alpha)_{\alpha > 0}$. Then the following are equivalent:*

(i) For all $u \in D(\mathcal{E})$ and $\alpha \ge 0$, $u \wedge \alpha \in D(\mathcal{E})$ and $\mathcal{E}(u \wedge \alpha, u - u \wedge \alpha) \ge 0$

(ii) For all $u \in D(\mathcal{E})$, $u^+ \wedge 1 \in D(\mathcal{E})$ and $\mathcal{E}(u^+ \wedge 1, u - u^+ \wedge 1) \ge 0$.

(iii) For all $u \in D(\mathcal{E})$, $u^+ \wedge 1 \in D(\mathcal{E})$ and $\mathcal{E}(u + u^+ \wedge 1, u - u^+ \wedge 1) \ge 0$.

(iv) $(G_\alpha)_{\alpha > 0}$ is sub-Markovian.

If $(G_\alpha)_{\alpha > 0}$ is replaced by its adjoint $(\hat{G}_\alpha)_{\alpha > 0}$ the analogous equivalences hold with the two respective entries of $\mathcal{E}(\cdot, \cdot)$ interchanged.

Proof. $(i) \Rightarrow (ii)$: Since $-u^+ = (-u) \wedge 0 \in D(\mathcal{E})$ by (i), we obtain $u^+ \in D(\mathcal{E})$. Hence $u^- \in D(\mathcal{E})$ and applying (i) again with $\alpha = 1$ we obtain that $u^+ \wedge 1 \in D(\mathcal{E})$ and

$$\begin{aligned}
\mathcal{E}(u^+ \wedge 1, u - u^+ \wedge 1) &= \mathcal{E}(u^+ \wedge 1, u^+ - u^+ \wedge 1) - \mathcal{E}(u^+ \wedge 1, u^-) \\
&\geq -\mathcal{E}((u \wedge 1)^+, (u \wedge 1)^-) \ .
\end{aligned}$$

Now (ii) follows since by (i)

$$(4.1) \quad \mathcal{E}(u^+, u^-) = \mathcal{E}(u^+, u^+ - u) = -\mathcal{E}((-u) \wedge 0, (-u) - (-u) \wedge 0) \leq 0 \ .$$

$(ii) \Rightarrow (iii)$: We have for all $u \in D(\mathcal{E})$ that

$$\mathcal{E}(u + u^+ \wedge 1, u - u^+ \wedge 1) = \mathcal{E}(u - u^+ \wedge 1, u - u^+ \wedge 1) + 2\mathcal{E}(u^+ \wedge 1, u - u^+ \wedge 1) \geq 0 \ .$$

$(iii) \Rightarrow (iv)$: Let $f \in L^2(E; m)$, $0 \leq f \leq 1$ and set $u := \alpha G_\alpha f$, then

$$\begin{aligned}
0 &\geq -\frac{1}{2}\Big[\mathcal{E}(u + u^+ \wedge 1, u - u^+ \wedge 1) + \mathcal{E}(u - u^+ \wedge 1, u - u^+ \wedge 1)\Big] \\
&= -\mathcal{E}(u, u - u^+ \wedge 1) = \alpha(u - f, u - u^+ \wedge 1) \\
&= \alpha\|u - u^+ \wedge 1\|^2 + \alpha(u^+ \wedge 1 - f, u - u^+ \wedge 1) \ .
\end{aligned}$$

But $u = (u-1)^+ + u \wedge 1$ and $u \wedge 1 = (u \wedge 1)^+ - (u \wedge 1)^- = u^+ \wedge 1 + u \wedge 0$, hence

$$(u^+ \wedge 1 - f, u - u^+ \wedge 1) = \int_{[u \geq 1]} (u^+ \wedge 1 - f)(u-1)^+ dm + \int_{[u \leq 0]} (u^+ \wedge 1 - f)(u \wedge 0) dm$$

and the right hand side is positive because $0 \leq f \leq 1$. Consequently, $\|u - u^+ \wedge 1\| = 0$, i.e., $0 \leq u \leq 1$.

$(iv) \Rightarrow (i)$: Let $u \in D(\mathcal{E})$, $\alpha \geq 0$. Since $u = (u - \alpha)^+ + u \wedge \alpha$ we have that for all $\beta > 0$

$$(4.2) \qquad \mathcal{E}^{(\beta)}(u \wedge \alpha, u - u \wedge \alpha) = \mathcal{E}^{(\beta)}(u \wedge \alpha, (u - \alpha)^+) \ .$$

But by (iv) and 4.2(i), $(u-\alpha)^+(1 - \beta G_\beta)(u \wedge \alpha) \geq (u-\alpha)^+(\alpha - \beta G_\beta(|u| \wedge \alpha)) \geq 0$; hence

$$(4.3) \qquad \mathcal{E}^{(\beta)}(u \wedge \alpha, (u - \alpha)^+) = \beta((1 - \beta G_\beta)(u \wedge \alpha), (u - \alpha)^+) \geq 0$$

and thus

$$\begin{aligned}
\mathcal{E}_1^{(\beta)}((u - \alpha)^+, (u - \alpha)^+) &= \mathcal{E}_1^{(\beta)}(u, (u - \alpha)^+) - \mathcal{E}^{(\beta)}(u \wedge \alpha, (u - \alpha)^+) \\
&\quad -(u \wedge \alpha, (u - \alpha)^+) \\
&\leq \mathcal{E}_1^{(\beta)}(u, (u - \alpha)^+) \\
&\leq (K + 1)\mathcal{E}_1(u, u)^{1/2}\mathcal{E}_1^{(\beta)}((u - \alpha)^+, (u - \alpha)^+)^{1/2}
\end{aligned}$$

where the last step follows from 2.11(iii). Consequently,

$$\sup_{\beta > 0} \mathcal{E}^{(\beta)}((u - \alpha)^+, (u - \alpha)^+) < \infty$$

and thus by 2.13(i), $(u - \alpha)^+ \in D(\mathcal{E})$, hence $u \wedge \alpha = u - (u - \alpha)^+ \in D(\mathcal{E})$. Now (4.2), (4.3) and 2.13 (iii) imply $\mathcal{E}(u \wedge \alpha, u - u \wedge \alpha) \geq 0$. $\qquad \square$

Definition 4.5. A coercive closed form $(\mathcal{E}, D(\mathcal{E}))$ on $L^2(E; m)$ is called a *Dirichlet form* if for all $u \in D(\mathcal{E})$, one has that

$$(4.4) \qquad \begin{aligned} u^+ \wedge 1 \in D(\mathcal{E}) \quad &\text{and} \quad \mathcal{E}(u + u^+ \wedge 1, u - u^+ \wedge 1) \geq 0 \\ &\text{and} \quad \mathcal{E}(u - u^+ \wedge 1, u + u^+ \wedge 1) \geq 0 \;. \end{aligned}$$

If $(\mathcal{E}, D(\mathcal{E}))$ is in addition symmetric in which case (4.4) is equivalent with

$$(4.5) \qquad \mathcal{E}(u^+ \wedge 1, u^+ \wedge 1) \leq \mathcal{E}(u, u) \;,$$

it is called a *symmetric Dirichlet form.*

Remark. (i) If $(\mathcal{E}, D(\mathcal{E}))$ is a Dirichlet form, the pair $(\mathcal{E}, D(\mathcal{E}))$ is also sometimes called a *Dirichlet space.*
(ii) A coercive closed form satisfying only one of the two inequalities in (4.4) is sometimes called a $\frac{1}{2}$-*Dirichlet form* as e.g. in the final diagram of this chapter (cf. p. 39 below).

Exercise 4.6. Prove that if a coercive closed form $(\mathcal{E}, D(\mathcal{E}))$ on $L^2(E; m)$ is a Dirichlet form then $(\tilde{\mathcal{E}}, D(\mathcal{E}))$ is a symmetric Dirichlet form.

$u^+ \wedge 1$ is called the *unit contraction* of u. The following shows that there is a "smoothed" version of (4.4).

Proposition 4.7. *Let $(\mathcal{E}, D(\mathcal{E}))$ be a coercive closed form on $L^2(E; m)$.*

(i) Let $u \in D(\mathcal{E})$ and assume that

$$(4.6) \qquad \begin{aligned} &\text{for every } \epsilon > 0 \text{ there exists } \varphi_\epsilon : \mathbb{R} \to [-\epsilon, 1 + \epsilon] \text{ such that} \\ &\varphi_\epsilon(t) = t \text{ for all } t \in [0, 1], \; 0 \leq \varphi_\epsilon(t_2) - \varphi_\epsilon(t_1) \leq t_2 - t_1 \text{ if } t_1 \leq \\ &t_2, \; \varphi_\epsilon \circ u \in D(\mathcal{E}), \text{ and } \liminf_{\epsilon \to 0} \mathcal{E}(u \pm \varphi_\epsilon \circ u, u \mp \varphi_\epsilon \circ u) \geq 0. \end{aligned}$$

Then (4.4) holds.

(ii) $(\mathcal{E}, D(\mathcal{E}))$ is a Dirichlet form if and only if (4.6) holds for all $u \in D(\mathcal{E})$.

Proof. (i): Observe that by adding the two inequalities in (4.6) we obtain that

$$\limsup_{\epsilon \to 0} \mathcal{E}(\varphi_\epsilon \circ u, \varphi_\epsilon \circ u) \leq \mathcal{E}(u, u) \;.$$

Since clearly, $\varphi_\epsilon \circ u \xrightarrow[\epsilon \to 0]{} u^+ \wedge 1$ in $L^2(E; m)$, this implies that $\varphi_{\epsilon_n} \circ u \xrightarrow[n \to \infty]{} u^+ \wedge 1$ weakly in $(D(\mathcal{E}), \tilde{\mathcal{E}}_1)$ for some subsequence $\epsilon_n \downarrow 0$ and $\mathcal{E}(u^+ \wedge 1, u^+ \wedge 1) \leq \liminf_{n \to \infty} \mathcal{E}(\varphi_{\epsilon_n} \circ u, \varphi_{\epsilon_n} \circ u)$ (by 2.12). Hence by (4.6)

$$\begin{aligned} \mathcal{E}(u \pm u^+ \wedge 1, u \mp u^+ \wedge 1) \;\geq\; & \mathcal{E}(u, u) \mp \lim_{n \to \infty} \mathcal{E}(u, \varphi_{\epsilon_n} \circ u) \\ & \pm \lim_{n \to \infty} \mathcal{E}(\varphi_{\epsilon_n} \circ u, u) - \liminf_{n \to \infty} \mathcal{E}(\varphi_{\epsilon_n} \circ u, \varphi_{\epsilon_n} \circ u) \\ =\; & \limsup_{n \to \infty} \mathcal{E}(u \pm \varphi_{\epsilon_n} \circ u, u \mp \varphi_{\epsilon_n} \circ u) \\ \geq\; & 0 \;. \end{aligned}$$

(ii): It is obvious that the condition is necessary since we can take $\varphi_\epsilon(t) := (t \vee 0) \wedge 1$ for all $\epsilon > 0$. The sufficiency follows by (i). $\qquad \square$

Exercise 4.8. Prove that 4.7 remains true if the last part in (4.6) is replaced by

$$(4.7) \qquad \begin{aligned} \varphi_\epsilon \circ u \in D(\mathcal{E}) \quad &\text{and} \quad \liminf_{\epsilon \to 0} \mathcal{E}(\varphi_\epsilon \circ u, u - \varphi_\epsilon \circ u) \geq 0 \\ &\text{and} \quad \liminf_{\epsilon \to 0} \mathcal{E}(u - \varphi_\epsilon \circ u, \varphi_\epsilon \circ u) \geq 0 \ . \end{aligned}$$

Lemma 4.9. *Let $(\mathcal{E}, D(\mathcal{E}))$ be a coercive closed form on $L^2(E; m)$ and $B : L^2(E; m) \to L^2(E; m)$ a continuous map such that for some $D \subset D(\mathcal{E})$, D dense w.r.t. $\tilde{\mathcal{E}}_1^{1/2}$, and all $u \in D$*

$$(4.8) \qquad B(u) \in D(\mathcal{E}) \ and \ \mathcal{E}(u \pm B(u), u \mp B(u)) \geq 0$$

or

$$(4.9) \quad B(u) \in D(\mathcal{E}) \ and \ \mathcal{E}(B(u), u - B(u)) \geq 0 \ , \quad \mathcal{E}(u - B(u), B(u)) \geq 0$$

then (4.8) resp. (4.9) hold for all $u \in D(\mathcal{E})$.

Proof. Let $u \in D(\mathcal{E})$ and $u_n \in D$, $n \in \mathbb{N}$, such that $u_n \xrightarrow[n \to \infty]{} u$ in $D(\mathcal{E})$. Adding the two inequalities in (4.8) (resp. by (4.9) and the weak sector condition (2.3)) we obtain that

$$\sup_{n \in \mathbb{N}} \mathcal{E}(B(u_n), B(u_n)) < \infty \ .$$

Since $B(u_n) \xrightarrow[n \to \infty]{} B(u)$ in $L^2(E; m)$, this implies that $B(u_n) \xrightarrow[n \to \infty]{} B(u)$ weakly in $D(\mathcal{E})$ and $\mathcal{E}(B(u), B(u)) \leq \liminf_{n \to \infty} \mathcal{E}(B(u_n), B(u_n))$ (cf. 2.12). Now the assertion follows similarly as in the proof of 4.7 resp. 4.8. $\square$

As an immediate consequence of 4.9 and 4.7 resp. 4.8 we obtain

Proposition 4.10. *A coercive closed form on $L^2(E; m)$ is a Dirichlet form if and only if (4.4), (4.6) or (4.7) hold for all u in a dense subset of $D(\mathcal{E})$.*

Proposition 4.11. *Let $(\mathcal{E}, D(\mathcal{E}))$ be a Dirichlet form on $L^2(E; m)$. Let $u_1, \ldots, u_n \in D(\mathcal{E})$ and $u \in L^2(E; m)$ such that (for some m-versions) $|u(x)| \leq \sum_{k=1}^{n} |u_k(x)|$ and $|u(x) - u(y)| \leq \sum_{k=1}^{n} |u_k(x) - u_k(y)|$ for all $x, y \in E$, then $u \in D(\mathcal{E})$ and $\mathcal{E}(u, u)^{1/2} \leq \sum_{k=1}^{n} \mathcal{E}(u_k, u_k)^{1/2}$. In particular, $D(\mathcal{E})$ is stable under taking $|\cdot|$, $\wedge$, $\vee$.*

Proof. Observe that we can replace $\mathcal{E}$ in the assertion by $\tilde{\mathcal{E}}$. Hence we only have to prove the symmetric case which will be done below (cf. 4.13). $\square$

We shall actually prove below that a symmetric closed form on $L^2(E; m)$ is a Dirichlet form if and only if it has the property in 4.11 (see 4.13 below). We first recall that a function $T : \mathbb{R} \to \mathbb{R}$ with $T(0) = 0$ and $|T(s) - T(t)| \leq |s - t|$ for all $s, t \in \mathbb{R}$, is called a *normal contraction*. Note that $T(t) := (t \vee 0) \wedge 1$ (see above) is a special case.

Theorem 4.12. *Let $(\mathcal{E}, D(\mathcal{E}))$ be a symmetric closed form on $L^2(E; m)$. Then the following are equivalent:*

(i) $(\mathcal{E}, D(\mathcal{E}))$ is a Dirichlet form.

(ii) For every normal contraction $T : \mathbb{R} \to \mathbb{R}$, one has $T(u) \in D(\mathcal{E})$ and

$$\mathcal{E}(T(u), T(u)) \leq \mathcal{E}(u, u) \ \text{for all } u \in D(\mathcal{E}) \ .$$

(iii) Given $n \in \mathbb{N}$ and $T : \mathbb{R}^n \to \mathbb{R}$ such that

$$|T(x)| \leq \sum_{k=1}^{n} |x_k| \ \text{and} \ |T(x) - T(y)| \leq \sum_{k=1}^{n} |x_k - y_k|$$

for all $x = (x_1, \ldots, x_n)$, $y = (y_1, \ldots, y_n) \in \mathbb{R}^n$, then for all $u_1, \ldots, u_n \in D(\mathcal{E})$, one has $T(u_1, \ldots, u_n) \in D(\mathcal{E})$ and

$$\mathcal{E}(T(u_1, \ldots, u_n), T(u_1, \ldots, u_n))^{1/2} \leq \sum_{k=1}^{n} \mathcal{E}(u_k, u_k)^{1/2} \ .$$

Proof. Clearly, $(iii) \Rightarrow (ii) \Rightarrow (i)$. Hence, it remains to show $(i) \Rightarrow (iii)$. So, assume $(\mathcal{E}, D(\mathcal{E}))$ is a Dirichlet form and denote its associated sub-Markovian strongly continuous contraction resolvent by $(G_\alpha)_{\alpha > 0}$. By 2.13 (i), (iii) it suffices to prove that for all $f_1, \ldots, f_n \in L^2(E; m)$, $\alpha > 0$,

$$(4.10) \quad ((1 - \alpha G_\alpha)T(f_1, \ldots, f_n), T(f_1, \ldots, f_n))^{1/2} \leq \sum_{k=1}^{n}((1 - \alpha G_\alpha)f_k, f_k)^{1/2} \ .$$

Of course, it is enough to prove (4.10) for f_k bounded, and we may even assume that

$$(4.11) \qquad\qquad f_k := \sum_{i=1}^{N} \alpha_{ki} 1_{A_i}$$

for $N \in \mathbb{N}$, $\alpha_{ki} \in \mathbb{R}$, $A_i \in \mathcal{B}$ pairwise disjoint with $m(A_i) < \infty$. If for $1 \leq i, j \leq N$

$$b_{ij} := ((1 - \alpha G_\alpha)1_{A_i}, 1_{A_j})$$

then (4.10) is equivalent with

$$(4.12) \ (\sum_{i,j=1}^{N} T(\alpha_{1i}, \ldots, \alpha_{ni})T(\alpha_{1j}, \ldots, \alpha_{nj})b_{ij})^{1/2} \leq \sum_{k=1}^{n}(\sum_{i,j=1}^{N} \alpha_{ki}\alpha_{kj}b_{ij})^{1/2}.$$

Let $\lambda_i := (1_{A_i}, 1_{A_i})$ and $a_{ij} := (\alpha G_\alpha 1_{A_i}, 1_{A_j})$, then $b_{ij} = \lambda_i \delta_{ij} - a_{ij}$, $1 \leq i, j \leq N$, and hence for all $z_1, \ldots, z_N \in \mathbb{R}$ since $a_{ij} = a_{ji}$ by the symmetry of G_α

$$\sum_{i,j=1}^{N} z_i z_j b_{ij} = \sum_{i<j} a_{ij}(z_i - z_j)^2 + \sum_{j=1}^{N} m_j z_j^2$$

with $m_j := \lambda_j - \sum_{i=1}^{N} a_{ij}$. Rewriting both sides of (4.12) correspondingly we see that (4.12) holds by Minkowski's inequality and the assumption on T if $a_{ij} \geq 0$ and $m_j \geq 0$. But $a_{ij} \geq 0$, since αG_α is sub-Markovian, and if $A := \bigcup_{i=1}^{N} A_i$, then

$$\sum_{i=1}^{n} a_{ij} = \alpha \int 1_A G_\alpha(1_{A_j}) dm = \alpha \int G_\alpha(1_A) 1_{A_j} dm$$

$$\leq \int 1_{A_j} dm = \lambda_j \ .$$

$\square$

Corollary 4.13. *Let $(\mathcal{E}, D(\mathcal{E}))$ be a symmetric closed form on $L^2(E; m)$. Then $(\mathcal{E}, D(\mathcal{E}))$ is a Dirichlet form if and only if for all $u_1, \ldots, u_n \in D(\mathcal{E})$ and $u \in L^2(E; m)$ such that (for some m-versions) $|u(x) - u(y)| \leq \sum_{k=1}^{n} |u_k(x) - u_k(y)|$ and $|u(x)| \leq \sum_{k=1}^{n} |u_k(x)|$ for all $x, y \in E$, we have that $u \in D(\mathcal{E})$ and $\mathcal{E}(u, u)^{1/2} \leq \sum_{k=1}^{n} \mathcal{E}(u_k, u_k)^{1/2}$.*

Proof. As in the proof of 4.11 it is enough to prove that for all $u_1, \ldots, u_n$, u as above but with $u_1, \ldots, u_n$ only in $L^2(E; m)$

$$(4.13) \qquad ((1 - \alpha G_\alpha)u, u)^{1/2} \leq \sum_{k=1}^{n} ((1 - \alpha G_\alpha)u_k, u_k)^{1/2} \ .$$

Again, it suffices to prove (4.13) for $u_1, \ldots, u_n$ bounded. If for $l \in \mathbb{N}$, $K_l := \{\sum_{k=1}^{n} |u_k|^2 > \frac{1}{l}\}$ then $1_{K_l} u \uparrow u$ and $1_{K_l} u_k \uparrow u_k$, $1 \leq k \leq n$, as $l \to \infty$. Furthermore, $m(K_l) < \infty$, $l \in \mathbb{N}$. Hence we can assume that

$$(4.14) \qquad m\left(\{u \neq 0\} \cup \bigcup_{k=1}^{n} \{u_k \neq 0\}\right) < \infty \ .$$

Define $T : (u_1, \ldots, u_n)(E) \to \mathbb{R}$ by $T(y_1, \ldots, y_n) = u(x)$ where $x \in E$ such that $(u_1(x), \ldots, u_n(x)) = (y_1, \ldots, y_n)$. Then by assumption T is well-defined and fulfills the assumptions on T in 4.12 (iii), but restricted to $C := (u_1, \ldots, u_n)(E)$. Hence, (4.13) becomes (4.10) and the same arguments as those following (4.11) in the proof of 4.12 imply the assertion if we can prove that we can approximate u_k, $1 \leq k \leq n$, in $L^2(E; m)$ by functions of the form

$$\sum_{i=1}^{N} \alpha_{ki} 1_{A_i}$$

for $N \to \infty$, with $(\alpha_{1i}, \ldots, \alpha_{ni}) \in C$. But this is an easy exercise using (4.14) (and a slight modification of the standard way of approximation by simple functions, cf. e.g. [B 78,11.6].) $\square$

Remark 4.14. (i) If $n = 1$ in 4.13, u is called a *normal contraction* of u_1.
(ii) Note that if $1 \in D(\mathcal{E})$ such that $\mathcal{E}(u, 1) = 0$ for all $u \in D(\mathcal{E})$, then 4.13 remains true for $u \in L^2(E; m)$ and $u_1, \ldots, u_n \in D(\mathcal{E})$ such that (for some m-versions) $|u(x) - u(y)| \leq \sum_{k=1}^{n} |u_k(x) - u_k(y)|$ for all $x, y \in E$.

Corollary 4.15. *Let $(\mathcal{E}, \mathcal{D}(\mathcal{E}))$ be a Dirichlet form on $L^2(E; m)$ and $u, v \in \mathcal{D}(\mathcal{E})$, u, v bounded. Then $u \cdot v \in \mathcal{D}(\mathcal{E})$ and*

$$\mathcal{E}(u \cdot v, u \cdot v)^{1/2} \leq \|u\|_\infty \mathcal{E}(v, v)^{1/2} + \|v\|_\infty \mathcal{E}(u, u)^{1/2}$$

(where $\| \ \|_\infty$ denotes the supremum norm).

Proof. Set $u_1 := v \cdot \|u\|_\infty$, $u_2 := u \cdot \|v\|_\infty$. Then $u_1, u_2 \in \mathcal{D}(\mathcal{E})$,

$$|(u \cdot v)(x)| \leq |u_1(x)| + |u_2(x)|$$

and

$$
\begin{aligned}
|(u \cdot v)(x) - (u \cdot v)(y)| &\leq\ |u(x)||v(x) - v(y)| + |v(y)||u(x) - u(y)| \\
&\leq\ |u_1(x) - u_1(y)| + |u_2(x) - u_2(y)|
\end{aligned}
$$

for all $x, y \in E$. Hence the assertion follows by 4.13. $\square$

Exercise 4.16. Let $(\mathcal{E}, (\mathcal{D}(\mathcal{E}))$ be a Dirichlet form on $L^2(E; m)$ and $u, v \in \mathcal{D}(\mathcal{E})$ with $u, v \geq 0$ and $u + v \geq c$ for some $c \in\ ]0, \infty[$. Show that $\frac{u}{u+v} \in \mathcal{D}(\mathcal{E})$ and that

$$\mathcal{E}\left(\frac{u}{u + v}, \frac{u}{u + v}\right)^{1/2} \leq c^{-1}\left(\mathcal{E}(u, u)^{1/2} + \mathcal{E}(v, v)^{1/2}\right).$$

We close this section with a result which will become very useful later.

Proposition 4.17. *Let $(\mathcal{E}, D(\mathcal{E}))$ be a coercive closed form on $L^2(E; m)$ and let $u \in D(\mathcal{E})$. Then:*

(i) $(u \wedge n) \vee (-n) \xrightarrow[n \to \infty]{} u$ in $D(\mathcal{E})$,

(ii) $(u \wedge \varepsilon) \vee (-\varepsilon) \xrightarrow[\varepsilon \to 0]{} 0$ in $D(\mathcal{E})$.

Proof. We may assume that $\mathcal{E} = \tilde{\mathcal{E}}$. Since $u_n := (u \wedge n) \vee (-n)$ is a normal contraction of u we have that

$$\mathcal{E}_1(u_n, u_n) \leq \mathcal{E}_1(u, u)$$

and hence

$$\mathcal{E}_1(u - u_n, u - u_n) \leq 2(\mathcal{E}_1(u, u) - \mathcal{E}_1(u, u_n))$$

for all $n \in \mathbb{N}$. Hence it suffices to prove that $\mathcal{E}_1(v, u_n) \xrightarrow[n \to \infty]{} \mathcal{E}_1(v, u)$ for all $v \in D(\mathcal{E})$, which obviously holds by 2.12. This proves (i), and (ii) can be proved similarly. $\square$

The following diagram summarizes the main achievements of this chapter and completes Diagram 2 on p. 27 for $\mathcal{H} := L^2(E; m)$. The new correspondences follow by 4.3 and 4.4.

Diagram 3

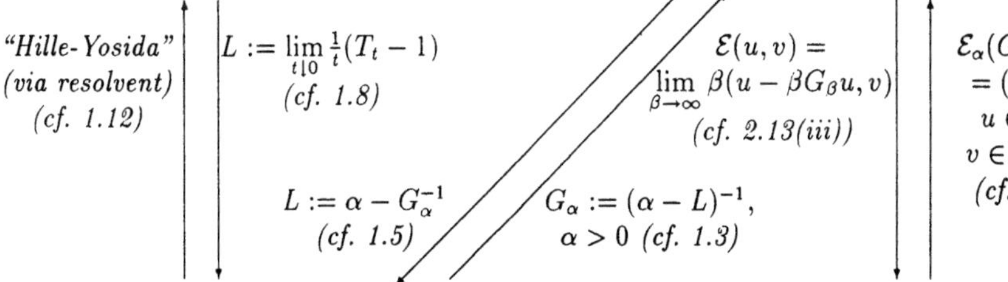

$$G_\alpha = \int_0^\infty e^{-\alpha t} T_t \, dt$$
(cf. 1.10)

$$T_t = \lim_{\alpha \to \infty} e^{t\alpha(\alpha G_\alpha - 1)}, \ t > 0$$
"Hille-Yosida" (cf. 1.12)

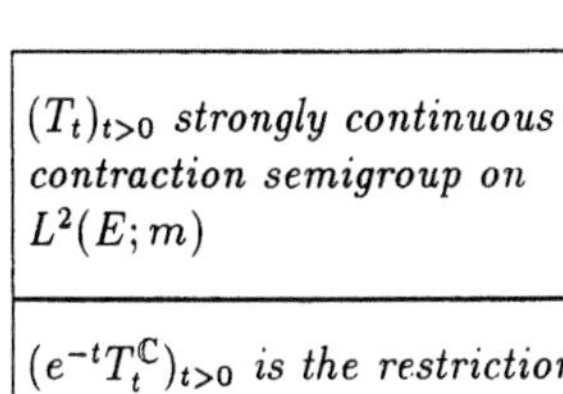

$(T_t)_{t>0}$ *strongly continuous contraction semigroup on* $L^2(E;m)$

$(e^{-t}T_t^{\mathbb{C}})_{t>0}$ *is the restriction of a holomorphic contraction semigroup on* $L^2(E \to \mathbb{C}; m)$

$(T_t)_{t>0}$ *sub-Markovian*

$(G_\alpha)_{\alpha>0}$ *strongly continuous contraction resolvent on* $L^2(E;m)$

$|(G_1 u, v)| \leq$ const $\cdot (G_1 u, u)^{1/2}(G_1 v, v)^{1/2}$

$(G_\alpha)_{\alpha>0}$ *sub-Markovian*

"Hille-Yosida" (via resolvent) (cf. 1.12)

$L := \lim_{t \downarrow 0} \frac{1}{t}(T_t - 1)$
(cf. 1.8)

$L := \alpha - G_\alpha^{-1}$
(cf. 1.5)

$\mathcal{E}(u,v) = \lim_{\beta \to \infty} \beta(u - \beta G_\beta u, v)$
(cf. 2.13(iii))

$G_\alpha := (\alpha - L)^{-1}, \ \alpha > 0$ *(cf. 1.3)*

$\mathcal{E}_\alpha(G_\alpha u, v) = (u, v), \ u \in \mathcal{H}, \ v \in D(\mathcal{E})$
(cf. 2.8)

$(L(D(L)))$ *densely defined, (closed) linear operator on* $L^2(E;m)$ *s.t.*
  (i) $]0, \infty[\subset \rho(L)$
  (ii) $\|\alpha(\alpha - L)^{-1}\| \leq 1$
  (iii) $|((1-L)u, v)| \leq$ const $\cdot ((1-L)u, u)^{1/2} \cdot ((1-L)v, v)^{1/2}$

L *Dirichlet, i.e.,*
$(Lu, (u-1)^+) \leq 0 \ \forall u \in D(L)$

$D(L) := \{u \in D(\mathcal{E}) \mid \exists Lu \in \mathcal{H} \ s.t. \ \mathcal{E}(u,v) = (-Lu, v) \ \forall v \in D(\mathcal{E})\}$ *(cf. 2.16)*

$\mathcal{E}(u,v) := (-Lu, v), \ u, v \in D(L)$ *& completition (cf. 2.15)*

$(\mathcal{E}, D(\mathcal{E}))$ *coercive closed form on* $L^2(E;m)$ *(cf. 2.4)*

$(\mathcal{E}, D(\mathcal{E}))$ $\frac{1}{2}$*-Dirichlet, i.e.,*
$u \in D(\mathcal{E}) \Rightarrow u^+ \wedge 1 \in D(\mathcal{E})$ &
$\mathcal{E}(u + u^+ \wedge 1, u - u^+ \wedge 1) \geq 0$

5   Notes/References

Section 1: The material of this section is entirely standard and can be found in many textbooks. Our proof of the Hille-Yosida theorem is close to that in [ReS 75, Chap. X, Sect.8].

Section 2: Apart from the fact that we only use the weak sector condition our presentation of this section up to Theorem 2.13 originates from [O 88, Sect.1.1] and [DM 88, Chap. XIII, Sect.4], but locally we made major modifications. A further standard reference is, of course, [K 66, Chap. VI, §§1,2]. However, we are not aware of a reference that includes all proofs to obtain the final diagram on p. 27. Though Corollary 2.21 appears to be new in this (real) form, it is quite a straightforward consequence of the beautiful work by J.T. Cannon (see [C 75, 1.1-1.4] and also [Go 85, Theorem 5.9]).

Section 3: This section is again quite standard and we refer e.g. to [K 66, Chapt. VI, §1]. The proof of Proposition 3.7 is taken from [ABR 89, see Theorem 1.1].

Section 4: The contents of this section in its present form can be considered as the final outcome of a long development to which many people contributed. Below we only mention those who influenced us the most. The proof of Proposition 4.3 is taken from [BH 86, Theorem 1.1]. Though it is presented there only in the symmetric case (that is, L is self-adjoint) it directly carries over to the general case. Theorem 4.4 including its proof can be found in [O 88, see Theorem 1.1.4]. In the case where $\mathcal{E}$ is symmetric, Proposition 4.7 (apart from the "$\lim\inf_{\epsilon \to 0}$") is contained in [F 80, Theorem 1.4.1]. The proof there , however, relies heavily on the symmetry. Our proof is entirely different and more elementary. It uses an idea we learnt from H. Brasche (cf. [ABR 89, Theorem 3.2]). The proof of Theorem 4.12 is a modification of one part of the proof of Theorem XIII.51 in [ReS 78]. A further extension of this method then leads to Corollary 4.13. In fact, for every normal contraction T the map $u \mapsto T(u)$ is a continuous map from $D(\mathcal{E})$ to $D(\mathcal{E})$. This result is due to A. Ancona (cf. [An 72, 76]). Finally, 4.15 and 4.17 are contained in [F 80, Theorem 1.4.2].

Chapter II

Examples

In this chapter we present examples of Dirichlet forms. We try to keep close to situations very likely to be encountered in applications, i.e., we consider in the respective sections, cases where one of the following is given: 1. some linear operator; 2. some bilinear form on finite or 3. infinite dimensional state space; 4. a semigroup of kernels; 5. a resolvent of kernels. In each case we show under what conditions and how to obtain the corresponding Dirichlet form. In this chapter for E we shall take various topological spaces. If we do not specify the σ-algebra $\mathcal{B}$ explicitly, it is understood to be the corresponding Borel-σ-algebra $\mathcal{B}(E)$. We denote by $\mathcal{B}_b$, $\mathcal{B}^+$ the bounded respectively positive $\mathcal{B}$-measurable functions on E and set $\mathcal{B}_b^+ := \mathcal{B}_b \cap \mathcal{B}^+$.

1 Starting point: operator

a) (a_{ij})-case

Let $E := U \subset \mathbb{R}^d$, $d \geq 1$, U open, and $m = dx :=$ *Lebesgue measure* on U. Let $a_{ij} : U \to \mathbb{R}$, $1 \leq i,j \leq d$, such that

(1.1) $\qquad a_{ij} = a_{ji}$ for all $1 \leq i,j \leq d$ and $\sum_{i,j=1}^{d} a_{ij}(x)\xi_i\xi_j \geq 0$ for all
$$\xi_1,\ldots,\xi_d \in \mathbb{R},\ dx\text{-a.e. } x \in U\ .$$

(1.2) $\qquad a_{ij} \in L^2_{loc}(U,dx),\ \dfrac{\partial}{\partial x_i}a_{ij} \in L^2_{loc}(U;dx),\ 1 \leq i,j \leq d,$

where the derivatives are taken in the sense of Schwartz distributions. Define the linear operator S on $L^2(U;dx)$ by

(1.3) $\qquad Su = \sum_{i,j=1}^{d} \dfrac{\partial}{\partial x_i}\left(a_{ij}\dfrac{\partial}{\partial x_j}\right)u\ ,\quad u \in D(S) := C_0^\infty(U)\ .$

Here $C_0^\infty(U)$ denotes the set of all infinitely differentiable functions on U with compact support. Note that (1.2) is necessary to have that $Su \in L^2(U; dx)$ for every $u \in C_0^\infty(U)$. Define the positive definite bilinear form

$$\mathcal{E}(u, v) := (-Su, v) = \sum_{i,j=1}^{d} \int \frac{\partial u}{\partial x_i} \frac{\partial v}{\partial x_j} a_{ij} dx \ , \quad u, v \in C_0^\infty(U) \ .$$

Then by I.3.3, $(\mathcal{E}, C_0^\infty(U))$ is closable on $L^2(U; dx)$. Since $C_0^\infty(U)$ is dense in $L^2(U; dx)$, its closure $(\mathcal{E}, D(\mathcal{E}))$ is a symmetric closed form on $L^2(U; dx)$, which is in fact a Dirichlet form since it is a special case of Subsection 2c) below.

b) Classical case with "minimal" domain

Consider a) with $a_{ij} := \frac{1}{2}\delta_{ij}$, i.e., $S = \frac{1}{2}\Delta$ with domain $C_0^\infty(U)$. We denote the corresponding symmetric closed form by $\mathbb{D}$, and its domain by $H_0^{1,2}(U)$ (since the completion of $C_0^\infty(U)$ w.r.t. $\mathbb{D}_1^{1/2}$ is by definition the $(1, 2)$-*Sobolev space on U with Dirichlet boundary conditions*).

Exercise 1.1. Recall that Δ is defined on all of $L^2(U; dx)$ in the sense of Schwartz distributions. Prove that $L := \frac{1}{2}\Delta$ with domain $D(L) := \{u \in H_0^{1,2}(U) | \Delta u \in L^2(U; dx)\}$ is the generator corresponding to $(\mathbb{D}, H_0^{1,2}(U))$ on $L^2(U; dx)$.

c) Classical case with "maximal" domain

Define

$$H^{1,2}(U) := \left\{ u \in L^2(U; dx) \, | \, \frac{\partial u}{\partial x_i} \in L^2(U; dx) \, , \, 1 \leq i \leq d \right\}$$

with derivatives in the Schwartz distribution sense (i.e., $H^{1,2}(U)$ is the $(1,2)$-*Sobolev space on U with Neumann-boundary conditions*). Define

$$\mathbb{D}(u, v) := \frac{1}{2} \sum_{i=1}^{d} \int \frac{\partial u}{\partial x_i} \frac{\partial v}{\partial x_i} \, dx \ ; \ u, v \in H^{1,2}(U) \ .$$

Exercise 1.2. Prove that $(\mathbb{D}, H^{1,2}(U))$ is a symmetric closed form on $L^2(U; dx)$ which extends $(\mathbb{D}, H_0^{1,2}(U))$.

Also $(\mathbb{D}, H^{1,2}(U))$ is a Dirichlet form. For the proof we refer to [F 80, Example 1.2.3]. The notions "minimal" and "maximal" in the sub-headings b) and c) can be justified. We refer to [F 80, §2.3], [AKR 90] and [T 91]. Note that in general $H^{1,2}(U) \neq H_0^{1,2}(U)$, e.g. if U is a ball then $1 \in H^{1,2}(U)$, but $1 \notin H_0^{1,2}(U)$. This is different if $U = \mathbb{R}^d$.

Exercise 1.3. Prove that $H_0^{1,2}(\mathbb{R}^d) = H^{1,2}(\mathbb{R}^d)$.

d) Powers of the Laplacian

Let $E = \mathbb{R}^d$, $m = dx$ and let " $\hat{\ }$ " resp. " $\check{\ }$ " denote *Fourier transform*, i.e.,

$$\hat{f}(x) = (2\pi)^{-d/2} \int \exp[i < x, y >_{\mathbb{R}^d}] f(y) dy \,,$$

resp. its inverse. Define for $\alpha > 0$

$$(-\Delta)^\alpha u := \left(|x|^{2\alpha} \hat{u}\right)^{\check{\ }} \ (\in L^2(\mathbb{R}^d; dx)) \,; \ u \in C_0^\infty(\mathbb{R}^d) \,.$$

Then $(-\Delta)^\alpha$ is a symmetric linear operator on $L^2(\mathbb{R}^d; dx)$ with dense domain $C_0^\infty(\mathbb{R}^d)$. Hence the form

$$\mathbb{D}^{(\alpha)}(u, v) := \frac{1}{2} \int \hat{u} \, \overline{\hat{v}} \, |x|^{2\alpha} \, dx \,, \ u, v \in C_0^\infty(\mathbb{R}^d) \,,$$

is closable by I.3.3, where "−" means complex conjugation. Its closure $(\mathbb{D}^{(\alpha)}, H^{\alpha,2}(\mathbb{R}^d))$ is hence a symmetric closed form on $L^2(\mathbb{R}^d; dx)$. By Subsection 4c) below if $0 < \alpha \leq 1$, it is a Dirichlet form. But this also follows by Subsection 2c) below from the following fact whose proof can be found in [Wl 82,p.97]. If $0 < \alpha < 1$, then $u \in H^{\alpha,2}(\mathbb{R}^d)$ if and only if

$$\iint \frac{|u(x) - u(y)|^2}{|x - y|^{2\alpha + d}} \, dxdy < \infty$$

and for $u, v \in H^{\alpha,2}(\mathbb{R}^d)$

$$\mathcal{E}(u, v) = c_{\alpha,d} \iint \frac{(u(x) - u(y))(v(x) - v(y))}{|x - y|^{2\alpha + d}} \, dxdy$$

for some constant $c_{\alpha,d} > 0$ (independent of u, v).
For a more detailed study of Dirichlet forms coming from more general pseudo-differential operators we refer to[HoJ 92], [J 92] and [J 91].

2 Starting point: bilinear form – finite dimensional case

a) Diagonal case

Let $E := U \subset \mathbb{R}^d$, U open, $m := \sigma \cdot dx$ for some $\sigma \in L^1_{loc}(U; dx)$, $\sigma \geq 0 \ dx - a.e.$ such that (cf. Remark 2.6 below)

$$(2.1) \qquad\qquad \int_V \sigma \, dx > 0 \ \text{ for all } \ V \subset U \,, V \text{ open} \,.$$

Let $\underline{\rho} := (\rho_1, \ldots, \rho_d)$ with $\rho_i \in L^1_{loc}(U; dx)$, $\rho_i \geq 0 \ dx - a.e.$ and define for $u, v \in C_0^\infty(U)$

$$(2.2) \qquad \mathcal{E}_{\underline{\rho}}(u,v) := \sum_{i=1}^{d} \int \frac{\partial u}{\partial x_i} \frac{\partial v}{\partial x_i} \rho_i \, dx \ .$$

Then $\left(\mathcal{E}_{\underline{\rho}}, C_0^\infty(U)\right)$ is a densely defined symmetric positive definite bilinear form on $L^2(U; \sigma \cdot dx)$. We want to give conditions on ρ_i, σ so that $(\mathcal{E}_{\underline{\rho}}, C_0^\infty(U))$ is closable on $L^2(U; \sigma \cdot dx)$.
Define for $\rho \in \mathcal{B}^+(U)$

$$(2.3) \quad R(\rho) := \left\{ x \in U \mid \int_{\{y \in U \,\mid\, |x-y| \le \varepsilon\}} \rho^{-1}(y) dy < \infty \ \text{ for some } \ \varepsilon > 0 \right\} \ .$$

Here we use the convention that $\frac{a}{0} := (\mathrm{sign}\,a) \cdot \infty$. Note that $R(\rho)$ is open and that $\rho > 0 \ dx - a.e.$ on $R(\rho)$. Obviously, $R(\rho)$ is the largest open set $V \subset U$ such that $\rho^{-1} \in L^1_{loc}(V; dx)$. Consider the following condition on ρ

$$(2.4) \qquad \rho = 0 \ dx - a.e. \text{ on } \ U \setminus R(\rho) \ .$$

Remark 2.1. Let $\rho : U \to [0, \infty[$ be Borel-measurable; then (2.4) is equivalent with

$$(2.5) \qquad \begin{array}{l} \text{for } dx\text{-a.e. } x \in \{\rho > 0\} \\ \int_{\{y \in U \,\mid\, |x-y| \le \varepsilon\}} \rho^{-1}(y) \, dy < \infty \text{ for some } \varepsilon > 0 \ . \end{array}$$

In particular, if ρ is lower semicontinuous or more generally if

$$(2.6) \qquad \begin{array}{l} \text{for } dx\text{-a.e. } x \in \{\rho > 0\} \\ \mathrm{ess\,inf}\{\rho(y) \mid |y - x| \le \varepsilon\} > 0 \text{ for some } \varepsilon > 0 \ , \end{array}$$

then ρ satisfies (2.5).

Lemma 2.2. *Let $\rho \in \mathcal{B}^+(U)$ satisfying (2.4). Then*

$$L^2(U; \rho \cdot dx) \subset L^1_{loc}(R(\rho); dx)$$

continuously.

Proof. Let $u \in L^2(U; \rho \cdot dx)$ and $K \subset R(\rho)$ compact. Then by the Cauchy-Schwarz inequality

$$\int_K |u| dx = \int_K |u| \rho \ \rho^{-1} dx \le \left(\int u^2 \rho \ dx \right)^{1/2} \left(\int_K \rho^{-1} dx \right)^{1/2} \ .$$

But

$$\int_K \rho^{-1} \, dx < \infty$$

since $K \subset R(\rho)$. $\qquad\qquad\square$

Suppose that σ and $\rho_1, \ldots, \rho_d$ satisfy (2.4) and that $dx(R(\rho_i) \setminus R(\sigma)) = 0$. Then $(\mathcal{E}_{\underline{\rho}}, C_0^\infty(U))$ is closable on $L^2(U; \sigma \cdot dx)$. To prove this, fix $1 \le i \le d$ and let $u_n \in$

$C_0^\infty(U)$, $n \in \mathbb{N}$, such that $(u_n)_{n \in \mathbb{N}}$ is $\mathcal{E}_\rho$-Cauchy and $u_n \xrightarrow[n \to \infty]{} 0$ in $L^2(U; \sigma \cdot dx)$, then $u_n \xrightarrow[n \to \infty]{} 0$ in $L^1_{loc}(R(\sigma); dx)$ by 2.2. Since $\frac{\partial u_n}{\partial x_i} \xrightarrow[n \to \infty]{} f_i$ in $L^2(U; \rho_i \cdot dx)$ for some $f_i \in L^2(U; \rho_i \cdot dx)$ we have that $\frac{\partial u_n}{\partial x_i} \xrightarrow[n \to \infty]{} f_i$ in $L^1_{loc}(R(\rho_i); dx)$ by 2.2. Hence, if $v \in C_0^\infty(R(\rho_i) \cap R(\sigma))$ then

$$\int f_i v\, dx = \lim_{n \to \infty} \int_{R(\rho_i)} \frac{\partial u_n}{\partial x_i} v\, dx = -\lim_{n \to \infty} \int_{R(\sigma)} u_n \frac{\partial v}{\partial x_i}\, dx = 0 \ .$$

Hence $f_i = 0$ dx-a.e. on $R(\rho_i) \cap R(\sigma)$, hence $f_i = 0$ $(\rho_i \cdot dx) - a.e.$ Consequently,

$$\lim_{n \to \infty} \mathcal{E}_\rho(u_n, u_n) = 0 \ .$$

Remark 2.3. The above arguments can also be used to prove the closability of

$$\mathcal{E}(u, v) = \int \langle \nabla u, \nabla v \rangle_x\, \rho\, dx \ ; \quad u, v \in C_0^\infty(U)$$

on $L^2(U; \sigma \cdot dx)$ if U is a Riemannian manifold with volume element dx and inner product $\langle\, ,\, \rangle_x$ on the tangent space at x (cf. [ABR 89, Theorem 4.2]).

In the above case the closure of $(\mathcal{E}_\rho, C_0^\infty(U))$ on $L^2(U; \sigma \cdot dx)$ is a symmetric closed form which is a Dirichlet form by Subsection 2c) below.

b) (a_{ij})-case

Let $E := U$ be as in Subsection a) and let $a_{ij} \in L^1_{loc}(U; dx)$, $a_{ij} = a_{ji}$, $1 \leq i, j \leq d$. Assume that there exist $\sigma, \rho_i \in L^1_{loc}(U; dx)$, $\rho_i, \sigma > 0$ dx-a.e. such that

$$(2.7) \qquad (\mathcal{E}_\rho, C_0^\infty(U)) \text{ is closable on } L^2(U; \sigma \cdot dx)$$

(cf. Subsection 2a)) and such that

$$(2.8) \qquad \begin{aligned} &\text{for } dx\text{-a.e. } x \in U \\ &\textstyle\sum_{i,j=1}^d a_{ij}(x)\xi_i\xi_j \geq \sum_{i=1}^d \rho_i(x)\xi_i^2 \text{ for all } \xi_1, \ldots, \xi_d \in \mathbb{R} \ . \end{aligned}$$

Then the form

$$\mathcal{E}(u, v) := \sum_{i,j=1}^d \int \frac{\partial u}{\partial x_i} \frac{\partial v}{\partial x_j} a_{ij}\, dx \ ; \quad u, v \in C_0^\infty(U) \ ,$$

is closable on $L^2(U; \sigma \cdot dx)$. Indeed, if $u_n \in C_0^\infty(U)$, $n \in \mathbb{N}$, such that $(u_n)_{n \in \mathbb{N}}$ is $\mathcal{E}$-Cauchy and $u_n \xrightarrow[n \to \infty]{} 0$ in $L^2(U; \sigma \cdot dx)$ then by (2.8), $(u_n)_{n \in \mathbb{N}}$ is $\mathcal{E}_\rho$-Cauchy, hence $\lim_{n \to \infty} \mathcal{E}_\rho(u_n, u_n) = 0$ by (2.7) and we can find a subsequence $(u_{n_k})_{k \in \mathbb{N}}$ such that $\frac{\partial}{\partial x_i} u_{n_k} \xrightarrow[k \to \infty]{} 0$ dx-a.e. for all $1 \leq i \leq d$. Hence by Fatou's Lemma and (2.8)

$$\begin{aligned} \mathcal{E}(u_n, u_n) &= \int \lim_{k \to \infty} \sum_{i,j=1}^d \left(\frac{\partial}{\partial x_i}(u_n - u_{n_k}) \frac{\partial}{\partial x_j}(u_n - u_{n_k}) a_{ij} \right) dx \\ &\leq \liminf_{k \to \infty} \mathcal{E}(u_n - u_{n_k}, u_n - u_{n_k}) \end{aligned}$$

which is arbitrarily small if n is large. The closure of $(\mathcal{E}, C_0^\infty(U))$ is a Dirichlet form by Subsection 2c) below.

Exercise 2.4. Prove that $(\mathcal{E}, C_0^\infty(U))$ is in particular closable on $L^2(U; \sigma \cdot dx)$ if for every $K \subset U$, K compact, there exists a constant $c_K > 0$ such that

$$(2.9) \qquad \begin{aligned} &\text{for } dx\text{-a.e. } x \in K \\ &\textstyle\sum_{i,j=1}^d a_{ij}(x)\xi_i\xi_j \geq c_K \|\xi\|_{\mathbb{R}^d}^2 \text{ for all } \xi = (\xi_1, \ldots, \xi_d) \in \mathbb{R}^d \end{aligned}$$

(*local (strict) ellipticity*).

c)   General regular symmetric case

Let $E := U$ be as in Subsection a) and m a positive Radon measure on U such that $\mathrm{supp}[m] = U$. Define for $u, v \in C_0^\infty(U)$

$$(2.10) \qquad \begin{aligned} \mathcal{E}(u, v) \ &:= \ \sum_{i,j=1}^d \int \frac{\partial u}{\partial x_i} \frac{\partial v}{\partial x_j}\, d\nu_{ij} \\ &+ \int_{U \times U \setminus \Delta} (u(x) - u(y))(v(x) - v(y))J(dx, dy) \\ &+ \int uv\, dk \ . \end{aligned}$$

Here k is a positive Radon measure on U and J is a symmetric positive Radon measure on $U \times U \setminus \Delta$, where $\Delta := \{(x, x) | x \in U\}$, such that for all $u \in C_0^\infty(U)$

$$(2.11) \qquad \int |u(x) - u(y)|^2 J(dxdy) < \infty \ .$$

For $1 \leq i, j \leq d$, ν_{ij} is a Radon measure on U such that

$$(2.12) \qquad \begin{aligned} &\text{for every } K \subset U, K \text{ compact}, \nu_{ij}(K) = \nu_{ji}(K) \text{ and} \\ &\textstyle\sum_{i,j=1}^d \xi_i\xi_j\nu_{ij}(K) \geq 0 \text{ for all } \xi_1, \ldots, \xi_d \in \mathbb{R}^d \ . \end{aligned}$$

Then $(\mathcal{E}, C_0^\infty(U))$ is a densely defined symmetric positive definite bilinear form on $L^2(U; m)$.

Exercise 2.5. Prove that (2.11) is equivalent with

$$(2.13) \qquad \begin{aligned} &\text{for all compact sets } K \text{ and open sets } U_1 \text{ with } K \subset U_1 \subset U, \\ &\textstyle\int_{K \times K \setminus \Delta} |x - y|^2 J(dxdy) < \infty \ , \ J(K, U \setminus U_1) < \infty \ . \end{aligned}$$

Remark 2.6. Note that the condition $\mathrm{supp}[m] = U$ (i.e., (2.1) if $m = \sigma \cdot dx$) is not necessary. It just ensures that $u \in C_0^\infty(U)$ is the unique representative of its class in $L^2(U; m)$, so that $\mathcal{E}$ is well-defined by (2.10). It would be enough to assume that definition (2.10) "respects m-classes" (cf. [AR 91] for examples).

Suppose that $(\mathcal{E}, C_0^\infty(U))$ is closable on $L^2(U; m)$ and let $(\mathcal{E}, D(\mathcal{E}))$ be its closure, then $(\mathcal{E}, D(\mathcal{E}))$ is a Dirichlet form. In order to prove this we need the following

Exercise 2.7. Show that for each $\varepsilon > 0$ there exists an infinitely differentiable function $\varphi_\varepsilon : \mathbb{R} \to [-\varepsilon, 1+\varepsilon]$ such that $\varphi_\varepsilon(t) = t$ for $t \in [0,1]$, $0 \leq \varphi_\varepsilon(t) - \varphi_\varepsilon(s) \leq t - s$ for all $t, s \in \mathbb{R}$, $t \geq s$, $\varphi_\varepsilon(t) = 1 + \varepsilon$ for $t \in [1 + 2\varepsilon, \infty[$ and $\varphi_\varepsilon(t) = -\varepsilon$ for $t \in]-\infty, -2\varepsilon]$.

Clearly, if $u \in C_0^\infty(U)$ and φ_ε is as in 2.7 then $\varphi_\varepsilon(u) \in C_0^\infty(U)$ and by the chain rule

$$
\begin{aligned}
\mathcal{E}(\varphi_\varepsilon(u), \varphi_\varepsilon(u)) &= \sum_{i,j=1}^d \int |\varphi_\varepsilon'(u)|^2 \frac{\partial u}{\partial x_i} \frac{\partial u}{\partial x_j}\, d\nu_{ij} \\
&\quad + \int_{U \times U \setminus \Delta} |\varphi_\varepsilon(u)(x) - \varphi_\varepsilon(u)(y)|^2\, J(dx\,dy) + \int \varphi_\varepsilon(u)^2\, dk \\
&\leq \mathcal{E}(u, u) \ .
\end{aligned}
$$

By I.4.10 it follows that $(\mathcal{E}, D(\mathcal{E}))$ is a symmetric Dirichlet form. Observe that all examples considered above (except those in Subsection 1c)) are of type (2.10), hence their closures are all Dirichlet forms. This is not a coincidence since conversely we have the following

Theorem 2.8. (Beurling-Deny formula) *Suppose $(\mathcal{E}, D(\mathcal{E}))$ is a symmetric Dirichlet form on $L^2(U; m)$ such that $C_0^\infty(U) \subset D(\mathcal{E})$. Then $(\mathcal{E}, C_0^\infty(U))$ can be expressed uniquely as in (2.10).*

For the proof we refer to[F 80, Section 2.2]. The three summands in (2.10) are called *diffusion part*, *jump part* and *killing part* respectively.

d)   Non-symmetric cases

Let $d \geq 3$ and $E := U$, $U \subset \mathbb{R}^d$, open, and $m = dx$ on U. Recall that by Sobolev's lemma (cf. e.g. [Da 89, Theorem 1.7.1]) if $\lambda := \frac{2(d-1)}{(d-2)d^{1/2}}$, then

$$
(2.14) \qquad \|u\|_q \leq \sqrt{2}\lambda \mathbb{D}(u, u)^{1/2} \text{ for all } u \in C_0^\infty(U)
$$

where $\frac{1}{q} + \frac{1}{d} = \frac{1}{2}$ and $\| \ \|_q$ denotes the usual norm in $L^q(U; dx)$. Let $a_{ij}, b_i, d_i, c \in L^1_{loc}(U; dx)$, $1 \leq i, j \leq d$, such that

$$
(2.15) \qquad
\begin{aligned}
&\text{for } dx\text{-a.e. } x \in U \\
&\textstyle\sum_{i,j=1}^d a_{ij}(x)\xi_i\xi_j \geq 0 \text{ for all } \xi = (\xi_1, \ldots, \xi_d) \in \mathbb{R}^d
\end{aligned}
$$

(i.e., *ellipticity*) and

$$c\,dx - \sum_{i=1}^{d} \frac{\partial d_i}{\partial x_i} \geq 0 \text{ and } c\,dx - \sum_{i=1}^{d} \frac{\partial b_i}{\partial x_i} \geq 0 \text{ (in the sense of Schwartz}$$

(2.16)

$$\text{distributions, i.e., } \int \left(cu + \sum_{i=1}^{d} d_i \frac{\partial u}{\partial x_i} \right) dx, \int \left(cu + \sum_{i=1}^{d} b_i \frac{\partial u}{\partial x_i} \right) dx \geq 0$$

$$\text{for all } u \in C_0^\infty(U) \text{ with } u \geq 0).$$

Define for $u, v \in C_0^\infty(U)$

(2.17)
$$\begin{aligned}
\mathcal{E}(u,v) &= \sum_{i,j=1}^{d} \int \frac{\partial u}{\partial x_i} \frac{\partial v}{\partial x_j} a_{ij}\,dx + \sum_{i=1}^{d} \int u \frac{\partial v}{\partial x_i} d_i\,dx \\
&\quad + \sum_{i=1}^{d} \int \frac{\partial u}{\partial x_i} v b_i\,dx + \int uv\,c\,dx .
\end{aligned}$$

Then $(\mathcal{E}, C_0^\infty(U))$ is a densely defined bilinear form on $L^2(U; dx)$ which is positive definite since for all $u \in C_0^\infty(U)$

$$\mathcal{E}(u,u) = \sum_{i,j=1}^{d} \int \frac{\partial u}{\partial x_i} \frac{\partial u}{\partial x_j} a_{ij}\,dx + \frac{1}{2} \int \left[\sum_{i=1}^{d} \frac{\partial u^2}{\partial x_i}(d_i + b_i) + u^2 2c \right] dx \geq 0$$

by (2.15), (2.16).

Remark 2.9. (i) Since $\sum_{i,j} a_{ij}\xi_i\xi_j = \frac{1}{2}\sum_{i,j}(a_{ij} + a_{ji})\xi_i\xi_j$, (2.15) is only a condition on the symmetric part of $A := (a_{ij})_{i,j}$.
(ii) Condition (2.16) is the usual condition to ensure that the "formal generator" and "formal cogenerator" satisfy the weak maximum principle. We refer to[GT 83, Sect.8.1] for details.
(iii) Of course, (2.16) just means that both Schwartz distributions $c\,dx - \sum_{i=1}^{d} \frac{\partial d_i}{\partial x_i}$ and $c\,dx - \sum_{i=1}^{d} \frac{\partial b_i}{\partial x_i}$ are positive Radon measures on U.

Now let us suppose that $(\mathcal{E}, C_0^\infty(U))$ is closable and satisfies the weak sector condition I.(2.3). Then its closure $(\mathcal{E}, D(\mathcal{E}))$ is a coercive closed form on $L^2(U; dx)$ in the sense of 2.4 by I.3.2 (iii). In this case $(\mathcal{E}, D(\mathcal{E}))$ is always a Dirichlet form. To prove this by I.4.10 it is enough to show that for all $\varepsilon > 0$

(2.18)
$$\varphi_\varepsilon \circ u \in D(\mathcal{E}) \ , \ \liminf_{\varepsilon \to 0} \mathcal{E}(\varphi_\varepsilon \circ u, u - \varphi_\varepsilon \circ u) \geq 0 \text{ and}$$
$$\liminf_{\varepsilon \to 0} \mathcal{E}(u - \varphi_\varepsilon \circ u, \varphi_\varepsilon \circ u) \geq 0 \text{ for all } u \in C_0^\infty(U)$$

where φ_ε is as in 2.7. So, let $u \in C_0^\infty(U)$ then clearly $\varphi_\varepsilon \circ u \in C_0^\infty(U) \subset D(\mathcal{E})$. We only show the first inequality, the second can be proved similarly. By the superadditivity of $\liminf$ we can consider the second order part and the remainder in (2.17) separately. For $\varepsilon > 0$

$$\sum_{i,j=1}^{d} \int \frac{\partial}{\partial x_i}(\varphi_\varepsilon \circ u)\frac{\partial}{\partial x_j}(u - \varphi_\varepsilon \circ u)a_{ij}dx$$

$$= \int (\varphi_\varepsilon'(u))(1 - \varphi_\varepsilon'(u)) \sum_{i,j=1}^{d} \frac{\partial u}{\partial x_i}\frac{\partial u}{\partial x_j}a_{ij}\,dx \geq 0$$

by (2.15) and since $0 \leq \varphi_\varepsilon' \leq 1$. Furthermore,

$$\int \sum_{i=1}^{d} \left[\varphi_\varepsilon(u)\frac{\partial(u - \varphi_\varepsilon(u))}{\partial x_i}d_i + \frac{\partial\varphi_\varepsilon(u)}{\partial x_i}(u - \varphi_\varepsilon(u))b_i\right]dx$$

$$+ \int \varphi_\varepsilon(u)(u - \varphi_\varepsilon(u))c\,dx$$

$$= \int \left[\sum_{i=1}^{d} \frac{\partial(\varphi_\varepsilon(u)(u - \varphi_\varepsilon(u)))}{\partial x_i}d_i + \varphi_\varepsilon(u)(u - \varphi_\varepsilon(u))c\right]dx$$

$$+ \int \varphi_\varepsilon'(u)(u - \varphi_\varepsilon(u)) \sum_{i=1}^{d} \frac{\partial u}{\partial x_i}(b_i - d_i)\,dx\ .$$

The first summand is positive by (2.16) since $\varphi_\varepsilon(u)(u - \varphi_\varepsilon(u)) \geq 0$; the second converges to zero as $\varepsilon \downarrow 0$, since by construction of φ_ε in 2.7

$$\varphi_\varepsilon'(u)(u - \varphi_\varepsilon(u)) \leq (1_{[-2\varepsilon,1+2\varepsilon]} \circ u)(u - \varphi_\varepsilon(u)) \xrightarrow[\varepsilon\searrow 0]{} 0\ .$$

Thus (2.18) is proved.

Now we shall describe conditions ensuring that $(\mathcal{E}, C_0^\infty(U))$ is closable and satisfies the weak sector condition I.(2.3).

Case 1. $(b_i = d_i = c = 0)$ We first consider

$$(2.19) \qquad \mathcal{E}_A(u,v) := \sum_{i,j=1}^{d} \int \frac{\partial u}{\partial x_i}\frac{\partial v}{\partial x_j}a_{ij}\,dx\ ,\quad u,v \in C_0^\infty(U)\ .$$

Define $\tilde{a}_{ij} := \frac{1}{2}(a_{ij} + a_{ji})$ and $\breve{a}_{ij} := \frac{1}{2}(a_{ij} - a_{ji})$, $1 \leq i,j \leq d$. Suppose that the following conditions hold

$$(2.20) \qquad \begin{array}{l}\text{there exists } \nu \in]0, \infty[\text{ such that} \\ \sum_{i,j=1}^{d} \tilde{a}_{ij}\xi_i\xi_j \geq \nu\|\xi\|_{\mathbb{R}^d}^2 \text{ for all } \xi = (\xi_1,\ldots,\xi_d) \in \mathbb{R}^d\end{array}$$

and

$$(2.21) \qquad |\breve{a}_{ij}| \leq M \in]0, \infty[\text{ for all } 1 \leq i,j \leq d\ .$$

Then $(\mathcal{E}_A, C_0^\infty(U))$ is closable and its closure $(\mathcal{E}_A, D(\mathcal{E}_A))$ is a coercive closed form (in the sense of 2.4). Indeed, $(\mathcal{E}_A, C_0^\infty(U))$ is closable, since

$$\tilde{\mathcal{E}}_A(u,v) = \sum_{i,j=1}^{d} \int \frac{\partial u}{\partial x_i}\frac{\partial v}{\partial x_j}\tilde{a}_{ij}\,dx\ ;\quad u,v \in C_0^\infty(U)\ ,$$

is closable by (2.20) and the results in Subsection 2b) (cf. e.g. 2.4). Furthermore, by Cauchy-Schwarz's inequality, (2.21) and (2.20)

$$
\begin{aligned}
|\check{\mathcal{E}}_A(u,v)| \;&\leq\; M \int \left[\sum_{i=1}^{d}\left(\frac{\partial u}{\partial x_i}\right)^2\right]^{1/2}\left[\sum_{i=1}^{d}\left(\frac{\partial v}{\partial x_i}\right)^2\right]^{1/2} dx \\
&\leq\; M \left(\sum_{i=1}^{d}\int(\frac{\partial u}{\partial x_i})^2\,dx\right)^{1/2}\left(\sum_{i=1}^{d}\int(\frac{\partial v}{\partial x_i})^2\,dx\right)^{1/2} \\
&\leq\; \frac{M}{\nu}\mathcal{E}_A(u,u)^{1/2}\mathcal{E}_A(v,v)^{1/2}
\end{aligned}
$$
(2.22)

for all $u,v \in C_0^\infty(U)$. By I.2.1 (ii),(iv) and I.3.2 (iii) it follows that $(\mathcal{E}_A, D(\mathcal{E}_A))$ is a coercive closed form.

Case 2. We first note that by (2.16) for all $u \in C_0^\infty(U)$

$$
\begin{aligned}
\mathcal{E}(u,u) \;&=\; \sum_{i,j=1}^{d}\int \frac{\partial u}{\partial x_i}\frac{\partial u}{\partial x_j}a_{ij}\,dx + \frac{1}{2}\int\left[\sum_{i=1}^{d}\frac{\partial u^2}{\partial x_i}(d_i + b_i) + u^2 2c\right]dx \\
&\geq\; \mathcal{E}_A(u,u)\,.
\end{aligned}
$$
(2.23)

We suppose that (2.20), (2.21) hold and assume, in addition, that

(2.24) $$b_i - d_i \in L^d(U;dx) \cup L^\infty(U;dx)\,,\; 1 \leq i \leq d\,.$$

Let $u,v \in C_0^\infty(U)$, then by Hölder's inequality and (2.20)

$$
\begin{aligned}
\sum_{i=1}^{d}\int &\left|u\frac{\partial v}{\partial x_i} - v\frac{\partial u}{\partial x_i}\right||d_i - b_i|\,dx \\
&\leq\; \left(\sum_{i=1}^{d}\|u(d_i - b_i)\|_2^2\right)^{1/2}\sqrt{2}\mathbb{D}(v,v)^{1/2} \\
&\quad +\left(\sum_{i=1}^{d}\|v(d_i - b_i)\|_2^2\right)^{1/2}\sqrt{2}\mathbb{D}(u,u)^{1/2} \\
&\leq\; \left(\sum_{i=1}^{d}\|u(d_i - b_i)\|_2^2\right)^{1/2}\frac{1}{\sqrt{\nu}}\mathcal{E}_A(v,v)^{1/2} \\
&\quad +\left(\sum_{i=1}^{d}\|v(d_i - b_i)\|_2^2\right)^{1/2}\frac{1}{\sqrt{\nu}}\mathcal{E}_A(u,u)^{1/2}\,.
\end{aligned}
$$
(2.25)

If $d_i - b_i \in L^\infty(U;dx)$, then for all $w \in C_0^\infty(U)$

$$\|w(d_i - b_i)\|_2^2 \leq \|d_i - b_i\|_\infty^2\|w\|_2^2 \leq \|d_i - b_i\|_\infty^2\mathcal{E}_{A,1}(w,w)$$

and if $d_i - b_i \in L^d(U;dx)$ then by Hölder's inequality, (2.14) and (2.20)

$$\|w(d_i - b_i)\|_2 \;\leq\; \|w\|_q\|d_i - b_i\|_d$$

$$\leq \quad \sqrt{2}\lambda \mathbb{D}(w,w)^{1/2}\|d_i - b_i\|_d$$

$$\leq \quad \frac{\lambda}{\sqrt{\nu}}\mathcal{E}_{A,1}(w,w)^{1/2}\|d_i - b_i\|_d \; .$$

In any case the left hand side of (2.25) is up to a constant dominated by

$$\mathcal{E}_{A,1}(u,u)^{1/2}\mathcal{E}_{A,1}(v,v)^{1/2} \; .$$

Combining this with (2.22), (2.23) we conclude that there exists $K' \in]0,\infty[$ such that

$$|\check{\mathcal{E}}(u,v)| \leq K' \, \mathcal{E}_1(u,u)^{1/2}\mathcal{E}_1(v,v)^{1/2} \; \text{ for all } u,v \in C_0^\infty(U) \; ,$$

hence by I.2.1(iv) there exists $K \in]0,\infty[$ such that

$$(2.26) \qquad |\mathcal{E}_1(u,v)| \leq K \, \mathcal{E}_1(u,u)^{1/2}\mathcal{E}_1(v,v)^{1/2} \; \text{ for all } u,v \in C_0^\infty(U) \; ,$$

i.e., $(\mathcal{E}, C_0^\infty(U))$ satisfies the weak sector condition I.(2.3). Now suppose, in addition, that for the positive Radon measure $\mu := 2c\,dx - \sum_{i=1}^d \frac{\partial(d_i+b_i)}{\partial x_i}$ (cf. 2.9 (iii)) we have:

$$\text{For any } u_n \in C_0^\infty(U), \; n \in \mathbb{N}, \text{ with } u_n \xrightarrow[n\to\infty]{} 0 \text{ and } \frac{\partial u_n}{\partial x_i} \xrightarrow[n\to\infty]{} 0,$$

$$(2.27) \qquad 1 \leq i \leq d, \text{ in } L^2(U;dx) \text{ there exists a subsequence } (u_{n_k})_{k\in\mathbb{N}}$$

$$\text{with } u_{n_k} \xrightarrow[k\to\infty]{} 0 \; \mu\text{-a.e.} \; .$$

Then $(\mathcal{E}, C_0^\infty(U))$ is closable. Indeed, if $u_n \in C_0^\infty(U)$, $n \in \mathbb{N}$, such that $u_n \to 0$ in $L^2(U;dx)$ as $n \to \infty$, and $(u_n)_{n\in\mathbb{N}}$ is $\mathcal{E}$-Cauchy, it follows by (2.23) and Case 1 that $\lim_{n\to\infty} \mathcal{E}_A(u_n,u_n) = 0$, hence by (2.20), $\frac{\partial u_n}{\partial x_i} \to 0$, $1 \leq i \leq d$, in $L^2(U;dx)$ as $n \to \infty$. If $(u_{n_k})_{k\in\mathbb{N}}$ is as in (2.27), Fatou's Lemma implies that for all $n \in \mathbb{N}$

$$\mathcal{E}(u_n,u_n) \leq \mathcal{E}_A(u_n,u_n) + \frac{1}{2}\liminf_{k\to\infty} \int (u_n - u_{n_k})^2 d\mu$$

(cf.(2.23)). Hence

$$\mathcal{E}(u_n,u_n) \leq \liminf_{k\to\infty} \mathcal{E}(u_n - u_{n_k}, u_n - u_{n_k})$$

which can be made arbitrarily small for large enough n. Clearly, (2.27) holds e.g. if μ is absolutely continuous w.r.t. dx (but see also Remark 2.13 (ii) below).

Lemma 2.10. *If $b_i + d_i \in L_{loc}^d(U;dx)$, $1 \leq i \leq d$, and $c \in L_{loc}^{d/2}(U;dx)$ then (2.27) holds.*

Proof. Let $u_n \in C_0^\infty(U)$, $n \in \mathbb{N}$, with $u_n \xrightarrow[n\to\infty]{} 0$ and $\frac{\partial u_n}{\partial x_i} \xrightarrow[n\to\infty]{} 0$, $1 \leq i \leq d$, in $L^2(U;dx)$. Replacing u_n by $u_n v$ for any $v \in C_0^\infty(U)$, $v \geq 0$ on U, we may assume that $\text{supp}\,u_n \subset K$ for some compact set $K \subset U$ and all $n \in \mathbb{N}$. By Hölder's inequality and (2.20) we obtain that

$$\frac{1}{2}\int u_n^2 d\mu \;=\; \sum_{i=1}^{d}\int u_n \frac{\partial u_n}{\partial x_i}(d_i + b_i)dx + \int u_n^2 c\, dx$$

$$\leq \left[\sum_{i=1}^{d}\|u_n 1_K (d_i + b_i)\|_2^2\right]^{1/2} \sqrt{2}\mathbb{D}(u_n, u_n)^{1/2} + \|u_n^2 1_K c\|_1 \;.$$

By Hölder's inequality and (2.14) we obtain that

$$\|u_n 1_K (d_i + b_i)\|_2 \leq \sqrt{2}\lambda \mathbb{D}(u_n, u_n)^{1/2}\|1_K (d_i + b_i)\|_d \;.$$

Similarly, by Hölder's inequality and (2.14)

$$\|u_n^2 1_K c\|_1 \leq \|u_n\|_q^2 \|1_K c\|_{d/2} \leq 2\lambda^2 \mathbb{D}(u_n, u_n)\|1_K c\|_{d/2} \;.$$

Hence, (2.27) holds. $\hfill\square$

As a consequence of 2.10 we obtain

Proposition 2.11. *Let $(\mathcal{E}, C_0^\infty(U))$ be as in (2.17). Assume that (2.16), (2.20) and (2.21) hold and that*

$$(2.28)\qquad \begin{aligned} &c \in L_{loc}^{d/2}(U; dx), \ b_i, d_i \in L_{loc}^{d}(U; dx) \ and \\ &d_i - b_i \in L^{d}(U; dx) \cup L^{\infty}(U; dx), \ 1 \leq i \leq d \;. \end{aligned}$$

Then $(\mathcal{E}, C_0^\infty(U))$ is closable and its closure is a Dirichlet form on $L^2(U; dx)$.

Exercise 2.12. Consider the situation of 2.11, but assume instead of (2.28) that even

$$b_i, d_i \in L^{d}(U; dx) \cup L^{\infty}(U; dx) \text{ and } c \in L^{d/2}(U; dx) \cup L^{\infty}(U; dx) \;.$$

Reprove the assertion of 2.11 using I.3.5.

Remark 2.13. (i) Condition (2.20) can be relaxed. Under suitable assumptions on b_i, d_i and c, one has Dirichlet forms of the above type in sub-elliptic cases (cf.[RS 91b]).
(ii) By III.2.3 and III.3.5 below (2.27) is fulfilled if μ does not charge $\mathbb{D}$-exceptional sets.

Exercise 2.14. Let $(\mathcal{E}, C_0^\infty(U))$ be as in (2.17) such that (2.16), (2.20) and (2.21) are satisfied. Set $m := \left(\sum_{i=1}^{d} b_i^2 + d_i^2 + c + 1\right) dx$, where we assume that $c \in L_{loc}^1(U; dx)$ and $b_i, d_i \in L_{loc}^2(U; dx)$ for all $1 \leq i \leq d$. Prove that $(\mathcal{E}, C_0^\infty(U))$ is closable on $L^2(U; m)$ and that its closure $(\mathcal{E}, D(\mathcal{E}))$ is a Dirichlet form on $L^2(U; m)$.

Except those in Subsection 1c), the examples we have considered so far are all *regular* in the sense of [F 80] (cf. Chap.IV, Subsection 4a) below), i.e., $C_0(U) \cap D(\mathcal{E})$ is dense both in $D(\mathcal{E})$ w.r.t. $\tilde{\mathcal{E}}_1^{1/2}$ and in $C_0(U)$ w.r.t. the uniform norm where $C_0(U)$ denotes the set of all continuous functions on U with compact support. In the next subsection we shall give an example of a Dirichlet form where $D(\mathcal{E}) \cap C(U) = \{0\}$.

e)   Non-regular case

Let $E = \mathbb{R}^d$ for some $d \geq 2$, and let $\{x_j \mid j \in \mathbb{N}\}$ be a dense subset of E, and $(\alpha_j)_{j \in \mathbb{N}}$ be an arbitrary sequence of real numbers. We consider a function V defined by

$$(2.29) \qquad V(x) = \sum_{j=1}^{\infty} c_j |x - x_j|^{\alpha_j} \,, \quad x \in E \,,$$

where $(c_j)_{j \in \mathbb{N}}$ is a sequence of strictly positive numbers. We define

$$(2.30) \qquad \begin{aligned} &\mathcal{E}^V(u, v) := \frac{1}{2} \int_{\mathbb{R}^d} \langle \nabla u, \nabla v \rangle_{\mathbb{R}^d} \, dx + \int uv \, V \, dx \\ &\text{for } u, v \in D(\mathcal{E}^V) := H^{1,2}(\mathbb{R}^d) \cap L^2(\mathbb{R}^d; V \cdot dx) \end{aligned}$$

(see Subsection 1c) for the definition of $H^{1,2}(\mathbb{R}^d)$).

Proposition 2.15. *For each $(x_j)_{j \in \mathbb{N}}$ and $(\alpha_j)_{j \in \mathbb{N}}$ specified as above, we can choose a sequence of strictly positive numbers $(c_j)_{j \in \mathbb{N}}$ so that $(\mathcal{E}^V, D(\mathcal{E}^V))$ defined by (2.30) becomes a symmetric Dirichlet form on $L^2(R^d; dx)$.*

For the proof of the above proposition we refer to [AM 88, Proposition 1.3] and[AM 91d, Proposition 3.1]. Suppose that there exists a natural number j_0 such that $\alpha_j \leq -d$ for all $j \geq j_0$, then $\int_U V dx = \infty$ for every non-empty open set U, i.e., $V \cdot dx$ is *nowhere Radon*. Suppose that $\alpha_j = -j$ for all $j \geq 1$, then we have even $\int_U |V|^p dx = \infty$ for any non-empty open set $U \subset \mathbb{R}^d$ and any $p > 0$, i.e., V is *nowhere L^p-integrable*. Let now $\alpha_j \leq -d$ for all $j \geq 1$. By 2.15 we can choose $(c_j)_{j \in \mathbb{N}}$ so that $(\mathcal{E}^V, D(\mathcal{E}^V))$ is a Dirichlet form. In this case there is no continuous function (except the null function) in the domain $D(\mathcal{E}^V)$. Since $(\mathcal{E}^V, D(\mathcal{E}^V))$ is a symmetric Dirichlet form, the generator L^V of $(\mathcal{E}^V, D(\mathcal{E}^V))$ is a self-adjoint operator on $L^2(\mathbb{R}^d; dx)$. One can show that L^V is a realization of the Schrödinger operator $\frac{1}{2}\Delta - V$. More precisely, we have the following result.

Proposition 2.16. *Suppose that $(\mathcal{E}^V, D(\mathcal{E}^V))$ is the Dirichlet form defined by (2.30) and L^V is its generator. Then*

(i) $D(L^V) \subset H^{1,2}(\mathbb{R}^d) \cap L^2(\mathbb{R}^d; V \cdot dx) \cap L^1_{loc}(\mathbb{R}^d; V \cdot dx)$

(ii) $L^V u = (\frac{1}{2}\Delta - V)u$ *in the following distributional sense:*

$$\int_{\mathbb{R}^d} (L^V u) v \, dx = \int_{\mathbb{R}^d} u (\frac{1}{2}\Delta v - Vv) \, dx \quad \text{for all } v \in C_0^\infty(\mathbb{R}^d) \,.$$

We refer to [AM 91a, Proposition 6.1] for the proof. $(\mathcal{E}^V, D(\mathcal{E}^V))$ defined by (2.30) is called the *perturbation of* $(\mathbb{D}, H^{1,2}(\mathbb{R}^d))$ *by* $V \cdot dx$. We shall discuss more general results on the perturbation of Dirichlet forms in Chap.IV, Subsection 4c) below.

3   Starting point: bilinear form – infinite dimensional case

a)   Classical Dirichlet forms on infinite dimensional space

Let E be a locally convex topological real vector space which is a *(topological) Souslin space*, that is, the continuous image of a separable complete metric space (e.g. a separable real Banach space). In particular, $\mathcal{B}(E) = \sigma(E')$ by [Ba 70, Exposé n° 8, N° 7, Corollaire]. Let $m := \mu$ be a finite positive measure on $\mathcal{B}(E)$ such that $\operatorname{supp}\mu = E$ (i.e., $\mu(U) > 0$ for every $\emptyset \neq U \subset E$, U open). Let E' denote the dual of E and $_{E'}\langle\,,\,\rangle_E : E' \times E \to \mathbb{R}$ the corresponding dualization. Define a linear space of functions on E by

$$(3.1) \quad \mathcal{F}C_b^\infty := \{ f(l_1, \ldots, l_m) \mid m \in \mathbb{N}\,,\, f \in C_b^\infty(\mathbb{R}^m)\,,\, l_1, \ldots, l_m \in E' \}\,.$$

Here $C_b^\infty(\mathbb{R}^m)$ denotes the set of all infinitely differentiable (real-valued) functions on $\mathbb{R}^m$ with all partial derivatives bounded. By the Hahn-Banach theorem (cf. [Ch 69b]) $\mathcal{F}C_b^\infty$ separates the points of E; hence, since $\mathcal{B}(E) = \sigma(E')$,

$$(3.2) \qquad\qquad \mathcal{F}C_b^\infty \text{ is dense in } L^2(E; \mu)\,,$$

by monotone class arguments (cf. [Sh 88, A.0.6]). Define for $u \in \mathcal{F}C_b^\infty$ and $k \in E$

$$\frac{\partial u}{\partial k}(z) := \frac{d}{ds} u(z + sk)|_{s=0}\,,\, z \in E\,.$$

Observe that if $u = f(l_1, \ldots, l_m)$, then

$$(3.3) \qquad \frac{\partial u}{\partial k} = \sum_{i=1}^m \frac{\partial f}{\partial x_i}(l_1, \ldots, l_m)\,_{E'}\langle l_i, k\rangle_E \in \mathcal{F}C_b^\infty\,.$$

Define for $k \in E$

$$(3.4) \qquad \mathcal{E}_k(u, v) := \int \frac{\partial u}{\partial k}\frac{\partial v}{\partial k}\,d\mu\,;\quad u, v \in \mathcal{F}C_b^\infty\,,$$

then $(\mathcal{E}_k, \mathcal{F}C_b^\infty)$ is a densely defined positive definite symmetric bilinear form on $L^2(E; \mu)$.

Definition 3.1. $k \in E$ is called μ-*admissible* if $(\mathcal{E}_k, \mathcal{F}C_b^\infty)$ is closable on $L^2(E; \mu)$.

For necessary and sufficient conditions for μ-admissibility we refer to [AR 90a, Sect. 3]. Now we shall see that $(\mathcal{E}_k, \mathcal{F}C_b^\infty)$ is closable if one has an integration by parts formula w.r.t. k.

Definition 3.2. $k \in E$ is called *well-μ-admissible* if there exists $\beta_k \in L^2(E; \mu)$ such that for all $u, v \in \mathcal{F}C_b^\infty$

$$(3.5) \qquad \int \frac{\partial u}{\partial k} v\,d\mu = -\int u\frac{\partial v}{\partial k}\,d\mu - \int uv\beta_k\,d\mu\,.$$

Exercise 3.3. Prove that the set of all well-μ-admissible elements form a linear subspace W of E and that $k \mapsto \beta_k$ is linear from W to $L^2(E; \mu)$.

Note that if $E = \mathbb{R}^d$ and $\mu = \rho \cdot dx$ for some nice function $\rho : \mathbb{R}^d \to [0, \infty[$ then $\beta_k = \frac{\partial}{\partial k} \ln \rho$. For a characterization of the well-μ-admissible elements in E we refer to [AKR 90].

Proposition 3.4. *If $k \in E$ is well-μ-admissible then it is μ-admissible.*

Proof. Define
$$S_k u := \frac{\partial}{\partial k}\left(\frac{\partial u}{\partial k}\right) + \beta_k \frac{\partial u}{\partial k} \, , \quad u \in \mathcal{F}C_b^\infty \, ,$$
then $(S_k, \mathcal{F}C_b^\infty)$ is a symmetric linear operator on $L^2(E; \mu)$ such that
$$\mathcal{E}_k(u, v) = (-S_k u, v) \ \text{ for all } u, v \in \mathcal{F}C_b^\infty \, .$$

Hence the assertion follows by I.3.3. $\square$

Fix a countable subset $K_0 \subset E$ such that

(3.6)
$$\sum_{k \in K_0} {}_{E'}\langle l, k \rangle_E^2 < \infty \ \text{ for all } l \in E' \, .$$

Note that (3.6) holds automatically if K_0 is finite. Define

(3.7)
$$\mathcal{E}_{K_0}(u, v) := \sum_{k \in K_0} \mathcal{E}_k(u, v) \ ; \ u, v \in \mathcal{F}C_b^\infty \, .$$

Observe that the sum in (3.7) converges since by (3.6) (and (3.3))

$$\sum_{k \in K_0} \mathcal{E}_k(u, u) \leq \left(\sum_{i=1}^m \int \left(\frac{\partial f}{\partial x_i}(l_1, \ldots, l_m) \right)^2 d\mu \right) \left(\sum_{i=1}^m \sum_{k \in K_0} {}_{E'}\langle l_i, k \rangle_E^2 \right) < \infty$$

if $u = f(l_1, \ldots, l_m) \in \mathcal{F}C_b^\infty$.

Proposition 3.5. *Suppose K_0 is a (finite or) countable subset of E consisting of μ-admissible elements satisfying (3.6). Then $(\mathcal{E}_{K_0}, \mathcal{F}C_b^\infty)$ given by (3.7) is closable on $L^2(E; \mu)$ and its closure is a symmetric Dirichlet form.*

Proof. The closability follows by I.3.7. Let φ_ε, $\varepsilon > 0$, be as in 2.7, then $\varphi_\varepsilon(u) \in \mathcal{F}C_b^\infty$ for every $u \in \mathcal{F}C_b^\infty$ and for each $k \in K_0$

$$\mathcal{E}_k(\varphi_\varepsilon(u), \varphi_\varepsilon(u)) = \int \varphi_\varepsilon'(u)^2 \left(\frac{\partial u}{\partial k} \right)^2 d\mu \leq \mathcal{E}_k(u, u) \, .$$

Hence the last assertion follows by I.4.10. $\square$

The Dirichlet forms arising in 3.5 are so-called *classical Dirichlet forms* introduced in[AR 90a]. But what we have done so far easily generalizes to the more general situation in the following

Exercise 3.6. Let K_0 be as in 3.5 and let $A := (a_{kk'})_{k,k' \in K_0}$ where each $a_{kk'}$ is a $\mathcal{B}(E)$-measurable function on E such that $a_{kk'} = a_{k'k}$ for all $k, k' \in K_0$. Let

$$\|A\|(z) := \left(\sum_{k,k' \in K_0} a_{kk'}(z)^2 \right)^{1/2} , \ z \in E ,$$

and assume that

$$\int \|A\| d\mu < \infty .$$

Suppose furthermore, that

$$\sum_{k,k' \in K_0} a_{kk'} \xi_k \xi_{k'} \geq 0 \ \mu\text{-a.e.}$$

for all $(\xi_k)_{k \in K_0} \in \ell^2(K_0)$. Prove:

(i) $\sum_{k,k' \in K_0} \int |a_{kk'}| \left|\frac{\partial u}{\partial k}\right| \left|\frac{\partial v}{\partial k'}\right| d\mu < \infty$ for all $u, v \in \mathcal{F}C_b^\infty$; in particular,

$$\mathcal{E}_A(u,v) = \sum_{k,k' \in K_0} \int a_{kk'} \frac{\partial u}{\partial k} \frac{\partial v}{\partial k'} d\mu ; \ u, v \in \mathcal{F}C_b^\infty$$

is well-defined, positive definite, and symmetric.

(ii) If there exists a constant $c > 0$ such that

$$\sum_{k,k' \in K_0} a_{kk'} \xi_k \xi_{k'} \geq c \sum_{k \in K_0} \xi_k^2 \ \mu\text{-a.e.}$$

for all $(\xi_k)_{k \in K_0} \in \ell^2(K_0)$, then $(\mathcal{E}_A, \mathcal{F}C_b^\infty)$ is closable on $L^2(E; \mu)$. (Hint: Use the same argument as in Subsection 2b) above. For more general closability criteria in terms of A, see [AR 91, Sect.3].)

(iii) If $(\mathcal{E}_A, \mathcal{F}C_b^\infty)$ is closable on $L^2(E; \mu)$, then its closure $(\mathcal{E}_A, D(\mathcal{E}_A))$ is a symmetric Dirichlet form on $L^2(E; \mu)$.

b) Gradient Dirichlet forms on infinite dimensional space

This subsection can be considered as a "coordinate free" version of the previous one. But we need a bit more structure. Assume that there exists a separable real Hilbert space $(H, \langle , \rangle_H)$ densely and continuously embedded into E. Identifying H with its dual H' we have that

$$(3.8) \qquad\qquad E' \subset H \subset E \text{ densely and continuously}$$

and $_{E'}\langle , \rangle_E$ restricted to $E' \times H$ coincides with $\langle , \rangle_H$. Here E' is endowed with the strong topology (or equivalently with the operator norm $\| \ \|_{E'}$ if E is a Banach space). H should be thought of as a tangent space to E at each point. Observe that by (3.3) for $u \in \mathcal{F}C_b^\infty$ and $z \in E$ fixed, $h \mapsto \frac{\partial u}{\partial h}(z)$ is a continuous linear functional on H. Define $\nabla u(z) \in H$ by

$$(3.9) \qquad \langle \nabla u(z), h \rangle_H = \frac{\partial u}{\partial h}(z) \ , \ h \in H \ .$$

Define

$$(3.10) \qquad \mathcal{E}(u,v) := \int \langle \nabla u, \nabla v \rangle_H \, d\mu \ ; \ u, v \in \mathcal{F}C_b^\infty \ ,$$

then $(\mathcal{E}, \mathcal{F}C_b^\infty)$ is a densely defined positive definite symmetric bilinear form on $L^2(E; \mu)$. The existence of the integral in (3.10) follows by

Remark 3.7. Clearly, $(\mathcal{E}, \mathcal{F}C_b^\infty)$ is a special case of (3.7). Since for every orthonormal basis K_0 of H, (3.6) holds and

$$\mathcal{E}(u,v) = \mathcal{E}_{K_0}(u,v) \ \text{ for all } u, v \in \mathcal{F}C_b^\infty \ .$$

3.7 and 3.5 imply:

Proposition 3.8. *Suppose there exists an orthonormal basis of H consisting of μ-admissible elements in E, then $(\mathcal{E}, \mathcal{F}C_b^\infty)$ given by (3.10) is closable on $L^2(E; \mu)$ and its closure is a symmetric Dirichlet form.*

Exercise 3.9. Formulate and prove the "coordinate free" version of 3.6. (Hint: See [AR 89b, Sect.3].)

c) Abstract Wiener spaces

In this section we look at the very special case where E, H, μ form an *abstract Wiener space* (E, H, μ) (cf.[Gr 65], [Kuo 75]) i.e., E is a separable real Banach space, H a separable real Hilbert space continuously and densely embedded into E, i.e.,

$$E' \subset H' \equiv H \subset E \text{ continuously and densely,}$$

and μ is a Gaussian measure on $\mathcal{B}(E)$ with covariance $\langle \, , \, \rangle_H$, i.e., each $l \in E'$ is $N(0, \|l\|_H^2)$-distributed under μ and $\operatorname{supp} \mu = E$. Consider the linear map

$$(3.11) \qquad l \mapsto {}_{E'}\langle l, \cdot \rangle_E \ , \ l \in E'$$

from E' to $L^2(E; \mu)$ and note that this is an isometry if E' is equipped with the norm inherited from H. Hence this map extends uniquely to an isometry

$$(3.12) \qquad h \mapsto X_h \ , \ h \in H \ ,$$

from H to $L^2(E; \mu)$ (i.e., in particular, $X_l = {}_{E'}\langle l, \cdot \rangle_E$ if $l \in E' \subset H$).

Example 3.10. Let $E := \{ f \in C([0,1], \mathbb{R}^d) \mid f(0) = 0 \}$ equipped with the uniform norm, $H := \{ f = (f_1, \ldots, f_d) \in E \mid f_i \text{ is absolutely continuous and } \int_0^1 f_i'(s)^2 \, ds < \infty, \ 1 \leq i \leq d \}$ with inner product $\langle f, g \rangle_H = \sum_{i=1}^d \int_0^1 f_i'(s) g_i'(s) \, ds$ and $\mu := $ Wiener measure on $\mathcal{B}(E)$. Then (E, H, μ) is an abstract Wiener space (cf. [Kuo 75]).

Theorem 3.11. *Each $h \in H$ is well-μ-admissible with $\beta_h = -X_h$.*

We need the following

Lemma 3.12. *Let $h \in E$ and $T_h(z) := z + h$, $z \in E$. Let $\mu \circ T_h^{-1}$ denote the image measure of μ under T_h. Then $\mu \circ T_h^{-1}$ is absolutely continuous w.r.t. μ if and only if $h \in H$. In this case*

$$(3.13) \qquad \frac{d\mu \circ T_h^{-1}}{d\mu} = \exp\left(X_h - \frac{1}{2}\|h\|_H^2 \right) .$$

Proof. Assume $h \in H \setminus \{0\}$ and let $l \in E'(\subset H \subset E)$. If

$$\alpha := \langle l, h\rangle_H / \langle h, h\rangle_H$$

then $\langle l - \alpha h, h\rangle_H = 0$. Hence, since μ is Gaussian with covariance $\langle\, ,\, \rangle_H$,

$$\int_E \exp\left[i_{E'}\langle l, z\rangle_E\right] \exp\left[X_h(z) - \frac{1}{2}\|h\|_H^2 \right] \mu(dz)$$

$$= \exp\left[-\frac{1}{2}\|h\|_H^2\right] \int_E \exp\left[i\, X_{l-\alpha h} + (1 + i\alpha)X_h\right] d\mu$$

$$= \exp\left[-\frac{1}{2}\|h\|_H^2\right] \int_E \exp\left[i\, X_{l-\alpha h}\right] d\mu \int_E \exp\left[(1 + i\alpha)X_h\right] d\mu$$

$$= \exp\left[-\frac{1}{2}\left(\|h\|_H^2 + \|l - \alpha h\|_H^2 - (1 + i\alpha)^2\|h\|_H^2 \right)\right]$$

$$= \exp\left[-\frac{1}{2}\langle l, l\rangle_H + i\langle l, h\rangle_H\right]$$

$$= \int \exp\left[i_{E'}\langle l, z\rangle_E\right] (\mu \circ T_h^{-1})(dz) .$$

In the second step above we used that $X_{l-\alpha h}$ and X_h are independent (under μ) since they are jointly Gaussian and

$$\int X_{l-\alpha h} X_h \, d\mu = \langle l - \alpha h, h\rangle_H = 0 .$$

Conversely, suppose $\mu \circ T_h^{-1} = \rho_h \cdot \mu$ for some $\rho_h \in \mathcal{B}^+(E)$. In order to show that $h \in H$ we have to show that $l \mapsto {}_{E'}\langle l, h\rangle_E$, $l \in E'$, is continuous w.r.t. $\|\ \|_H$. But if $l_n \in E'$, $n \in \mathbb{N}$, such that $l_n \underset{n\to\infty}{\longrightarrow} 0$ w.r.t. $\|\ \|_H$ then $l_n \underset{n\to\infty}{\longrightarrow} 0$ in $L^2(E; \mu)$, in particular in μ-measure and hence in $(\rho_h \cdot \mu)$-measure. Since $\rho_h \cdot \mu$ is Gaussian, it follows that $l_n \underset{n\to\infty}{\longrightarrow} 0$ in $L^1(E; \rho_h \cdot \mu)$ (cf. e.g. [Ro 82, p.42]) and hence

$$\limsup_{n\to\infty} |{}_{E'}\langle l_n, h\rangle_E| = \limsup_{n\to\infty} \left| \int {}_{E'}\langle l_n, z\rangle_E\, \rho_h(z)\mu(dz) \right| = 0$$

□

Proof of Theorem 3.11. By 3.12 and Lebesgue's dominated convergence theorem we obtain that for all $u \in \mathcal{F}C_b^\infty$

$$
\begin{aligned}
\int \frac{\partial u}{\partial h}\, d\mu \;&=\; \frac{d}{ds} \int u(z+sh)\mu(dz)\big|_{s=0} \\
&=\; \frac{d}{ds} \int u(z)\frac{d\mu \circ T_{sh}^{-1}}{d\mu}(z)\mu(dz)\big|_{s=0} \\
&=\; \int u \frac{d}{ds}\exp\left(X_{sh} - \frac{1}{2}s^2\|h\|_H^2\right)\big|_{s=0}\, d\mu \\
&=\; \int u\, X_h d\mu \; .
\end{aligned}
$$

By the product rule we obtain (3.5) and the assertion follows. $\square$

As a consequence we obtain by 3.4, 3.8

Corollary 3.13. *Define*

$$
(3.14) \qquad \mathcal{E}(u,v) := \int \langle \nabla u, \nabla v\rangle_H d\mu \; ; u,v \in \mathcal{F}C_b^\infty
$$

(cf. (3.9), (3.10)). Then $(\mathcal{E}, \mathcal{F}C_b^\infty)$ is closable on $L^2(E;\mu)$ and its closure $(\mathcal{E}, D(\mathcal{E}))$ is a symmetric Dirichlet form.

Remark 3.14. (i) $(\mathcal{E}, D(\mathcal{E}))$ in 3.13 is of fundamental importance in the Malliavin calculus (cf.[Ma 78]). If $(\overline{\nabla}, D(\overline{\nabla}))$ denotes the closure of $(\nabla, \mathcal{F}C_b^\infty)$ (which exists by 3.13, cf. I.3.2 (i)) then $D(\overline{\nabla}) = D(\mathcal{E})$ and

$$
(3.15) \qquad \mathcal{E}(u,v) = \int \langle \overline{\nabla}u, \overline{\nabla}v\rangle_H \, d\mu \; ; u,v \in D(\mathcal{E}) \; .
$$

$\overline{\nabla}$ is just the Malliavin gradient (cf.[W 84]).
(ii) H in (E, H, μ) is called the *(generalized) Cameron-Martin space*. Examples where the Hilbert space appearing in (3.14) is replaced by some other Hilbert space can be found in [AR 90a], [RZ 90] and [R 90a].

d)   Cases with densities

Suppose E, H, μ are as in Subsection 3b). Assume there exists an orthonormal basis K_0 of H consisting of well-μ-admissible elements in E. Then $(\mathcal{E}, \mathcal{F}C_b^\infty)$ and ∇ defined by (3.10) resp. (3.9) are closable by 3.4, 3.8. Let $(\mathcal{E}, D(\mathcal{E}))$ and $\overline{\nabla}$ denote their corresponding closures and let

$$
(3.16) \quad \varphi \in D(\mathcal{E}) \text{ such that } \varphi > 0 \text{ and } \varphi \cdot \beta_k \in L^2(E;\mu) \text{ for all } k \in K_0 \; .
$$

Then $(\mathcal{E}_\varphi, \mathcal{F}C_b^\infty)$ defined by

$$
(3.17) \qquad \mathcal{E}_\varphi(u,v) = \int \langle \nabla u, \nabla v\rangle_H\, \varphi^2\, d\mu \; ; \; u,v \in \mathcal{F}C_b^\infty \; ,
$$

is closable on $L^2(E; \varphi^2 \cdot \mu)$ and its closure $(\mathcal{E}_\varphi, D(\mathcal{E}_\varphi))$ is a symmetric Dirichlet form. Indeed, by 3.4, 3.8 we only have to prove that each $k \in K_0$ is well-$(\varphi^2 \cdot \mu)$-admissible. So, let $k \in K_0$ and define

$$\varphi_n := \varphi \wedge n \, , \quad n \in \mathbb{N} \, .$$

Then $\varphi_n^2 \in D(\mathcal{E})$ and since (3.5) extends to all $v \in D(\mathcal{E})$ we have that for all $u \in \mathcal{F}C_b^\infty$

$$(3.18) \qquad \int \frac{\partial u}{\partial k} \varphi_n^2 \, d\mu = -\int u \, 2\varphi_n \langle k, \overline{\nabla}\varphi_n \rangle_H \, d\mu - \int u \, \varphi_n^2 \beta_k d\mu \, .$$

But $\varphi_n \xrightarrow[n\to\infty]{} \varphi$ in $D(\mathcal{E})$ by I.4.17, hence $\left(\frac{\partial \varphi_n}{\partial k} \right)_{n\in\mathbb{N}}$ converges to $\langle k, \overline{\nabla}\varphi \rangle_H$ in $L^2(E;\mu)$. Consequently, by (3.18)

$$\int \frac{\partial u}{\partial k} \varphi^2 \, d\mu = -\int u \overline{\beta}_k \varphi^2 d\mu$$

with

$$(3.19) \qquad \overline{\beta}_k := \beta_k + 2\frac{\langle k, \overline{\nabla}\varphi \rangle_H}{\varphi} \, ,$$

and hence k is well-$(\varphi^2 \cdot \mu)$-admissible.

Remark 3.15. (i) Above, well-admissibility can be replaced by admissibility (cf. [AR 90b, Theorem 4.7]).
(ii) For more examples, e.g. where μ is not even absolutely continuous w.r.t. a Gaussian measure, we refer to [AR 89a,b, 90a,b, 91].

e)   Non-symmetric cases

Suppose E, H, μ are as in Subsection 3b) and let $(\mathcal{E}_\mu, \mathcal{F}C_b^\infty)$, defined by

$$\mathcal{E}_\mu(u, v) = \int \langle \nabla u, \nabla v \rangle_H \, d\mu \, ; \qquad u, v \in \mathcal{F}C_b^\infty$$

(cf. (3.9), (3.10)), be closable on $L^2(E;\mu)$. Let $\mathcal{L}_\infty(H)$ denote the set of all bounded linear operators on H with operator norm $\| \, \|$. Suppose $z \mapsto A(z)$, $z \in E$, is a map from E to $\mathcal{L}_\infty(H)$ such that $z \mapsto \langle A(z)h_1, h_2 \rangle_H$ is $\mathcal{B}(E)$-measurable for all $h_1, h_2 \in H$. Furthermore, assume that

$$(3.20) \qquad \begin{array}{l} \text{there exists } \alpha \in \,]0, \infty[\text{ such that} \\ \langle A(z)h, h \rangle_H \geq \alpha \|h\|_H^2 \text{ for all } h \in H \, . \end{array}$$

and that $\|\tilde{A}\|_\infty \in L^1(E;\mu)$ and $\|\check{A}\|_\infty \in L^\infty(E;\mu)$ where $\tilde{A} := \frac{1}{2}(A + \hat{A})$, $\check{A} := \frac{1}{2}(A - \hat{A})$ and $\hat{A}(z)$ denotes the adjoint of $A(z)$, $z \in E$. Let $c \in L^\infty(E;\mu)$ and $b, d \in L^\infty(E \to H;\mu)$ such that

$$(3.21) \qquad \begin{array}{l} \int (\langle d, \nabla u \rangle_H + cu) \, d\mu \, , \quad \int (\langle b, \nabla u \rangle_H + cu) \, d\mu \geq 0 \\ \text{for all } u \in \mathcal{F}C_b^\infty, \, u \geq 0 \, . \end{array}$$

(3.21) is e.g. fulfilled if d, b are in the domain of $\hat{\nabla}$, the adjoint of $\nabla : \mathcal{F}C_b^\infty \to L^2(E \to H; d\mu)$ and $\hat{\nabla}d + c \geq 0$, $\hat{\nabla}b + c \geq 0$. Define for $u, v \in \mathcal{F}C_b^\infty$

$$
\begin{aligned}
(3.22) \qquad \mathcal{E}(u,v) \;=\; & \int \langle A(z)\nabla u(z), \nabla v(z)\rangle_H \, \mu(dz) + \int u \,\langle d, \nabla v\rangle_H \, d\mu \\
& + \int \langle b, \nabla u\rangle_H \, v \, d\mu + \int u\, v\, c\, d\mu \; .
\end{aligned}
$$

Then $(\mathcal{E}, \mathcal{F}C_b^\infty)$ is a densely defined bilinear form on $L^2(E; \mu)$ which is positive definite since for all $u \in \mathcal{F}C_b^\infty$

$$
(3.23) \quad \mathcal{E}(u,u) = \int \langle A\nabla u, u\rangle_H \, d\mu + \frac{1}{2} \int ((d + b, \nabla u^2)_H + cu^2) d\mu \geq 0 \; .
$$

Furthermore, $(\mathcal{E}, \mathcal{F}C_b^\infty)$ is closable on $L^2(E; \mu)$ and its closure $(\mathcal{E}, D(\mathcal{E}))$ is a Dirichlet form on $L^2(E; \mu)$. Indeed, once we know that $(\mathcal{E}, \mathcal{F}C_b^\infty)$ is closable and satisfies the weak sector condition, the Dirichlet property for $(\mathcal{E}, D(\mathcal{E}))$ can be shown in exactly the same way as in Subsection 2d) simply using the product rule of ∇ on $\mathcal{F}C_b^\infty$. Hence we omit this part and only prove the closability and the weak sector condition, though also this part is quite similar to Subsection 2d). We start with the weak sector condition. Let $u, v \in \mathcal{F}C_b^\infty$, then

$$
\begin{aligned}
|\check{\mathcal{E}}(u,v)| \;\leq\; & \int (\|\check{A}\|\, \|\nabla u\|_H \|\nabla v\|_H + |u|\|d - b\|_H \|\nabla v\|_H \\
& + \|d - b\|_H \|\nabla u\|_H |v|) \, d\mu \\
\leq\; & K[(\int \|\nabla u\|_H^2 \, d\mu)^{1/2}(\int \|\nabla v\|_H^2 \, d\mu)^{1/2} \\
& + (\int u^2 \, d\mu)^{1/2}(\int \|\nabla v\|_H^2 \, d\mu)^{1/2} \\
& + (\int \|\nabla u\|_H^2 \, d\mu)^{1/2}(\int v^2 \, d\mu)^{1/2}
\end{aligned}
$$

where $K = \sup_{z \in E}(\|\check{A}(z)\|_H \vee \|d(z) - b(z)\|_H)$. Consequently, by (3.20), (3.21), (3.23)

$$
\begin{aligned}
|\check{\mathcal{E}}(u,v)| \;\leq\; & 3K\, \mathcal{E}_{\mu,1}(u,u)^{1/2}\, \mathcal{E}_{\mu,1}(v,v)^{1/2} \\
\leq\; & 3K\alpha^{-1}\mathcal{E}_1(u,u)^{1/2}\mathcal{E}_1(v,v)^{1/2} \; .
\end{aligned}
$$

Hence $(\mathcal{E}, \mathcal{F}C_b^\infty)$ satisfies the weak sector condition by I.2.1(iv).

To show the closability let $u_n \in \mathcal{F}C_b^\infty$, $n \in \mathbb{N}$, such that $u_n \to 0$ in $L^2(E; \mu)$ as $n \to \infty$ and $(u_n)_{n\in\mathbb{N}}$ is $\mathcal{E}$-Cauchy. By (3.20), (3.21), (3.23) $(u_n)_{n\in\mathbb{N}}$ is also $\mathcal{E}_\mu$-Cauchy, hence by the closability of $(\mathcal{E}_\mu, \mathcal{F}C_b^\infty)$ $\nabla u_n \xrightarrow[n\to\infty]{} 0$ in $L^2(E \to H; \mu)$ and $\|\nabla u_{n_k}\|_H \xrightarrow[k\to\infty]{} 0$ μ-a.e. for some subsequence $(n_k)_{k\in\mathbb{N}}$. Therefore, by Fatou's Lemma and (3.20), (3.21)

$$
\begin{aligned}
0 \;\leq\; & \int \langle A\nabla u_n, \nabla u_n\rangle_H \, d\mu \\
\leq\; & \liminf_{k\to\infty} \int \langle A\nabla(u_n - u_{n_k}), \nabla(u_n - u_{n_k})\rangle_H \, d\mu \\
\leq\; & \liminf_{k\to\infty} \mathcal{E}(u_n - u_{n_k}, u_n - u_{n_k})
\end{aligned}
$$

which is arbitrary small for n large. Furthermore,

$$\int u_n(\langle d + b, \nabla u_n \rangle_H + c\, u_n)\, d\mu \xrightarrow[n\to\infty]{} 0$$

and thus, $\mathcal{E}(u_n, u_n) \xrightarrow[n\to\infty]{} 0$. Consequently, $(\mathcal{E}, \mathcal{F}C_b^\infty)$ is closable on $L^2(E; \mu)$.

Remark 3.16. Let J be a symmetric finite positive measure on $(E \times E, \mathcal{B}(E) \otimes \mathcal{B}(E))$ such that $(\mathcal{E}_J, \mathcal{F}C_b^\infty)$ defined by

$$\mathcal{E}_J(u, v) = \iint (u(x) - u(y))(v(x) - v(y)) J(dx\, dy); \quad u, v \in \mathcal{F}C_b^\infty$$

is closable. This e.g. is obviously the case if $J = \mu \otimes \mu$. Then by I.3.7(i) and I.2.1(iv) it follows that $(\mathcal{E} + \mathcal{E}_J, \mathcal{F}C_b^\infty)$ is closable, where $(\mathcal{E}, \mathcal{F}C_b^\infty)$ is defined by (3.22), and that its closure is a coercive bilinear form which is easily checked to be a Dirichlet form (cf. Subsection 2c)). Thus it is easy to construct non-symmetric Dirichlet forms in (finite and) infinite dimensions which are not *local* (cf. Chap.V, Sect.1 below, in particular V.1.1).

4 Starting point: semigroup of kernels

Let $(E, \mathcal{B})$ be an arbitrary measurable space.

a) Kernels and L^2-operators

Let $(E', \mathcal{B}')$ be another measurable space and let π be a *kernel from* $(E', \mathcal{B}')$ *to* $(E, \mathcal{B})$, i.e., $\pi : E' \times \mathcal{B} \to [0, \infty[$ such that $\pi(z, \cdot)$ is a positive measure on $\mathcal{B}$ for each $z \in E'$ and $z \mapsto \pi(z, A)$ is $\mathcal{B}'$-measurable for every $A \in \mathcal{B}$. Define for $f : E \to \mathbb{R}$, $\mathcal{B}$-measurable, and $z \in E'$

$$(4.1) \qquad\qquad \pi f(z) := \pi(z, f) := \int f(y) \pi(z, dy)$$

if $\pi(z, f^+) \wedge \pi(z, f^-) < \infty$. π is called *sub-Markovian* if $\pi 1 \leq 1$ and *Markovian* if $\pi 1 = 1$. In this section we assume from now on that $(E', \mathcal{B}') = (E, \mathcal{B})$. In this case we simply say that π is a *kernel on* $(E, \mathcal{B})$. Let m be a positive measure on $(E, \mathcal{B})$. m is called π-*supermedian* if

$$\int \pi f\, dm \leq \int f\, dm \quad \text{for all } f \in \mathcal{B}^+ .$$

Two kernels $\pi, \hat{\pi}$ on $(E, \mathcal{B})$ are said to be *in duality w.r.t.* m if

$$(4.2) \qquad\qquad \int \pi f\, g\, dm = \int f\, \hat{\pi} g\, dm \quad \text{for all } f, g \in \mathcal{B}^+ .$$

If $\pi = \hat{\pi}$ in (4.2) then π is called m-*symmetric*. Note that if there exists $\hat{\pi}$ satisfying (4.2) which is sub-Markovian then

$$\int \pi f\, dm = \int f\, \hat{\pi} 1\, dm \leq \int f\, dm \quad \text{for all } f \in \mathcal{B}^+ ,$$

i.e., m is π-supermedian. A (slightly weakened) converse to this statement is contained in the next proposition.

Proposition 4.1. *Let π be a sub-Markovian kernel on $(E, \mathcal{B})$ and let m be a positive measure on $(E, \mathcal{B})$ which is π-supermedian. Then there exists a bounded linear operator Π on $L^2(E; m)$ such that*

$$(4.3) \qquad \pi f \text{ is an } m\text{-version of } \Pi f \quad \text{for all } f \in \mathcal{B}_b \cap L^2(E; m) \ .$$

Both Π and its adjoint $\hat{\Pi}$ are sub-Markovian contractions on $L^2(E; m)$.

Remark. Let $f \in \mathcal{B}_b \cap L^2(E; m)$. In 4.1 and henceforth (cf. also the following lemma) if we say "πf is an m-version of Πf" for some sub-Markovian kernel π resp. a bounded linear operator Π on $L^2(E; m)$, we mean $\pi \tilde{f}$ is an m-version of Πf for any $\mathcal{B}$-measurable representative $\tilde{f}$ of the class $f \in L^2(E; m)$. In particular, $\pi g = \pi \tilde{f}$ m-a.e. if $g = \tilde{f}$ m-a.e. .

Proof of 4.1. Let $f \in \mathcal{B}_b \cap L^2(E; m)$, then by Cauchy-Schwarz's inequality

$$(4.4) \qquad \int (\pi f)^2 \, dm \leq \int \pi(f^2) \, dm \leq \int f^2 \, dm \ .$$

Hence $f = g$ m-a.e. implies $\pi f = \pi g$ m-a.e. for all $g \in \mathcal{B}_b$, and we can define Πf to be the m-class in $L^2(E; m)$ corresponding to πf. Since the bounded functions are dense in $L^2(E; m)$, Π extends by (4.4) to a unique bounded operator on $L^2(E; m)$ which is clearly a sub-Markovian contraction on $L^2(E; m)$. For its adjoint $\hat{\Pi}$ we have for any $f \in L^2(E; m)$, $0 \leq f \leq 1$, that

$$(4.5) \qquad \int g \, \hat{\Pi} f \, dm = \int \Pi g \, f \, dm \quad \text{for all } g \in L^2(E; m) \ .$$

But if $g \in L^2(E; m)$, $g \geq 0$, then $\Pi g \geq 0$ by I.4.2(i), hence

$$0 \leq \int \Pi g \, f \, dm \leq \int \pi g \, dm \leq \int g \, dm \ .$$

Consequently, (4.5) implies that $0 \leq \hat{\Pi} f \leq 1$, i.e., $\hat{\Pi}$ is sub-Markovian. $\hat{\Pi}$ is a contraction since it is the adjoint of a contraction (cf. A.1.1 below). $\qquad \square$

The following result will be useful.

Lemma 4.2. *Let π be a sub-Markovian kernel on $(E, \mathcal{B})$ and m a positive measure on $(E, \mathcal{B})$. Suppose there exists a bounded linear operator Π on $L^2(E; m)$ such that πf is an m-version of Πf for all $f \in \mathcal{B}_b \cap \mathcal{D}$ for some dense subset $\mathcal{D}$ of $L^2(E; m)$ with $f \wedge k \in \mathcal{D}$ for all $f \in \mathcal{D}$, $k \in \mathbb{N}$. Then πf is an m-version of Πf for all $f \in L^2(E; m)$.*

Proof. Let $f \in L^2(E; m)$. Since $f = f^+ - f^-$ we may assume that $f \geq 0$. Since $f \wedge n \uparrow f$ pointwise on E and $f \wedge n \to f$ in $L^2(E; m)$ as $n \to \infty$ we may also assume that $f \in \mathcal{B}_b(E)$. Let $k \in \mathbb{N}$, $k \geq \|f\|_\infty$, and $f_n \in \mathcal{D}$, $n \in \mathbb{N}$, such that $f_n \xrightarrow[n \to \infty]{} f$ in $L^2(E; m)$ and pointwise on E, hence $g_n := f_n \wedge k \xrightarrow[n \to \infty]{} f$ in $L^2(E; m)$ and pointwise on E. Consequently, $\Pi g_n \xrightarrow[n \to \infty]{} \Pi f$ in $L^2(E; m)$ and by Lebesgue's dominated convergence theorem $\pi g_n \xrightarrow[n \to \infty]{} \pi f$ pointwise on E. By assumption this implies that πf is an m-version of Πf. $\qquad \square$

Now let $(p_t)_{t>0}$ be a *semigroup of kernels on $(E, \mathcal{B})$* (i.e., $p_t(p_s f)(z) = p_{t+s} f(z)$ for all $f \in \mathcal{B}^+$, $z \in E$ and $t, s > 0$) which is *sub-Markovian*, i.e., each p_t is sub-Markovian.

Proposition 4.3. *Let $(p_t)_{t>0}$ be as above and let m be a positive measure on $(E, \mathcal{B})$ such that m is p_t-supermedian for all $t > 0$. Let T_t, $t > 0$, be the corresponding sub-Markovian operators on $L^2(E; m)$. Then $(T_t)_{t>0}$ is a semigroup of contractions on $L^2(E; m)$. Suppose, in addition, that the following condition holds:*

$$(4.6) \qquad \begin{array}{c} \textit{There exists a dense subset } \mathcal{D} \subset L^2(E; m) \textit{ such that} \\[4pt] p_t f \xrightarrow[t \to 0]{} f \textit{ in } m\textit{-measure for all } f \in \mathcal{D} \; . \end{array}$$

Then $(T_t)_{t>0}$ is strongly continuous.

Proof. The first part of the assertion is obvious by 4.1. To prove the second we first note that by (4.6)

$$(4.6)' \qquad p_t f \xrightarrow[t \to 0]{} f \text{ in } m\text{-measure for all } f \in L^2(E; m) \; .$$

To see   this we may assume that $f \geq 0$. Let $\varepsilon > 0$ and $A \in \mathcal{B}$ with $m(A) < \infty$, then by the Chebychev-Markov inequality for all $t > 0$ and $g \in \mathcal{D}$

$$\begin{aligned}
m\left[\{|p_t f - f| > \varepsilon\} \cap A\right] \\
\leq \; & m\left(\{|p_t(f - g)| > \tfrac{\varepsilon}{3}\} \cap A\right) + m\left(\{|p_t g - g| > \tfrac{\varepsilon}{3}\} \cap A\right) \\
& + m\left(\{|g - f| > \tfrac{\varepsilon}{3}\} \cap A\right) \\
\leq \; & (\tfrac{3}{\varepsilon})^2 \int (p_t(f - g))^2 \, dm + m\left(\{|p_t g - g| > \tfrac{\varepsilon}{3}\} \cap A\right) + (\tfrac{3}{\varepsilon})^2 \int |f - g|^2 \, dm \; .
\end{aligned}$$

The first summand is by (4.4) less or equal to the third which can be made arbitrarily small, and $(4.6)'$ now follows from (4.6). Furthermore, since each p_t is sub-Markovian, we have for all $f \in L^2(E; m)$, $f \geq 0$,

$$\begin{aligned}
\int (p_t f - f)^2 \, dm \; &= \; \int (p_t f)^2 \, dm + \int f^2 \, dm - 2 \int f \, p_t f \, dm \\
&\leq \; 2\left(\int f^2 \, dm - \int f \, p_t f \, dm\right) \\
&= \; 2\left(\int f^2 \, dm - \int f(p_t f - f)^+ \, dm - \int f(f \wedge p_t f) \, dm\right) \\
&\leq \; 2\left(\int f^2 \, dm - \int f(f \wedge p_t f) \, dm\right)
\end{aligned}$$

which converges to zero by $(4.6)'$ and Lebesgue's dominated convergence theorem. Since $f = f^+ - f^-$, $p_t f \to f$ in $L^2(E; m)$ as $t \to 0$ for all $f \in L^2(E; m)$. Now the last part of the assertion follows by 4.2. $\qquad \square$

Consider the situation of 4.3. Let L be the generator of $(T_t)_{t>0}$ on $L^2(E; m)$. We know that if $1 - L$ satisfies the sector condition I.(2.5) (which is e.g. the case if p_t is m-symmetric, hence T_t is symmetric on $L^2(E; m)$ for each $t > 0$, see also I.2.21) then there exists a corresponding coercive closed form $(\mathcal{E}, D(\mathcal{E}))$ on $L^2(E; m)$. $(\mathcal{E}, D(\mathcal{E}))$ is a Dirichlet form by I.4.3, I.4.4 since both $(T_t)_{t>0}$ and $(\hat{T}_t)_{t>0}$ are sub-Markovian by 4.1.

b)   Construction of semigroups via subordination

Let $(\mu_t)_{t>0}$ be a *vaguely continuous convolution semigroup* of measures on $(]0, \infty[, \mathcal{B}(]0, \infty[))$, i.e.,

$$(4.7) \qquad \mu_t(]0, \infty[) \leq 1 \quad \text{ for all } t > 0 .$$

$$(4.8) \qquad \mu_s * \mu_t = \mu_{s+t} \quad \text{ for all } s, t > 0 .$$

$$(4.9) \qquad \lim_{t \to 0} \mu_t = \varepsilon_0 \quad \text{ vaguely,}$$

where $*$ denotes convolution w.r.t. the additive structure on $\mathbb{R}$. Then for any semigroup $(p_t)_{t>0}$ of sub-Markovian kernels on $(E, \mathcal{B})$ we can define

$$(4.10) \qquad p_t^\mu f := \int_0^\infty p_s f \, \mu_t(ds) , \quad t > 0 , \quad f \in \mathcal{B}_b .$$

Then $(p_t^\mu)_{t>0}$ is a semigroup of sub-Markovian kernels on $(E, \mathcal{B})$. $(p_t^\mu)_{t>0}$ is called *subordinated to* $(p_t)_{t>0}$ *by means of* $(\mu_t)_{t>0}$. If m is a positive measure on $(E, \mathcal{B})$ which is p_t-supermedian for all $t > 0$ then it is p_t^μ-supermedian for each $t > 0$. The corresponding contraction semigroup $(T_t^\mu)_{t>0}$ on $L^2(E; m)$ (cf. 4.1, 4.3) is strongly continuous on $L^2(E; m)$ if $(p_t)_{t>0}$ satisfies (4.6). In this case the final remarks in Subsection 4a) above apply.

c)   Symmetric stable semigroup of order $\alpha \in]0, 1[$

As a special case we consider $(\eta_t^\alpha)_{t>0}$, i.e., the *one-sided stable semigroup of order $\alpha \in]0, 1]$*, as an example for $(\mu_t)_{t>0}$ above. $(\eta_t^\alpha)_{t>0}$ is defined by its Laplace transform $\mathcal{L}$ as follows

$$(4.11) \qquad \mathcal{L}\eta_t^\alpha(s) = \exp(-t \, s^\alpha) , \quad s > 0 .$$

If $(p_t)_{t>0}$ is the *Brownian semigroup* on $(\mathbb{R}^d, \mathcal{B}(\mathbb{R}^d))$, i.e., for $f \in \mathcal{B}(\mathbb{R}^d)_b$ and $t > 0$,

$$(4.12) \qquad p_t f(x) = \int f(y)(2\pi t)^{-d/2} \exp\left[-|x - y|^2/2t\right] dy , \quad x \in \mathbb{R}^d ,$$

then define for $\alpha \in]0, 1]$,

$$(4.13) \qquad p_t^\alpha f := \int_0^\infty p_s f \, \eta_t^\alpha(ds) , \quad t > 0 .$$

Note that $p_t = p_t^1$, $t > 0$. $(p_t^\alpha)_{t>0}$ is called *symmetric stable semigroup of order* $\alpha > 0$. Since $(p_t)_{t>0}$ is dx-symmetric, so is $(p_t^\alpha)_{t>0}$, and clearly $(p_t)_{t>0}$ satisfies (4.6) for $D = C_0^\infty(\mathbb{R}^d)$. Hence $(p_t^\alpha)_{t>0}$ gives rise to a symmetric Dirichlet form

$(\mathcal{E}, D(\mathcal{E}))$ on $L^2(\mathbb{R}^d; dx)$ which in fact coincides (up to a constant) with the closed form $(\mathbb{D}^{(\alpha)}, H^{\alpha,2}(\mathbb{R}^d))$ defined in Subsection 1d). To show this, note that if

$$\nu_t^\alpha := p_t^\alpha(0, \cdot) \, , \quad t > 0 \, ,$$

then $p_t^\alpha f = \nu_t^\alpha * f$ (where $*$ denotes convolution w.r.t. the additive structure on $\mathbb{R}^d$). Hence in terms of their Fourier transforms for all $f \in L^2(\mathbb{R}^d; dx)$,

$$(2\pi)^{-d/2}(T_t^\alpha f)\hat{\ } = \hat{\nu}_t^\alpha \cdot \hat{f} \, , \; t > 0 \, ,$$

where $\hat{\nu}_t^\alpha(x) = (2\pi)^{-d/2} \exp[-2^{-\alpha} t |x|^{2\alpha}]$, $x \in \mathbb{R}^d$, as easily follows from (4.11), (4.13), and T_t^α is the contraction on $L^2(\mathbb{R}^d; dx)$ corresponding to p_t^α. Now by Plancherel's theorem for all $u \in L^2(\mathbb{R}^d; dx)$

$$
\begin{aligned}
(4.14) \qquad \frac{1}{t}(u - T_t^\alpha u, u) \;&=\; \frac{1}{t} \int (\hat{u} - \hat{\nu}_t^\alpha \hat{u}) \bar{\hat{u}} \, dx \\
&=\; \int |\hat{u}|^2(x) \frac{1 - \exp[-2^{-\alpha} t |x|^{2\alpha}]}{t} \, dx.
\end{aligned}
$$

By the results of Chap.I, Sect. 2 (cf. the remarks following I.2.22) we know that $u \in D(\mathcal{E})$ if and only if $\sup_{t > 0} \frac{1}{t}(u - T_t^\alpha u, u) < \infty$ and that $\mathcal{E}(u, v) = \lim_{t \to 0} \frac{1}{t}(u - T_t^\alpha u, v)$ for $u, v \in D(\mathcal{E})$. Hence by (4.14)

$$D(\mathcal{E}) = \{ u \in L^2(\mathbb{R}^d; dx) \mid \int |\hat{u}|^2(x) |x|^{2\alpha} \, dx < \infty \}$$

and for $c := 2 \cdot 2^{-\alpha}$

$$\mathcal{E}(u, v) = \frac{c}{2} \int \hat{u}(x) \bar{\hat{v}}(x) |x|^{2\alpha} \, dx \, , \quad u, v \in D(\mathcal{E}) \, .$$

Hence $(\mathcal{E}, D(\mathcal{E})) = (c \, \mathbb{D}^\alpha, H^{\alpha,2}(\mathbb{R}^d))$ by the following

Exercise 4.4. Prove that $C_0^\infty(\mathbb{R}^d)$ is dense in $D(\mathcal{E})$ w.r.t. $\mathcal{E}_1^{1/2}$.

5   Starting point: resolvent of kernels

a)   Preliminaries

Let $(E, \mathcal{B})$ be a measurable space and $(R_\alpha)_{\alpha>0}$ a *sub-Markovian resolvent of kernels on* $(E, \mathcal{B})$, i.e., each αR_α is a sub-Markovian kernel on $(E, \mathcal{B})$ and for all $\alpha, \beta > 0$

$$(5.1) \qquad\qquad R_\alpha - R_\beta = (\beta - \alpha) R_\alpha R_\beta \, .$$

$(R_\alpha)_{\alpha>0}$ is called *Markovian* if $\alpha R_\alpha 1 = 1$ for all $\alpha > 0$.

Exercise 5.1. Let $(p_t)_{t>0}$ be a semigroup of sub-Markovian kernels on $(E, \mathcal{B})$. Assume that $(p_t)_{t>0}$ is *measurable*, i.e., $(t, z) \mapsto p_t f(z)$ is $\mathcal{B}(]0, \infty[) \otimes \mathcal{B}$-measurable for all $f \in \mathcal{B}_b$. Prove that if

$$R_\alpha f(z) := \int_0^\infty e^{-\alpha t} p_t f(z) dt \ , \quad \alpha > 0 \ , \quad f \in \mathcal{B}_b \ , \quad z \in E \ ,$$

then $(R_\alpha)_{\alpha>0}$ is a sub-Markovian resolvent of kernels on $(E, \mathcal{B})$.

Let $(R_\alpha)_{\alpha>0}$ be a sub-Markovian resolvent of kernels on $(E, \mathcal{B})$ and m a positive measure on $(E, \mathcal{B})$ such that m is αR_α-supermedian for every $\alpha > 0$. Let G_α, $\alpha > 0$, be the corresponding operators on $L^2(E; m)$ according to 4.1. Suppose we can find a dense subset $D \subset L^2(E; m)$ such that

$$(5.2) \qquad \alpha R_\alpha f \xrightarrow[\alpha \to \infty]{} f \text{ in } m\text{-measure for all } f \in D \ .$$

Exercise 5.2. Show that (by (5.2)), $(G_\alpha)_{\alpha>0}$ is a strongly continuous contraction resolvent on $L^2(E; m)$.

We now assume in addition that

$$(5.3) \qquad \begin{array}{c} \text{there exists } K > 0 \text{ such that for all } f, g \in L^2(E; m) \cap \mathcal{B}_b(E), \\ |(R_1 f, g)| \leq K (R_1 f, f)^{1/2} (R_1 g, g)^{1/2} \end{array}$$

(where again $(,)$ denotes the inner product in $L^2(E; m)$). Note that $(R_\alpha f, f) = (G_\alpha f, f) \geq 0$ by I.(1.5). Clearly, (5.3) implies that each G_α satisfies I.(2.5), hence by the results in Chap.I, Sect. 2 there exists a corresponding coercive closed form $(\mathcal{E}, D(\mathcal{E}))$ which is a Dirichlet form by I.4.3, I.4.4 since both resolvents $(G_\alpha)_{\alpha>0}$ and $(\hat{G}_\alpha)_{\alpha>0}$ are sub-Markovian by 4.1.

b)   Construction of resolvents of kernels

Let $(R_\alpha)_{\alpha>0}$ be a resolvent of kernels on $(E, \mathcal{B})$ and let $f \in \mathcal{B}^+$. It follows by (5.1) that $\alpha \mapsto R_\alpha f$, $\alpha > 0$, is decreasing. Define

$$(5.4) \qquad R_0 f := \sup_{\alpha>0} R_\alpha f = \lim_{\alpha \to 0} R_\alpha f \ .$$

Then $R_0 f(z) = R_0(z, f)$, $z \in E$, $f \in \mathcal{B}^+$, defines a kernel on $(E, \mathcal{B})$ called the *potential kernel* of $(R_\alpha)_{\alpha>0}$. Note that in the special situation of 5.1 one has that

$$(5.5) \qquad R_0 f(z) = \int_0^\infty p_t f(z) \, dt \ , \qquad f \in \mathcal{B}^+ \ , \quad z \in E \ .$$

In the general case R_0 satisfies for $f \in \mathcal{B}^+$

$$(5.6) \qquad R_0 f = \alpha R_\alpha R_0 f + R_\alpha f = \alpha R_0 R_\alpha f + R_\alpha f$$

and

$$(5.7) \qquad I + \alpha R_0 = \sum_{n=0}^{\infty} (\alpha R_\alpha)^n$$

where I is the identity kernel (i.e., $If = f$ for all $f \in \mathcal{B}^+$). (5.6) is obvious from (5.1) and (5.7) can be proved as follows. Iterating (5.1) we obtain for $f \in \mathcal{B}_b^+$ and $\beta > 0$

$$(5.8) \qquad f + \alpha R_\beta f = \sum_{n=0}^{N} (\alpha R_{\alpha+\beta})^n f + (\alpha R_{\alpha+\beta})^N (\alpha R_\beta) f \ .$$

Since $R_\beta f \uparrow R_0 f$ and $R_{\alpha+\beta} f \uparrow R_\alpha f$ as $\beta \downarrow 0$, it remains to show that $\lim_{N \to \infty} (\alpha R_{\alpha+\beta})^N R_\beta f = 0$. Note that by (5.8) $\left((\alpha R_{\alpha+\beta})^N R_\beta f \right)_{N \in \mathbb{N}}$ is decreasing. Set

$$g := \lim_{N \to \infty} (\alpha R_{\alpha+\beta})^N R_\beta f \ .$$

Then $\alpha R_{\alpha+\beta} g = g$ and applying (5.6) to g we obtain that $R_\beta g = R_\beta g + \alpha^{-1} g$, hence $g = 0$ since $R_\beta g < \infty$. Thus (5.7) is proved. Note that by (5.6) in the situation described at the end of Subsection 5a) we have that informally, $R_0 = (-L)^{-1}$ where L is the generator of $(G_\alpha)_{\alpha>0}$.

It is now possible to reconstruct $(R_\alpha)_{\alpha>0}$ from R_0 using (5.6). (Informally, (5.6) is equivalent to $R_\alpha = (I + \alpha R_0)^{-1} \circ R_0$). Hence in connection with Subsection 5a) we thus can construct Dirichlet forms starting with R_0. The reconstruction of $(R_\alpha)_{\alpha>0}$ requires some techniques which are outside the main theme of this book. Therefore, we only quote the important results and give references for the proofs. We start with

Proposition 5.3. *Let R_0 be the potential kernel of a sub-Markovian resolvent $(R_\alpha)_{\alpha>0}$ of kernels on $(E, \mathcal{B})$. Then R_0 satisfies the complete maximum principle, i.e., for all $f, g \in \mathcal{B}^+$*

$$R_0 f \leq R_0 g + 1 \ \text{ on } \ \{f > 0\} \ \text{ implies } R_0 f \leq R_0 g + 1 \ \text{ on } \quad E \ .$$

Proof. See [BlHa 86, II.7.1]. $\qquad\qquad\qquad\qquad\qquad\qquad\qquad\qquad\qquad\qquad$ $\square$

Theorem 5.4. (Hunt) *Let R_0 be a kernel on $(E, \mathcal{B})$ satisfying the complete maximum principle such that $R_0 1$ is bounded. Then for $\alpha > 0$, $I + \alpha R_0 : \mathcal{B}_b \to \mathcal{B}_b$ is invertible and if*

$$R_\alpha := (I + \alpha R_0)^{-1} \circ R_0 \ ,$$

then $(R_\alpha)_{\alpha>0}$ is a sub-Markovian resolvent of kernels on $(E, \mathcal{B})$ with potential kernel R_0.

Proof. See [BlHa 86, II.7.2-7.7]. $\qquad\qquad\qquad\qquad\qquad\qquad\qquad\qquad\qquad\qquad$ $\square$

Remark 5.5. There is a generalization of 5.4 due to F. Hirsch [Hi 74] and J.C. Taylor [Ta 75] giving necessary and sufficient conditions for the existence of a sub-Markovian resolvent of kernels on $(E, \mathcal{B})$ with a given potential kernel R_0 if R_0 satisfies the complete maximum principle and is *proper*, i.e., there exists $E_n \in \mathcal{B}$, $n \in \mathbb{N}$, with $E_n \uparrow E$ and each $R_0 1_{E_n}$ is bounded.

Exercise 5.6. Let $E = \mathbb{R}^d$, $d \geq 3$, $\mathcal{B} = \mathcal{B}(\mathbb{R}^d)$ and $\alpha \in]0,1]$. Let

$$g^{(\alpha)}(x,y) := |x - y|^{2\alpha - d} \; ; \;\; x, y \in \mathbb{R}^d \; .$$

Define a kernel $R_0^{(\alpha)}$ on $(\mathbb{R}^d, \mathcal{B}(\mathbb{R}^d))$ by

$$(5.9) \qquad R_0^{(\alpha)} f(x) := \int f(y) g^{(\alpha)}(x,y)\, dy \; , \;\; f \in \mathcal{B}(\mathbb{R}^d)^+.$$

Define for $\beta > 0$,

$$R_\beta^{(\alpha)} f := \int_0^\infty e^{-t\beta} p_t^\alpha f\, dt \; , \;\; f \in \mathcal{B}(\mathbb{R}^d)^+,$$

where $(p_t^\alpha)_{t>0}$ is given by (4.13). Then by 5.1, $\left(R_\beta^{(\alpha)}\right)_{\beta>0}$ is an m-symmetric sub-Markovian resolvent of kernels on $(\mathbb{R}^d, \mathcal{B}(\mathbb{R}^d))$. Prove that its potential kernel is up to a constant given by (5.9). (Hint: Use that the Fourier transform of $g(\cdot, 0)$ is up to a constant given by $x \mapsto |x|^{-2\alpha}$).

Remark 5.7. It is also possible to verify directly that $R_0^{(\alpha)}$ in (5.9) satisfies the complete maximum principle (cf. [L 72]) and then apply the Hirsch-Taylor Theorem to obtain $\left(R_\beta^{(\alpha)}\right)_{\beta>0}$.

6   Notes/References

Section 1. This section is standard. Almost all of it is contained in [F 80].
Section 2. Parts a) and b) are based on [RW 85, Sections 3,4] and [AR 90a]. In the latter reference conditions for the closability of symmetric forms of this type are given which are both sufficient and necessary. In particular, a related conjecture of M. Fukushima is proved (cf. [AR 90a, Sect. 5a)]). Part c) is taken from [F 80, Example 1.2.1]. Part d) is new to some extent. It extends classical results which can be found in [Bl 71, Sect.10],[HH 68] and [St 65] and contains a considerably shorter proof for the Dirichlet property based on the "φ_ε-criterion" in I.4.8. Part e) is taken from [AM 88, 91a,d].
Section 3. Parts $a) - c)$ are essentially contained in [AR 90a]. For the proof of Lemma 3.12 we refer to [R 89, A.2]. Part d) is contained in [RZ 90] to which we also refer for the so-called "uniqueness problem" (cf. [R 90b] and the references therein) which is important for this type of forms and not touched upon at all in this book. A more general closability result applicable in this case is given in [AR 90b, Theorem 4.7]. We also note that the Dirichlet forms appearing in white noise analysis (cf. [HPS 88]) are special cases of those studied in this section. The interested reader is referred to [AHPRS 90 a,b] and [PR 90] for details.
Section 4. Most of the material of this section is well-known (cf. [BlHa 86, Chap.V, Sect. 3] w.r.t. Part b) and [F 80, Example 1.4.1] w.r.t. Part c). Proposition 4.3, however, seems to be new under these weak assumptions.
Section 5. Part b) is mainly based on [BlHa 86, Chapt.II, Sect. 7].

Chapter III

Analytic Potential Theory of Dirichlet Forms

In this chapter we develop some analytic potential theory of Dirichlet forms. We try to keep the amount of material as small as possible but sufficient for understanding the probabilistic part of the theory contained in Chapters IV, V below. The corresponding probabilistic potential theory and the relation between the two is studied in Chapter V, Sect.5, as far as necessary for the purpose of this book. In Section 1 we consider excessive and reduced (or balayaged) functions. In Section 2 we look at the capacities corresponding to a Dirichlet form $(\mathcal{E}, D(\mathcal{E}))$ and introduce an "intrinsic" notion of $\mathcal{E}$-exceptional sets. Section 3 contains all on $\mathcal{E}$-quasi-continuity that we use later. In this chapter we consider the following situation: let E be a Hausdorff topological space. Let $\mathcal{B}(E)$ be the σ-algebra consisting of its Borel subsets and m be a σ-finite positive measure on $(E, \mathcal{B}(E))$. Let $(\mathcal{E}, D(\mathcal{E}))$ be a fixed Dirichlet form on $L^2(E; m)$ with associated generator L , semigroups $(T_t)_{t>0}$, $(\hat{T}_t)_{t>0}$ and resolvents $(G_\alpha)_{\alpha>0}$, $(\hat{G}_\alpha)_{\alpha>0}$ on $L^2(E; m)$. Let $K \geq 1$ be a continuity constant of $(\mathcal{E}, D(\mathcal{E}))$ and, as before, let $\tilde{\mathcal{E}}$ denote its symmetric part. Recall that $\mathcal{E}_\alpha := \mathcal{E} + \alpha(\,,\,)$, $\alpha > 0$, where $(\,,\,)$ is the usual inner product in $L^2(E; m)$ and that $\|\,\| := (\,,\,)^{1/2}$.

1 Excessive functions and balayage

Definition 1.1. Let $\alpha \in]0, \infty[$. $u \in L^2(E; m)$ is called α-*excessive* (resp. α-*coexcessive*) if $e^{-\alpha t} T_t u \leq u$ (resp. $e^{-\alpha t} \hat{T}_t u \leq u$) for all $t > 0$.

Proposition 1.2. *Let $u \in L^2(E; m)$ and $\alpha > 0$. If u is α-excessive then $u \geq 0$ and $\beta G_{\beta+\alpha} u \leq u$ for all $\beta > 0$. Moreover, for $u \in D(\mathcal{E})$ the following assertions are equivalent:*

(i) u is α-excessive.

(ii) $\beta G_{\beta+\alpha} u \leq u$ for all $\beta > 0$.

(iii) $\mathcal{E}_\alpha(u, v) \geq 0$ for all $v \in D(\mathcal{E})$, $v \geq 0$.

Corresponding statements hold for α-coexcessive functions.

Proof. Since $\|e^{-\alpha t} T_t u\| \leq e^{-\alpha t} \|u\|$ for $t > 0$, we can find a sequence $t_n \in]0, \infty[$ with $t_n \uparrow \infty$ as $n \to \infty$ such that $e^{-\alpha t_n} T_{t_n} u \underset{n \to \infty}{\longrightarrow} 0$ m-a.e. . If u is α-excessive it follows that $u \geq \lim\limits_{n \to \infty} e^{-\alpha t_n} T_{t_n} u = 0$. The second statement is obvious by I.(1.4). Now we assume that $u \in D(\mathcal{E})$ and complete the proof of the equivalence of (i)-(iii).

$(ii) \Rightarrow (iii)$: We have by I.2.13 (iii) and the resolvent equation that for all $v \in D(\mathcal{E})$, $v \geq 0$,

$$
\begin{aligned}
\mathcal{E}_\alpha(u, v) &= \lim_{\beta \to \infty} \left[\beta(u - \beta G_\beta u, v) + \alpha(u, v) \right] \\
&= \lim_{\beta \to \infty} \left[\beta(u - \beta G_{\beta+\alpha} u, v) + \alpha(u - \beta^2 G_\beta G_{\beta+\alpha} u, v) \right] \\
&\geq \lim_{\beta \to \infty} \alpha(u - \beta G_\beta u, v)
\end{aligned}
$$

where we used (ii) in the last step. Using the strong continuity of $(G_\alpha)_{\alpha>0}$ we obtain (iii).

$(iii) \Rightarrow (i)$: Let $v \in L^2(E; m)$, $v \geq 0$, then

$$
\hat{G}_\alpha v - e^{-\alpha t} \hat{T}_t \hat{G}_\alpha v = \int_0^t e^{-\alpha s} \hat{T}_s v \, ds \geq 0 \; .
$$

Consequently, by (iii)

$$
(u - e^{-\alpha t} T_t u, v) = (u, v - e^{-\alpha t} \hat{T}_t v) = \mathcal{E}_\alpha(u, \hat{G}_\alpha v - e^{-\alpha t} \hat{T}_t \hat{G}_\alpha v) \geq 0 \; ,
$$

i.e., $u - e^{-\alpha t} T_t u \geq 0$ and (i) is proved. $\qquad\square$

Exercise 1.3. (i) Let $\alpha \in]0, \infty[$ and $f \in L^2(E; m)$, $f \geq 0$. Prove that $G_\alpha f$ is α-excessive resp. $\hat{G}_\alpha f$ is α-coexcessive. Hence prove that the linear span of all α-excessive resp. α-coexcessive functions is dense in $D(\mathcal{E})$. Furthermore, show that $G_\alpha f > 0$ if $f > 0$.

(ii) Let $u \in L^2(E; m)$ be α-excessive. Assume that there exists $v \in D(\mathcal{E})$, v α-coexcessive, such that $u \leq v$. Prove that $u \in D(\mathcal{E})$. (Hint: Use I.2.13(i) and 1.2(ii)).

(iii) If $u \in D(\mathcal{E})$ is α-excessive then $G_\beta u$ is α-excessive for all $\beta > 0$.

(iv) If $u, v \in L^2(E; m)$ are α-excessive (resp. α-coexcessive), then $u \wedge v$ is α-excessive (resp. α-coexcessive).

(v) If $u \in L^2(E; m)$ is α-excessive then u is β-excessive for all $\beta > \alpha$.

(vi) If $u \in D(\mathcal{E})$ is α-excessive then $\beta \mapsto \beta G_{\beta+\alpha} u$ is increasing.

Define for $U \subset E$, U open,

$$(1.1) \qquad D(\mathcal{E})_{U^c} := \{u \in D(\mathcal{E}) \mid u = 0 \; m - a.e. \text{ on } \quad U\}$$

where $U^c := E \setminus U$. Note that $D(\mathcal{E})_{U^c}$ is a closed subspace of $D(\mathcal{E})$. Of course, in (1.1) as in 1.4-1.7 below we can replace the open set U by an arbitrary subset A of E. We confine ourselves to open sets in order to avoid notational confusion with other literature (cf. e.g. [F 80]).

Lemma 1.4. *Let $U \subset E$, U open, and $u \in D(\mathcal{E})$. Then there exists a unique $u'_U \in D(\mathcal{E})$ satisfying*

  (i) $u'_U = u \;\; m$-a.e. on U

  (ii) $\mathcal{E}_1(u'_U, w) = 0$ for all $w \in D(\mathcal{E})_{U^c}$.

Proof. Applying I.2.7 to $J(w) := \mathcal{E}_1(u, w)$, $w \in D(\mathcal{E})$, and $C := D(\mathcal{E})_{U^c}$ we obtain a unique $v \in D(\mathcal{E})_{U^c}$ such that $\mathcal{E}_1(v, w) = \mathcal{E}_1(u, w)$ for all $w \in D(\mathcal{E})_{U^c}$. Define $u'_U := u - v$, then u'_U satisfies (i), (ii). The uniqueness of u'_U follows from the uniqueness of v. $\qquad\qquad\square$

Proposition 1.5. *Let h be a function on E. Define for $U \subset E$, U open,*

$$\mathcal{L}_{h,U} := \{w \in D(\mathcal{E}) \mid w \geq h \; m\text{-a.e. on } U\} \; .$$

Suppose that $\mathcal{L}_{h,U} \neq \emptyset$. Then:

  (i) There exist unique $h_U, \hat{h}_U \in \mathcal{L}_{h,U}$ such that for all $w \in \mathcal{L}_{h,U}$

$$\mathcal{E}_1(h_U, w) \geq \mathcal{E}_1(h_U, h_U) \text{ and } \mathcal{E}_1(w, \hat{h}_U) \geq \mathcal{E}_1(\hat{h}_U, \hat{h}_U) \; .$$

  (ii) $\mathcal{E}_1(h_U, w) \geq 0$ for all $w \in D(\mathcal{E})$ with $w \geq 0$ m-a.e. on U. In particular, h_U is 1-excessive and $\mathcal{E}_1(h_U, w) = 0$ for all $w \in D(\mathcal{E})_{U^c}$.

  (iii) h_U is the smallest function u on E such that $u \wedge h_U$ is a 1-excessive function in $D(\mathcal{E})$ and $u \geq h$ m-a.e. on U. In particular, $(0 \leq)h_U \leq h(m$-a.e. on E) if and only if $h \wedge h_U$ is a 1-excessive function in $D(\mathcal{E})$. In this case $h_U = h$ m-a.e. on U.

Suppose that $V \subset U \subset E$, V open. Then:

  (iv) $\mathcal{L}_{h,V} \supset \mathcal{L}_{h,U}(\neq \emptyset)$, $h_V \leq h_U$, and

$$\mathcal{E}_1(h_V, h_V) \leq K^2 \mathcal{E}_1(h_U, h_U) \; .$$

  (v) If $h \wedge h_U$ is a 1-excessive function in $D(\mathcal{E})$, then $(h_U)_V = h_V$.

  (vi) $(h_V)_U = h_V$.

Interchanging the two entries of $\mathcal{E}$ and replacing excessive by coexcessive we have statements for $\hat{h}_U$ (and $\hat{h}_V$) corresponding to (ii)-(vi).

Proof. It suffices to give the proof for h_U.

(i): $\mathcal{L}_{h,U}$ is a non-empty closed convex subset of $D(\mathcal{E})$. Hence by I.2.6 (applied to $J \equiv 0$) there exists a unique $h_U \in \mathcal{L}_{h,U}$ such that $\mathcal{E}_1(h_U, w - h_U) \geq 0$ for all $w \in \mathcal{L}_{h,U}$ and (i) is proved.

(ii): $\mathcal{E}_1(h_U, w) = \mathcal{E}_1(h_U, (w + h_U) - h_U) \geq 0$ for all $w \in D(\mathcal{E})$, $w \geq 0$ m-a.e. on U, since $w + h_U \geq h_U \geq h$ m-a.e. on U.

(iii): By definition and (ii) h_U has the stated property. Let $u : E \to \mathbb{R}$ be such that $u \wedge h_U \in D(\mathcal{E})$, $u \wedge h_U$ is 1-excessive and $u \geq h$ m-a.e. on U. Since by 1.2(iii)

$$\mathcal{E}_1(u \wedge h_U , h_U - u \wedge h_U) \geq 0$$

we have that

$$0 \leq \mathcal{E}_1(h_U - u \wedge h_U , h_U - u \wedge h_U) \leq \mathcal{E}_1(h_U , h_U - u \wedge h_U)$$

which is negative by (i). Hence $h_U = u \wedge h_U$, i.e., $h_U \leq u$.

(iv): The first part is obvious by (iii); the second follows from

$$\mathcal{E}_1(h_V, h_V) \leq \mathcal{E}_1(h_V, h_U) \leq K \ \mathcal{E}_1(h_V, h_V)^{1/2} \mathcal{E}_1(h_U, h_U)^{1/2} \ .$$

(v): Let $w \in \mathcal{L}_{h,V}$, then by (iii), $w \geq h = h_U$ m-a.e. on V. Hence

$$\mathcal{E}_1((h_U)_V, w) \geq \mathcal{E}_1((h_U)_V, (h_U)_V) \ .$$

But $(h_U)_V \geq h_U \geq h$ m-a.e. on V, i.e., $(h_U)_V \in \mathcal{L}_{h,V}$, hence (by the uniqueness of h_V in (i)) $(h_U)_V = h_V$.

(vi): By (ii), (iii) (and their "coexcessive versions") for all $w \in D(\mathcal{E})$, w 1-coexcessive,

$$\mathcal{E}_1((h_V)_U, w) = \mathcal{E}_1((h_V)_U, \hat{w}_U) = \mathcal{E}_1(h_V, \hat{w}_U) = \mathcal{E}_1(h_V, w).$$

Hence (vi) follows by 1.3(i). $\qquad\qquad\qquad\qquad\qquad\qquad\qquad\qquad\qquad$ $\square$

Remark 1.6. (i) If $h \in D(\mathcal{E})$ then $\mathcal{L}_{h,U} \neq \emptyset$ for every $U \subset E$, U open. Suppose in addition, that $h \wedge h_U$ is 1-excessive (or equivalently $h_U \leq h$ m-a.e. by 1.5 (iii), which is always the case if h is 1-excessive), then $h_U = h'_U$ (defined as in 1.4) and in case $E = U$, $h_E = h$.

(ii) Let $h \equiv 1$ and let $U \subset E$, U open with $\mathcal{L}_{h,U} \neq \emptyset$. Then $h \wedge h_U, h \wedge \hat{h}_U \in D(\mathcal{E})$ and they are 1-excessive, 1-coexcessive respectively, i.e., $h_U, \hat{h}_U \leq h$ by 1.5 (iii).

(iii) Let $U \subset E$, U open. Since h_U is 1-excessive, it follows by 1.5 (iii) that h_U is the smallest 1-excessive function $u \in D(\mathcal{E})$ such that $u \geq h$ m-a.e. on U. In particular, if $g : E \to \mathbb{R}$, with $\mathcal{L}_{g,U} \neq \emptyset$ and $g \geq h$ m-a.e. on U, then $g_U \geq h_U$, $\hat{g}_U \geq \hat{h}_U$ (m-a.e. on E).

Because of 1.5 (iii) (or 1.6 (iii)) for $U \subset E$, U open, h_U is called the (1-)*reduced* (or *balayaged*) *function* and $\hat{h}_U$ the (1-)*coreduced* (or *cobalayaged*) *function* of h on U.

Exercise 1.7. Let h be a function on E and $U \subset E$, U open, such that $\mathcal{L}_{h,U} \neq \emptyset$. Prove that if $h_U, \hat{h}_U \leq h$, then

$$\mathcal{E}_1(h_U, h_U) = \mathcal{E}_1(h_U, \hat{h}_U) = \mathcal{E}_1(\hat{h}_U, \hat{h}_U) \ .$$

2 $\mathcal{E}$-exceptional sets and capacities

Definition 2.1.   (i) An increasing sequence $(F_k)_{k\in\mathbb{N}}$ of closed subsets of E is called an $\mathcal{E}$-*nest* if $\bigcup_{k\geq 1} D(\mathcal{E})_{F_k}$ is dense in $D(\mathcal{E})$ (w.r.t. $\tilde{\mathcal{E}}_1^{1/2}$).

 (ii) A subset $N \subset E$ is called $\mathcal{E}$-*exceptional* if $N \subset \bigcap_{k\geq 1} F_k^c$ for some $\mathcal{E}$-nest $(F_k)_{k\in\mathbb{N}}$. We say that a property of points in E holds $\mathcal{E}$-*quasi-everywhere* (abbreviated $\mathcal{E}$-*q.e.*), if the property holds outside some $\mathcal{E}$-exceptional set.

Remark 2.2. Note that 2.1 only depends on the symmetric part of $(\mathcal{E}, D(\mathcal{E}))$.

Exercise 2.3. Show that every Borel $\mathcal{E}$-exceptional set has m-measure zero.

The description of "small sets" by 2.1 is essentially sufficient to formulate all subsequent results in this book. For the proofs and practical purposes, however, 2.1 is not very handy. In this section we therefore introduce and study capacities which can be used to characterize $\mathcal{E}$-nests and whose zero sets are exactly the $\mathcal{E}$-exceptional sets (cf. Theorem 2.11 below). But these capacities are not unique.

Definition 2.4. Let h, g be functions on E such that one of the following two conditions holds

 (i) $h, g \in D(\mathcal{E})$, h 1-excessive, g 1-coexcessive

 (ii) $h = g$ and $h_U \wedge h$, $\hat{h}_U \wedge h$ are 1-excessive resp. 1-coexcessive functions in $D(\mathcal{E})$ for all $U \subset E$, U open with $\mathcal{L}_{h,U} \neq \emptyset$.

Define for $U \subset E$, U open

$$(2.1) \qquad \mathrm{Cap}_{h,g}(U) := \begin{cases} \mathcal{E}_1(h_U, \hat{g}_U) & \text{if} \quad \mathcal{L}_{h,U} \neq \emptyset \neq \mathcal{L}_{g,U} \\ +\infty & \text{else} \end{cases}.$$

and for arbitrary $A \subset E$

$$(2.2) \qquad \mathrm{Cap}_{h,g}(A) := \inf\{\mathrm{Cap}_{h,g}(U) | A \subset U \subset E\ , U \text{ open}\}\ .$$

Remark 2.5. (i) It follows from Exercise 2.6 below that (2.1), (2.2) are consistent.

(ii) Assumptions (i), (ii) in 2.4 are used below to obtain Lemma 2.7. Note here that (i) or (ii) implies that $h, g \geq 0$ and that for $U \subset E$, U open, both $h_U = h$ and $\hat{g}_U = g$ m-a.e. on U by 1.5 (iii) provided $\mathcal{L}_{h,U} \neq \emptyset$ in case (ii). A typical example for (ii) is $h = g \equiv 1$ (cf. 1.6 (ii)).

(iii) It can occur in case 2.4 (ii) that $\mathrm{Cap}_{h,g}(A) = \infty$ for every $\emptyset \neq A \subset E$. In case 2.4 (i), however, $\mathrm{Cap}_{h,g}(A) \leq \mathrm{Cap}_{h,g}(E) < \infty$ for all $A \subset E$.

(iv) If $h', g' : E \to \mathbb{R}$ are as in 2.4 (i) or (ii) such that $h \leq h'$, $g \leq g'$ then $\mathrm{Cap}_{h,g} \leq \mathrm{Cap}_{h',g'}$ by 1.6 (iii).

From now on we fix h, g as in 2.4.

Exercise 2.6. (i) Let U, W be open subsets of E, $U \subset W$. Prove that $\mathrm{Cap}_{h,g}(U) \leq \mathrm{Cap}_{h,g}(W)$. Furthermore, show that if $\mathcal{L}_{h,U} \neq \emptyset \neq \mathcal{L}_{g,U}$ and $m(W \setminus U) = 0$ then $\mathcal{L}_{h,W} \neq \emptyset \neq \mathcal{L}_{g,W}$ and $\mathrm{Cap}_{h,g}(U) = \mathrm{Cap}_{h,g}(W)$.
(ii) Prove that if $h = g > 0$ then $\mathrm{Cap}_{h,g}(N) = 0$ implies $m(N) = 0$ for all $N \in \mathcal{B}(E)$. (See also 2.11 below.)

Lemma 2.7.   *(i) Let U, W be open subsets of E, then*

$$\mathrm{Cap}_{h,g}(U \cup W) + \mathrm{Cap}_{h,g}(U \cap W) \leq \mathrm{Cap}_{h,g}(U) + \mathrm{Cap}_{h,g}(W) \ ,$$

(strong subadditivity).

(ii) Let $(U_n)_{n \in \mathbb{N}}$ be an increasing sequence of open subsets of E, then

$$\mathrm{Cap}_{h,g}\big(\bigcup_{n \geq 1} U_n\big) = \lim_{n \to \infty} \mathrm{Cap}_{h,g}(U_n) \ .$$

Proof. (i): We may assume that $\mathrm{Cap}_{h,g}(U), \mathrm{Cap}_{h,g}(W) < \infty$. Then $h_U + h_W \geq h$ on $U \cup W$, hence $\mathcal{L}_{h,U \cup W} \neq \emptyset$. Since

$$h_{U \cup W} + h_{U \cap W} \leq h_U + h_W \quad \text{on } U \cup W \ ,$$

and since $\hat{g}_V = g = \hat{g}_{U \cup W}$ m-a.e. on V for all $V \subset U \cup W$, V open, we obtain that

$$\begin{aligned}
\mathrm{Cap}_{h,g}(U \cup W) + \mathrm{Cap}_{h,g}(U \cap W) &= \mathcal{E}_1(h_{U \cup W}, \hat{g}_{U \cup W}) + \mathcal{E}_1(h_{U \cap W}, \hat{g}_{U \cup W}) \\
&\leq \mathcal{E}_1(h_U + h_W, \hat{g}_{U \cup W}) \\
&= \mathcal{E}_1(h_U, \hat{g}_U) + \mathcal{E}_1(h_W, \hat{g}_W) \\
&= \mathrm{Cap}_{h,g}(U) + \mathrm{Cap}_{h,g}(W) \ .
\end{aligned}$$

(ii): Let $U := \bigcup_{n \geq 1} U_n$. By 2.6(i) we only have to prove that

$$\mathrm{Cap}_{h,g}(U) \leq \lim_{n \to \infty} \mathrm{Cap}_{h,g}(U_n)$$

and we may assume that $\sup_{n \geq 1} \mathrm{Cap}_{h,g}(U_n) < \infty$. Since $(h_{U_n})_{n \in \mathbb{N}}$ is increasing, $h_\infty := \lim_{n \to \infty} h_{U_n}$ exists m-a.e. . In case 2.4 (i) we have by 1.5 (iv) that

$$\mathcal{E}_1(h_{U_n}, h_{U_n}) \leq K^2 \mathcal{E}_1(h, h) \ ,$$

and in case 2.4 (ii) by 1.7 that

$$\mathcal{E}_1(h_{U_n}, h_{U_n}) = \mathrm{Cap}_{h,g}(U_n) \leq \sup_{n \geq 1} \mathrm{Cap}_{h,g}(U_n) < \infty \ .$$

In particular, $h_{U_n} \to h_\infty$ in $L^2(E; m)$ as $n \to \infty$; therefore, by I.2.12 $(h_{U_n})_{n \in \mathbb{N}}$ weakly converges to h_∞ in the Hilbert space $(D(\mathcal{E}), \tilde{\mathcal{E}}_1)$. Consequently, since $h_\infty \geq h$ m-a.e. on U, $\mathcal{L}_{h,U} \neq \emptyset$ and for all $w \in D(\mathcal{E})$ with $w \geq h$ m-a.e. on U

$$\mathcal{E}_1(h_\infty, w) = \lim_{n \to \infty} \mathcal{E}_1(h_{U_n}, w) \geq \liminf_{n \to \infty} \mathcal{E}_1(h_{U_n}, h_{U_n}) \geq \mathcal{E}_1(h_\infty, h_\infty)$$

(by I.2.12). Hence $h_\infty = h_U$ and

$$\begin{aligned}
\mathrm{Cap}_{h,g}(U) &= \mathcal{E}_1(h_U, \hat{g}_U) = \lim_{n\to\infty} \mathcal{E}_1(h_{U_n}, \hat{g}_U) \\
&= \lim_{n\to\infty} \mathcal{E}_1(h_{U_n}, \hat{g}_{U_n}) = \lim_{n\to\infty} \mathrm{Cap}_{h,g}(U_n) .
\end{aligned}$$

$\square$

Theorem 2.8. $\mathrm{Cap}_{h,g}$ *is a Choquet capacity on E, i.e., it has the following two properties:*

(i) If $(A_n)_{n\in\mathbb{N}}$ is an increasing sequence of subsets of E then

$$\mathrm{Cap}_{h,g}(\bigcup_{n\geq 1} A_n) = \sup_{n\geq 1} \mathrm{Cap}_{h,g}(A_n) .$$

(ii) If $(K_n)_{n\in\mathbb{N}}$ is a decreasing sequence of compact subsets of E then

$$\mathrm{Cap}_{h,g}(\bigcap_{n\geq 1} K_n) = \inf_{n\geq 1} \mathrm{Cap}_{h,g}(K_n) .$$

Furthermore, if $A_n \subset E$, $n \in \mathbb{N}$, then

$$\mathrm{Cap}_{h,g}(\bigcup_{n\geq 1} A_n) \leq \sum_{n\geq 1} \mathrm{Cap}_{h,g}(A_n) .$$

Proof. (i): Because of 2.6 (i), $\left(\mathrm{Cap}_{h,g}(A_n)\right)_{n\in\mathbb{N}}$ is increasing, so we only have to prove that

$$\mathrm{Cap}_{h,g}(\bigcup_{n\geq 1} A_n) \leq \sup_{n\geq 1} \mathrm{Cap}_{h,g}(A_n) .$$

So, let $\epsilon > 0$. We may assume that $\mathrm{Cap}_{h,g}(A_n) < \infty$ for all $n \in \mathbb{N}$. Hence there exist open sets $U_n \subset E$ such that $A_n \subset U_n$, $n \geq 1$, and $\mathrm{Cap}_{h,g}(U_n) \leq \mathrm{Cap}_{h,g}(A_n) + \epsilon \, 2^{-n}$. Assume that for $N \in \mathbb{N}$, $N \geq 2$,

$$\mathrm{Cap}_{h,g}(\bigcup_{n=1}^{N-1} U_n) \leq \mathrm{Cap}_{h,g}(A_{N-1}) + (1 - 2^{1-N})\epsilon$$

which certainly holds for $N = 2$. Then by 2.7(i)

$$\begin{aligned}
\mathrm{Cap}_{h,g}(\bigcup_{n=1}^{N} U_n) &\leq \mathrm{Cap}_{h,g}(A_{N-1}) + (1 - 2^{1-N})\epsilon + \mathrm{Cap}_{h,g}(U_N) \\
&\quad - \mathrm{Cap}_{h,g}((\bigcup_{n=1}^{N-1} U_n) \cap U_N) \\
&\leq \mathrm{Cap}_{h,g}(A_N) + (1 - 2^{-N})\epsilon .
\end{aligned}$$

Hence

$$\mathrm{Cap}_{h,g}(\bigcup_{n=1}^{N} U_n) \leq \sup_{n\geq 1} \mathrm{Cap}_{h,g}(A_n) + \epsilon \quad \text{for all } N \in \mathbb{N} .$$

Letting $N \to \infty$ and using 2.7 (ii) we obtain (i) since $\bigcup_{n\geq 1} A_n \subset \bigcup_{n\geq 1} U_n$.
(ii): We only have to show that $\mathrm{Cap}_{h,g}(\bigcap_{n\geq 1} K_n) \geq \inf_{n\geq 1} \mathrm{Cap}_{h,g}(K_n)$ and we may hence assume that $\mathrm{Cap}_{h,g}(\bigcap_{n\geq 1} K_n) < \infty$. Let $\epsilon > 0$ and $U \subset E$, U open with $\bigcap_{n\geq 1} K_n \subset U$ and

$$(2.3) \qquad \mathrm{Cap}_{h,g}(U) \leq \mathrm{Cap}_{h,g}\Big(\bigcap_{n\geq 1} K_n\Big) + \epsilon \, .$$

Then $K_n \subset U$ for all $n \geq n_0$ for some $n_0 \in \mathbb{N}$, hence

$$(2.4) \qquad \mathrm{Cap}_{h,g}(K_n) \leq \mathrm{Cap}_{h,g}(U) \quad \text{for all } n \geq n_0 \, .$$

(2.3), (2.4) imply the desired inequality.
The last part is an easy consequence of (i) and the (strong) subadditivity (cf. 2.7(i)). $\qquad\qquad\square$

Remark 2.9. (i) Suppose that E is a (topological) Souslin space and that $\mathrm{Cap}_{h,g}(E) < \infty$, then 2.8 implies that for every $A \in \mathcal{B}(E)$

$$\mathrm{Cap}_{h,g}(A) = \sup\{\mathrm{Cap}_{h,g}(K) | K \subset A \, , \, K \text{ compact}\}$$

(Choquet's Capacitability Theorem). This follows by [Bou 74, Chap.IX, Sect. 6, Def. 9, Théorème 6 and Proposition 10].
(ii) In a major part of the literature (cf. e.g. [O 88]) only the case $h \equiv 1 \equiv g$ is considered and E is assumed to be a locally compact separable metric space such that $D(\mathcal{E}) \cap C_0(E)$ is dense both in $D(\mathcal{E})$ w.r.t. $\tilde{\mathcal{E}}_1^{1/2}$ and in $C_0(E)$ ($:=$ all continuous functions on E with compact support) w.r.t. uniform norm (i.e., $(\mathcal{E}, D(\mathcal{E}))$ is *regular*, cf. Chap.IV, Subsection 4a) below). Then clearly, $\mathcal{L}_{1,U} \neq \emptyset$ and hence $\mathrm{Cap}_{1,1}(U) < \infty$ for every relatively compact open subset U of E.

Exercise 2.10. Suppose $\mathcal{E} = \tilde{\mathcal{E}}$, i.e., $\mathcal{E}$ is symmetric. Let $h : E \to \mathbb{R}$ and $U \subset E$, U open, such that $\mathcal{L}_{h,U} \neq \emptyset$. Prove that h_U is the unique function in $\mathcal{L}_{h,U}$ minimizing $w \mapsto \mathcal{E}_1(w,w)$ on $\mathcal{L}_{h,U}$. In particular,

$$\mathrm{Cap}_{1,1}(U) = \inf\{\mathcal{E}_1(w,w) \,|\, w \in D(\mathcal{E}) \, , \, w \geq 1 \ m\text{-a.e. on } U\}$$

(i.e., for $h \equiv g \equiv 1$ our definition of capacity coincides with the one in [F 80]).

Now we want to consider the special case where

$$h := G_1\varphi \, , \, g := \hat{G}_1\hat{\varphi}(\in D(\mathcal{E}))$$

for some $\varphi, \hat{\varphi} \in L^2(E;m)$, $\varphi, \hat{\varphi} > 0$. In this case using 2.1 we shall now see that the system of all sets $N \subset E$ with $\mathrm{Cap}_{h,g}(N) = 0$ is the same for all (h,g) of this type. Moreover, for such (h,g), $\mathrm{Cap}_{h,g}$ are all "equivalent" in a certain sense (cf. 2.11(i) below).

Theorem 2.11.   *(i) An increasing sequence $(F_k)_{k\in\mathbb{N}}$ of closed subsets of E is an $\mathcal{E}$-nest if and only if $\lim_{k\to\infty} \mathrm{Cap}_{h,g}(F_k^c) = 0$.*

(ii) A subset $N \subset E$ is $\mathcal{E}$-exceptional if and only if $\mathrm{Cap}_{h,g}(N) = 0$.

Proof. Clearly (ii) follows from (i). So, we only have to prove (i). Let $(F_k)_{k\in\mathbb{N}}$ be an increasing sequence of closed subsets of E. Since $\left(h_{F_k^c}\right)_{k\in\mathbb{N}}$ is decreasing by 1.5(iv), $h_\infty := \lim_{k\to\infty} h_{F_k^c}$ exists m-a.e. and $h_{F_k^c} \xrightarrow[k\to\infty]{} h_\infty$ in $L^2(E;m)$. By 1.5(iv) we know that

$$\mathcal{E}_1(h_{F_k^c}, h_{F_k^c}) \leq K^2 \mathcal{E}_1(h_{F_1^c}, h_{F_1^c}) \ \text{ for all } k \in \mathbb{N} \, .$$

Hence by I.2.12 we can conclude that $h_{F_k^c} \xrightarrow[k\to\infty]{} h_\infty$ weakly in $(D(\mathcal{E}), \tilde{\mathcal{E}}_1)$; in particular,

$$(2.5) \qquad \mathcal{E}_1(h_\infty, w) = \lim_{k\to\infty} \mathcal{E}_1(h_{F_k^c}, w) = 0 \ \text{ for all } w \in \bigcup_{k\geq 1} D(\mathcal{E})_{F_k}$$

and

$$(2.6) \quad \lim_{k\to\infty} \mathrm{Cap}_{h,g}(F_k^c) = \lim_{k\to\infty} \mathcal{E}_1(h_{F_k^c}, \hat{g}_{F_k^c}) = \lim_{k\to\infty} \mathcal{E}_1(h_{F_k^c}, g) = \mathcal{E}_1(h_\infty, g) \, .$$

Now suppose that $(F_k)_{k\in\mathbb{N}}$ is an $\mathcal{E}$-nest, then (2.5) implies that $\mathcal{E}_1(h_\infty, w) = 0$ for all $w \in D(\mathcal{E})$, hence $\mathcal{E}_1(h_\infty, h_\infty) = 0$, i.e., $h_\infty = 0$. Consequently, $\lim_{k\to\infty} \mathrm{Cap}_{h,g}(F_k^c) = 0$ by (2.6).

Suppose $\lim_{k\to\infty} \mathrm{Cap}_{h,g}(F_k^c) = 0$ and let $v \in D(\mathcal{E})$. To show that v is in the closure of $\bigcup_{k\geq 1} D(\mathcal{E})_{F_k}$ we may assume that v is 1-coexcessive (cf. 1.3(i)). Since $\left(\hat{v}_{F_k^c}\right)_{k\in\mathbb{N}}$ is decreasing we conclude as above by 1.5 (iv) and I.2.12 that there exists $\hat{v}_\infty \in D(\mathcal{E})$ such that $\hat{v}_\infty = \lim_{k\to\infty} \hat{v}_{F_k^c}$ m-a.e. and

$$\lim_{k\to\infty} \mathcal{E}_1(w, \hat{v}_{F_k^c}) = \mathcal{E}_1(w, \hat{v}_\infty) \ \text{ for all } w \in D(\mathcal{E}) \, .$$

Hence

$$(2.7) \quad \begin{aligned} \int \varphi\, \hat{v}_\infty \, dm &= \mathcal{E}_1(G_1\varphi, \hat{v}_\infty) = \mathcal{E}_1(h, \hat{v}_\infty) = \lim_{k\to\infty} \mathcal{E}_1(h, \hat{v}_{F_k^c}) \\ &= \lim_{k\to\infty} \mathcal{E}_1(h_{F_k^c}, v) = \mathcal{E}_1(h_\infty, v). \end{aligned}$$

By (2.6) and the assumption we have that

$$0 = \mathcal{E}_1(h_\infty, g) = \mathcal{E}_1(h_\infty, \hat{G}_1\hat{\varphi}) = \int h_\infty \hat{\varphi} \, dm$$

and therefore, $h_\infty = 0$ m-a.e. since $h_\infty \geq 0$ and $\hat{\varphi} > 0$. Because $\hat{v}_\infty \geq 0$ and $\varphi > 0$, (2.7) now implies that $\hat{v}_\infty = 0$. By I.2.12 the Cesaro mean $(w_n)_{n\in\mathbb{N}}$ of a subsequence of $\left(\hat{v}_{F_k^c}\right)_{k\in\mathbb{N}}$ converges to $\hat{v}_\infty$ hence to zero in $D(\mathcal{E})$. Clearly,

$$v - w_n \in \bigcup_{k\geq 1} D(\mathcal{E})_{F_k} \, ,$$

and the proof is complete. $\qquad\qquad\square$

Similar arguments that we used in the first half of the preceding proof lead to the following

Proposition 2.12. *Let $U \subset E$, U open, and let $(F_k)_{k\in\mathbb{N}}$ be an $\mathcal{E}$-nest. Let h be any 1-excessive function in $D(\mathcal{E})$. Then $h_{U\cup F_k^c} \to h_U$ in $D(\mathcal{E})$ as $k \to \infty$.*

Proof. Since $(h_{U\cup F_k^c})_{k\in\mathbb{N}}$ is decreasing by 1.5(iv), $h_\infty := \lim_{k\to\infty} h_{U\cup F_k^c}$ exists m-a.e. everywhere and $h_{U\cup F_k^c} \xrightarrow[k\to\infty]{} h_\infty$ in $L^2(E;m)$. Again by 1.5(iv) and by I.2.12 we obtain that $h_{U\cup F_k^c} \xrightarrow[k\to\infty]{} h_\infty$ weakly in $(D(\mathcal{E}), \tilde{\mathcal{E}}_1)$. If $U = \emptyset$ it follows as in the proof of 2.11 that $h_\infty = 0$, hence by 1.5(ii) (since $h = h_{F_k^c}$ m-a.e. on F_k^c for all $k \in \mathbb{N}$)

$$\limsup_{k\to\infty} \mathcal{E}_1(h_{F_k^c}, h_{F_k^c}) = \lim_{k\to\infty} \mathcal{E}_1(h_{F_k^c}, h) = 0 \;,$$

i.e., $h_{F_k^c} \xrightarrow[k\to\infty]{} 0$ in $D(\mathcal{E})$. If U is an arbitrary open subset in E we have that for all $k \in \mathbb{N}$

$$h_{U\cup F_k^c} \leq h_U + h_{F_k^c} \;,$$

hence $h_\infty \leq h_U$. But, clearly $h_\infty \geq h$ m-a.e. on U and h_∞ is 1-excessive. Consequently, $h_U = h_\infty$ by 1.6(iii). Now again by 1.5(ii) (since $h_{U\cup F_k^c} = h_U$ m-a.e. on U)

$$
\begin{aligned}
&\mathcal{E}_1(h_U - h_{U\cup F_k^c}, h_U - h_{U\cup F_k^c}) \\
&= \; \mathcal{E}_1(h_U, h_U) + \mathcal{E}_1(h_{U\cup F_k^c}, h_{U\cup F_k^c}) - \mathcal{E}_1(h_U, h_{U\cup F_k^c}) - \mathcal{E}_1(h_{U\cup F_k^c}, h_U) \\
&= \; \mathcal{E}_1(h_{U\cup F_k^c}, h_{U\cup F_k^c}) - \mathcal{E}_1(h_{U\cup F_k^c}, h_U) \\
&= \; \mathcal{E}_1(h_{U\cup F_k^c}, h) - \mathcal{E}_1(h_{U\cup F_k^c}, h_U)
\end{aligned}
$$

which converges to zero since $h_{U\cup F_k^c} \xrightarrow[k\to\infty]{} h_\infty = h_U$ weakly in $(D(\mathcal{E}), \tilde{\mathcal{E}}_1)$. $\square$

Exercise 2.13.   (i) If $1 \in D(\mathcal{E})$, then an increasing sequence of closed sets $(F_k)_{k\in\mathbb{N}}$ is an $\mathcal{E}$-nest and only if $\lim_{k\to\infty} \mathrm{Cap}_{1,1}(F_k^c) = 0$. (Hint: cf. the proof of V.2.15 below.)

(ii) Let $h \in D(\mathcal{E})$ and $(U_n)_{n\in\mathbb{N}}$ an increasing sequence of open subsets of E. If $U := \bigcup_{n\geq 1} U_n$, then $h_{U_n} \xrightarrow[n\to\infty]{} h_U$ in $D(\mathcal{E})$. (Hint: Recall that by the proof of 2.7(ii), $h_{U_n} \xrightarrow[n\to\infty]{} h_U$ weakly in $(D(\mathcal{E}), \tilde{\mathcal{E}}_1)$.)

Remark 2.14. (i) If we define h_A as in 1.5 for any $A \subset E$ and if we just assume $(E, \mathcal{B}(E))$ to be a measurable space and replace the system of open sets above by any class of sets in $\mathcal{B}(E)$ which contains $\emptyset$ and E, and is closed under taking countable unions and finite intersections, then all statements above except 2.8(ii) and 2.9(i) remain true.

(ii) In order to use 2.11 we have to ensure the existence of a $\varphi \in L^2(E;m)$ with $\varphi > 0$ m-a.e. . Therefore, we have assumed and shall do so from now on that m is σ-finite.

3   Quasi-continuity

In this section we fix $\varphi \in L^2(E;m)$, $0 < \varphi \leq 1$ m-a.e. and set

$$(3.1) \qquad\qquad h := G_1\varphi \;, \quad g := \hat{G}_1\varphi \;.$$

Remark 3.1. Note that $\mathrm{Cap}_{h,g}$ only depends on φ, but we shall not express this in the notation. The capacity associated with the symmetric part $(\tilde{\mathcal{E}}, D(\mathcal{E}))$ of $(\mathcal{E}, D(\mathcal{E}))$ we denote by $\tilde{\mathrm{Cap}}_h$ where $h := \tilde{G}_1 \varphi$ and $(\tilde{G}_\alpha)_{\alpha>0}$ is the strongly continuous contraction resolvent associated with $(\tilde{\mathcal{E}}, D(\mathcal{E}))$.

Given an $\mathcal{E}$-nest $(F_k)_{k\in\mathbb{N}}$ we define

(3.2)
$$C(\{F_k\}) := \{f : A \to \mathbb{R} \mid \bigcup_{k\geq 1} F_k \subset A \subset E , \\ f_{|F_k} \text{ is continuous for every } k \in \mathbb{N}\} .$$

Note that the notion introduced in the following definition only depends on the symmetric part of $(\mathcal{E}, D(\mathcal{E}))$ (cf. 2.2).

Definition 3.2. An $\mathcal{E}$-q.e. defined function f on E is called $\mathcal{E}$-*quasi-continuous* if there exists an $\mathcal{E}$-nest $(F_k)_{k\in\mathbb{N}}$ such that $f \in C(\{F_k\})$.

Proposition 3.3. *Let S be a countable family of $\mathcal{E}$-quasi-continuous functions on E. Then there exists an $\mathcal{E}$-nest $(F_k)_{k\in\mathbb{N}}$ such that $S \subset C(\{F_k\})$.*

Proof. Let $S = \{f_l \mid l \in \mathbb{N}\}$. For every $l \in \mathbb{N}$ there exists an $\mathcal{E}$-nest $(F_{lk})_{k\in\mathbb{N}}$ such that $f_l \in C(\{F_{lk}\})$ and $\mathrm{Cap}_{h,g}(F_{lk}^c) < \frac{1}{2^l k}$ (where we used 2.11). Let $F_k := \bigcap_{l\geq 1} F_{lk}$, $k \in \mathbb{N}$. Then each F_k is closed and

$$\mathrm{Cap}_{h,g}(F_k^c) \leq \sum_{l\geq 1} \mathrm{Cap}_{h,g}(F_{lk}^c) \leq \frac{1}{k} .$$

Hence $(F_k)_{k\in\mathbb{N}}$ is an $\mathcal{E}$-nest and obviously $S \subset C(\{F_k\})$. $\qquad\square$

Proposition 3.4. *Let $u \in D(\mathcal{E})$ such that it has an $\mathcal{E}$-quasi-continuous m-version $\tilde{u}$. Then for all $\lambda > 0$*

(i) $\mathrm{Cap}_{h,g}\{|\tilde{u}| > \lambda\} \leq \frac{K^2}{\lambda}\mathcal{E}_1(u,u)^{1/2}\mathcal{E}_1(g,g)^{1/2}$.

(ii) $\tilde{\mathrm{Cap}}_h\{|\tilde{u}| > \lambda\} \leq \frac{1}{\lambda^2}\mathcal{E}_1(u,u)$.

Proof. Let $(F_k)_{k\in\mathbb{N}}$ be an $\mathcal{E}$-nest such that $\tilde{u} \in C(\{F_k\})$. For $\lambda > 0$ we set $U_k := \{|\tilde{u}| > \lambda\} \cup F_k^c$, $k \in \mathbb{N}$. Then each U_k is open and if

$$u_k := \lambda^{-1}|\tilde{u}| + h_{F_k^c} , \quad k \in \mathbb{N} ,$$

then $u_k \in D(\mathcal{E})$ and $u_k \geq h$ m-a.e. on U_k since $h = G_1\varphi \leq 1$. Consequently, for every $k \in \mathbb{N}$ by 1.5

$$\begin{aligned}
\mathrm{Cap}_{h,g}(U_k) &= \mathcal{E}_1(h_{U_k}, \hat{g}_{U_k}) = \mathcal{E}_1(h, \hat{g}_{U_k}) \leq \mathcal{E}_1(u_k, \hat{g}_{U_k}) \\
&= \mathcal{E}_1(\lambda^{-1}|u|, \hat{g}_{U_k}) + \mathcal{E}_1(h_{F_k^c}, \hat{g}_{U_k}) \\
&= \mathcal{E}_1(\lambda^{-1}|u|, \hat{g}_{U_k}) + \mathcal{E}_1(h_{F_k^c}, \hat{g}_{F_k^c}) \\
&\leq K^2\lambda^{-1}\mathcal{E}_1(|u|, |u|)^{1/2}\mathcal{E}_1(g,g)^{1/2} + \mathrm{Cap}_{h,g}(F_k^c) .
\end{aligned}$$

Letting $k \to \infty$ we obtain (i), since $\mathrm{Cap}_{h,g}\{|\tilde{u}| > \lambda\} \leq \mathrm{Cap}_{h,g}(U_k)$ for all $k \in \mathbb{N}$. Similarly, if now for $U \subset E$, U open, h_U denotes the reduced function w.r.t. $\tilde{\mathcal{E}}$, then

$$
\begin{aligned}
\tilde{\mathrm{Cap}}_h(U_k) &= \tilde{\mathcal{E}}_1(h_{U_k}, h_{U_k}) \\
&\leq \tilde{\mathcal{E}}_1(\lambda^{-1}|u|, h_{U_k}) + \tilde{\mathcal{E}}_1(h_{F_k^c}, h_{U_k}) \\
&\leq \left[\lambda^{-1}\tilde{\mathcal{E}}_1(|u|, |u|)^{1/2} + \tilde{\mathrm{Cap}}_h(F_k^c)^{1/2}\right] \tilde{\mathrm{Cap}}_h(U_k)^{1/2}
\end{aligned}
$$

and (ii) follows as above. $\qquad\qquad\square$

3.4 will only be used to prove 3.5 below. Since the notion of $\mathcal{E}$-quasi-continuity only depends on $(\tilde{\mathcal{E}}, D(\mathcal{E}))$, it would have been enough to derive 3.4(ii).

Let f, f_n, $n \in \mathbb{N}$, be ($\mathcal{E}$-q.e. defined) functions on E. We say that $(f_n)_{n \in \mathbb{N}}$ converges to f *$\mathcal{E}$-quasi-uniformly* if there exists an $\mathcal{E}$-nest $(F_k)_{k \in \mathbb{N}}$ such that $f_n \underset{n \to \infty}{\longrightarrow} f$ uniformly on each F_k.

Proposition 3.5. *Let $u_n \in D(\mathcal{E})$, which have $\mathcal{E}$-quasi-continuous m-versions $\tilde{u}_n$, $n \in \mathbb{N}$, such that $u_n \underset{n \to \infty}{\longrightarrow} u \in D(\mathcal{E})$ w.r.t. $\tilde{\mathcal{E}}_1^{1/2}$. Then there exists a subsequence $(u_{n_k})_{k \in \mathbb{N}}$ and an $\mathcal{E}$-quasi-continuous m-version $\tilde{u}$ of u such that $(\tilde{u}_{n_k})_{k \in \mathbb{N}}$ converges $\mathcal{E}$-quasi-uniformly to $\tilde{u}$.*

Proof. By 3.4 we may choose a subsequence $(u_{n_i})_{i \in \mathbb{N}}$ such that

$$
\mathrm{Cap}_{h,g}\left\{\left|\tilde{u}_{n_{i+1}} - \tilde{u}_{n_i}\right| > 2^{-i}\right\} \leq 2^{-i} \text{ for all } i \in \mathbb{N}.
$$

Let

$$
A_k := \bigcup_{i \geq k}\left\{|\tilde{u}_{n_{i+1}} - \tilde{u}_{n_i}| > 2^{-i}\right\}, \quad k \in \mathbb{N}.
$$

Let $(F_k')_{k \in \mathbb{N}}$ be an $\mathcal{E}$-nest such that $\tilde{u}_n \in C(\{F_k'\})$ for each $n \in \mathbb{N}$, and define

$$
F_k := F_k' \cap A_k^c, \quad k \in \mathbb{N}.
$$

Then $(F_k)_{k \in \mathbb{N}}$ is an $\mathcal{E}$-nest and $(\tilde{u}_{n_i})_{i \in \mathbb{N}}$ converges uniformly on each F_k. Set

$$
\tilde{u}(z) := \begin{cases} \lim_{i \to \infty} \tilde{u}_{n_i}(z) & \text{if} \quad z \in \bigcup_{k \geq 1} F_k \\ 0 & \text{else.} \end{cases}
$$

Since $\tilde{\mathcal{E}}_1^{1/2}$-convergence implies the m-a.e. convergence of a subsequence and since $m\left(\bigcap_{k \geq 1} F_k^c\right) = 0$ by 2.3, $\tilde{u}$ is an m-version of u. $\qquad\square$

Proposition 3.6. *Suppose h in (3.1) has an $\mathcal{E}$-quasi-continuous m-version $\tilde{h}$ and let $(F_k)_{k \in \mathbb{N}}$ be an $\mathcal{E}$-nest such that $\tilde{h} \in C(\{F_k\})$. Let $(\delta_k)_{k \in \mathbb{N}}$ be a decreasing sequence of positive numbers such that $\lim_{k \to \infty} \delta_k = 0$. Then there exists an $\mathcal{E}$-nest $(\tilde{F}_k)_{k \in \mathbb{N}}$ such that $\tilde{F}_k \subset F_k$ and $\tilde{h} \geq \delta_k$ on $\tilde{F}_k$ for every $k \in \mathbb{N}$. In particular, $\{\tilde{h} = 0\}$ is $\mathcal{E}$-exceptional.*

Proof. Set
$$\tilde{F}_k := \{\tilde{h} \geq \delta_k\} \cap F_k \, ,$$
then $(\tilde{F}_k)_{k\in\mathbb{N}}$ is an increasing sequence of closed sets. Let
$$u_k := (h \wedge \delta_k) + h_{F_k^c} \, , \ k \in \mathbb{N} \, .$$
Then for each $k \in \mathbb{N}$, $u_k \in D(\mathcal{E})$ and $u_k \geq h$ m-a.e. on $\tilde{F}_k^c$, and by 1.5
$$
\begin{aligned}
\mathrm{Cap}_{h,g}(\tilde{F}_k^c) &= \mathcal{E}_1(h_{\tilde{F}_k^c}, \hat{g}_{\tilde{F}_k^c}) \leq \mathcal{E}_1(u_k, \hat{g}_{\tilde{F}_k^c}) \\
&= \mathcal{E}_1(h \wedge \delta_k, \hat{g}_{\tilde{F}_k^c}) + \mathcal{E}_1(h_{F_k^c}, \hat{g}_{\tilde{F}_k^c}) \\
&\leq K^2 \mathcal{E}_1(h \wedge \delta_k, h \wedge \delta_k)^{1/2} \mathcal{E}_1(g,g)^{1/2} + \mathrm{Cap}_{h,g}(F_k^c) \\
&\xrightarrow[k\to\infty]{} 0 \, ,
\end{aligned}
$$
since $\mathcal{E}_1(h \wedge \delta_k, h \wedge \delta_k) \to 0$ as $k \to \infty$ (cf. I.4.17(ii)). Hence $(\tilde{F}_k)_{k\in\mathbb{N}}$ has the required properties. $\qquad\square$

Definition 3.7. We say that an $\mathcal{E}$-nest $(F_k)_{k\in\mathbb{N}}$ is *regular* if for all $k \in \mathbb{N}$, $U \subset E$, U open, $m(U \cap F_k) = 0$ implies that $U \subset F_k^c$.

Proposition 3.8. *Let $(F_k)_{k\in\mathbb{N}}$ be an $\mathcal{E}$-nest. Suppose that the relative topology on each F_k is strongly Lindelöf (i.e., every open cover of any given open set has a countable subcover). Set $\tilde{F}_k := \mathrm{supp}\,[1_{F_k} \cdot m]$, then $(\tilde{F}_k)_{k\in\mathbb{N}}$ is a regular $\mathcal{E}$-nest, such that $\tilde{F}_k \subset F_k$ for all $k \in \mathbb{N}$.*

Proof. Fix $k \in \mathbb{N}$. Note that by definition $\tilde{F}_k$ is the smallest closed set F such that $m(F^c \cap F_k) = 0$ (which exists since F_k is strongly Lindelöf). In particular, $\tilde{F}_k \subset F_k$ and $m(F_k \setminus \tilde{F}_k) = m(\tilde{F}_k^c \cap F_k) = 0$. Hence $\mathrm{Cap}_{h,g}(F_k^c) = \mathrm{Cap}_{h,g}(\tilde{F}_k^c)$ by 2.6(i) and if $U \subset E$, U open with $m(U \cap \tilde{F}_k) = 0$ then $m(U \cap F_k) = 0$. Consequently, $U^c \supset \tilde{F}_k$, equivalently $U \subset \tilde{F}_k^c$, and $(\tilde{F}_k)_{k\in\mathbb{N}}$ is therefore a regular $\mathcal{E}$-nest. $\qquad\square$

Proposition 3.9. *Suppose $(F_k)_{k\in\mathbb{N}}$ is a regular $\mathcal{E}$-nest and $f \in C(\{F_k\})$. If $f \geq 0$ m-a.e. on an open set U then $f(z) \geq 0$ for all $z \in \bigcup_{k\geq 1} F_k \cap U$.*

Proof. Let $k \in \mathbb{N}$, then $(\{f < 0\} \cup F_k^c) \cap U$ is open and
$$m((\{f < 0\} \cup F_k^c) \cap U \cap F_k) = 0$$
since $f \geq 0$ m-a.e. on U. Hence $(\{f < 0\} \cup F_k^c) \cap U \subset F_k^c$, i.e., $\{f < 0\} \cap U \subset F_k^c$, equivalently
$$F_k \cap U \subset \{f \geq 0\} \, .$$
Consequently,
$$\bigcup_{k\geq 1} F_k \cap U \subset \{f \geq 0\} \, .$$
$$\qquad\square$$

We conclude this section with an important result on capacities which, however, will only be used in Chap.VI, Sect. 1 below.

Exercise 3.10. Suppose that the following condition holds.

(3.3)   There exists an $\mathcal{E}$-nest of closed sets which are strongly Lindelöf and each $u \in D(\mathcal{E})$ has an $\mathcal{E}$-quasi-continuous m-version denoted by $\tilde{u}$.

Let f be an $\mathcal{E}$-quasi-continuous function on E.

(i) Prove that if $f \geq 0$ m-a.e. on some open set $U \subset E$ then $f \geq 0$ $\mathcal{E}$-q.e. on U. In particular, $\mathcal{E}$-quasi-continuous m-versions of elements in $D(\mathcal{E})$ are unique $\mathcal{E}$-q.e. . (See IV.3.3(iii) below.)

(ii) Let $A \subset E$ and define

$$\mathcal{L}_{f,A} := \{w \in D(\mathcal{E}) \,|\, \tilde{w} \geq f \ \mathcal{E}\text{-q.e. on } \ A\} \ .$$

Assume $\mathcal{L}_{f,A} \neq \emptyset$. Prove that there exist unique $f_A, \hat{f}_A \in \mathcal{L}_{f,A}$ such that for all $w \in \mathcal{L}_{f,A}$

$$\mathcal{E}_1(f_A, w) \geq \mathcal{E}_1(f_A, f_A) \ \text{and} \ \mathcal{E}_1(w, \hat{f}_A) \geq \mathcal{E}_1(\hat{f}_A, \hat{f}_A)$$

and that this extends the notion of "reduced function on A" to arbitrary sets $A \subset E$. Furthermore, prove the correspondingly modified analogues of 1.5-1.7 with A, f replacing U, h respectively and "$\mathcal{E}$-q.e." replacing "m-a.e.".

Theorem 3.11. *Let h, g be as in (3.1). Suppose that (3.3) holds and let $A \subset E$. Then*

$$\mathrm{Cap}_{h,g}(A) = \mathcal{E}_1(h_A, \hat{g}_A) \ .$$

Proof. Let $U \subset E$, U open, with $A \subset U$. Then by 3.10 (ii) we have that $\tilde{h}_A \leq \tilde{h}_U$ $\mathcal{E}$-q.e. on E and $\hat{\tilde{g}}_U = \hat{\tilde{g}}_A$ $\mathcal{E} - q.e.$ on A, hence

$$\mathrm{Cap}_{h,g}(U) = \mathcal{E}_1(h_U, \hat{g}_U) \geq \mathcal{E}_1(h_A, \hat{g}_U) = \mathcal{E}_1(h_A, \hat{g}_A) \ .$$

Consequently, $\mathrm{Cap}_{h,g}(A) \geq \mathcal{E}_1(h_A, \hat{g}_A)$. To prove the dual inequality let $(U_n)_{n\in\mathbb{N}}$ be a decreasing sequence of open sets in E such that $A \subset U_n$, $n \in \mathbb{N}$, and

$$\mathrm{Cap}_{h,g}(U_n) \downarrow \mathrm{Cap}_{h,g}(A) \ \text{as } n \to \infty \ .$$

By 3.3 and 3.6 there exists an $\mathcal{E}$-nest $(F_k)_{k\in\mathbb{N}}$ such that $\tilde{h}, \tilde{h}_A, \tilde{g}, \hat{\tilde{g}}_A \in C(\{F_k\})$ and $\tilde{h} > 0$ on each F_k. Let $k \in \mathbb{N}$ and define

$$V_k := \left\{ \tilde{h}_A > (1 - \frac{1}{k})\tilde{h} \right\} \cup F_k^c \ .$$

Then V_k is open, $V_k \supset A \setminus N$ for some $\mathcal{E}$-exceptional set N and

$$h_A + \frac{1}{k}h + h_{F_k^c} \geq h \ m\text{-a.e. on } V_k \ .$$

Since $h, h_A, h_{F_k^c}$ are 1-excessive, we conclude by 1.6 (iii) and 1.5 (iv) that

$$(3.4) \qquad h_A + \frac{1}{k}h + h_{F_k^c} \geq h_{V_k} \geq h_{V_k \cap U_n} \text{ for all } n \in \mathbb{N} .$$

Since $h_{V_k \cap U_n}$ is m-a.e. decreasing as k or n increases and since

$$\mathcal{E}_1(h_{V_k \cap U_n}, h_{V_k \cap U_n}) \leq K^2 \mathcal{E}_1(h, h) \text{ for all } n, k \in \mathbb{N}$$

by 1.5 (iv), $(h_{V_k \cap U_n})_{n \in \mathbb{N}}$ converges to some 1-excessive $h_k \in D(\mathcal{E})$ and $(h_k)_{k \in \mathbb{N}}$ to some 1-excessive $h_\infty \in D(\mathcal{E})$ m-a.e. and also weakly in $(D(\mathcal{E}), \tilde{\mathcal{E}}_1)$ by I.2.12. 2.12 and (3.4) imply that

$$h_A \geq h_\infty \text{ } m\text{-a.e. .}$$

Similarly, we can find $\hat{g}_k$, $k \in \mathbb{N} \cup \{\infty\}$ with the corresponding dual properties. Now since all involved functions are 1-excessive, resp. 1-coexcessive and by 1.5(ii) it follows that

$$
\begin{aligned}
\mathcal{E}_1(h_A, \hat{g}_A) &\geq \mathcal{E}_1(h_\infty, \hat{g}_A) \geq \mathcal{E}_1(h_\infty, \hat{g}_\infty) \\
&= \lim_{\substack{k \to \infty}} \lim_{\substack{l \to \infty \\ l \geq k}} \lim_{\substack{n \to \infty}} \lim_{\substack{m \to \infty \\ m \geq n}} \mathcal{E}_1(h_{U_n \cap V_k}, \hat{g}_{U_m \cap V_l}) \\
&= \lim_{\substack{k \to \infty}} \lim_{\substack{l \to \infty \\ l \geq k}} \lim_{\substack{n \to \infty}} \lim_{\substack{m \to \infty \\ m \geq n}} \mathcal{E}_1(h_{U_m \cap V_l}, \hat{g}_{U_m \cap V_l}) \\
&= \liminf_{k \to \infty} \liminf_{n \to \infty} \mathrm{Cap}_{h,g}(U_n \cap V_k) \\
&\geq \mathrm{Cap}_{h,g}(A \setminus N) = \mathrm{Cap}_{h,g}(A) .
\end{aligned}
$$

$\square$

4   Notes/References

Though the concepts and results in this chapter are with respect to a Dirichlet form without any regularity assumption, basically the analytic potential theory developed in this chapter is a modification and/or extension of the one represented in [F 80, §3].

Section 1. Definition 1.1 is from [F 80, p. 69]. Proposition 1.2 is taken from the part of [F 80, Theorem 3.2.1] which does not rely on the regularity assumption. We adopted some arguments from [O 88, Theorem 2.3.1] for the proof of 1.2. Here we mention that there is also an extension of the equivalence between (i) and (ii) of [F 80, Theorem 3.2.1] (which depends on the regularity assumption) to quasi-regular Dirichlet forms (see [DoM 91]). Proposition 1.5 extends [F 80, Theorem 3.3.3] (see also [O 88, (2.3.13)-(2.3.15)]), though in a standard way.

Section 2. The concept of $\mathcal{E}$-nest was introduced in [AM 91c,f]. The notion of $\mathrm{Cap}_{h,g}$ is new. It covers the classical capacity Cap (i.e., $\mathrm{Cap}_{1,1}$) and the h-weighted capacity Cap_φ ($:= \mathrm{Cap}_{h,g}$ with $h = G_1\varphi$ and $g = \hat{G}_1\varphi$) of [AM 91c,f] as special cases (see also [R 85]). This non-symmetric notion of $\mathrm{Cap}_{h,g}$ is particularly convenient for introducing a new capacity $\mathrm{Cap}_{1,g}$ (cf. V.2.1) which is essential for Chap.V, Sect. 2 below. Theorem 2.11 extends [AM 91f, Proposition 2.5] to the non-symmetric case.

Section 3. Apart from the "non-symmetry" all results in this section are close to those in [F 80, §3] with two exceptions. The proof of Proposition 3.4 is different from that of [F 80, Lemma 3.1.5] because we do not assume the continuous functions to be dense in $D(\mathcal{E})$ w.r.t. $\tilde{\mathcal{E}}_1^{1/2}$. The proof of Theorem 3.11 is new and entirely different from previous ones (cf. [O 88, Theorem 2.3.7], [F 80, Theorem 3.3.1]).

Chapter IV

Markov Processes and Dirichlet Forms

In this chapter we develop the probabilistic part of the theory of Dirichlet forms. We start with some basics on Markov processes in Section 1. In particular, we recall some facts on Ray resolvents whose proofs are, however, postponed to the Appendix (cf. A. Sect.3). In Section 2 we explain the "(proper) association" of Dirichlet forms with (a pair of "nice") Markov processes. In Section 3 we introduce the notion of "quasi-regularity" and prove that a Dirichlet form on an arbitrary topological state space always possesses an associated "nice" Markov process provided it is quasi-regular. This extends M. Fukushima's fundamental existence theorem for regular (symmetric) Dirichlet forms on locally compact, separable metric state spaces (cf. [F 71, 80]). In Section 4 we present examples of quasi-regular Dirichlet forms including cases with infinite dimensional state spaces. In Section 5 we prove that the quasi-regularity of a Dirichlet form $(\mathcal{E}, D(\mathcal{E}))$ is also necessary for the existence of an associated "nice" Markov process $\mathbf{M}$. On the way, we study the essentials of the probabilistic potential theory of $\mathbf{M}$ and its relation with the analytic potential theory of $(\mathcal{E}, D(\mathcal{E}))$. In all of this chapter we assume E to be a Hausdorff topological space and $\mathcal{B}(E)$ denotes its Borel σ-algebra. In Sections 2, 3 and 5 we assume for convenience in addition, that $\mathcal{B}(E) = \sigma(C(E))$ where $C(E)$ denotes the set of all continuous functions on E.

1 Basics on Markov processes

Given a measurable space $(S, \mathcal{B})$, the *completion* of the σ-algebra $\mathcal{B}$ w.r.t. a probability measure P on $(S, \mathcal{B})$ is denoted by $\mathcal{B}^P$. An element B of $\mathcal{B}^* := \bigcap_{P \in \mathcal{P}(S)} \mathcal{B}^P$ is called a *universally measurable* set, where $\mathcal{P}(S)$ denotes the family of all probability measures on $(S, \mathcal{B})$. Given a sub-σ-algebra $\mathcal{C}$ of $\mathcal{B}$, we define

$\mathcal{C}^P$, the *completion of $\mathcal{C}$ in $\mathcal{B}^P$ w.r.t. $P \in \mathcal{P}(S)$*, by

(1.1) $\mathcal{C}^P = \{A \subset S \mid A \Delta B \subset N \text{ for some } B \in \mathcal{C} \, , \, N \in \mathcal{B} \, , \, P(N) = 0\} \, .$

If $\mathcal{N}_P := \{N \in \mathcal{B}^P \mid P(N) = 0\}$ then $\mathcal{C}^P = \sigma\{\mathcal{C}, \mathcal{N}_P\}$ i.e., the smallest σ-algebra containing $\mathcal{C}, \mathcal{N}_P$.

Now fix a measurable space $(\Omega, \mathcal{M})$. Let $\mathcal{M}_t$, $t \in [0, \infty]$, be sub-σ-algebras of $\mathcal{M}$. $(\mathcal{M}_t) := (\mathcal{M}_t)_{t\in[0,\infty]}$ is called a *filtration* (on $(\Omega, \mathcal{M})$) if $\mathcal{M}_s \subset \mathcal{M}_t$ for $s \leq t$ and

(1.2)
$$\mathcal{M}_\infty = \bigvee_{t\in[0,\infty[} \mathcal{M}_t \, ,$$

i.e., $\mathcal{M}_\infty$ is the smallest σ-algebra containing all $\mathcal{M}_t$, $t \in [0, \infty[$. Given a filtration $(\mathcal{M}_t)$ we set

(1.3)
$$\mathcal{M}_{t+} := \bigcap_{s>t} \mathcal{M}_s \, .$$

$(\mathcal{M}_t)$ is called *right continuous* if $\mathcal{M}_{t+} = \mathcal{M}_t$ for all $t \geq 0$.

Exercise 1.1. Show that $(\mathcal{M}_{t+})^P = (\mathcal{M}_t^P)_+$ for all $t \geq 0$ and any probability measure P on $(\Omega, \mathcal{M})$.

Remark 1.2. By convention a filtration $(\mathcal{M}_t)$ on a probability space $(\Omega, \mathcal{M}, P)$ is said to satisfy the *usual conditions* if $\mathcal{M}_t = \mathcal{M}_{t+}^P$ for all $t \geq 0$ (cf. [DM 78, IV.48]).

We adjoin an extra point Δ (the *cemetery*) to E as an isolated point to obtain a Hausdorff topological space $E_\Delta := E \cup \{\Delta\}$ with Borel σ-algebra $\mathcal{B}(E_\Delta)(= \mathcal{B}(E) \cup \{B \cup \{\Delta\} \mid B \in \mathcal{B}(E)\})$. Any function $f : E \to \mathbb{R}$ is considered as a function on E_Δ by putting $f(\Delta) = 0$. However, if E is locally compact then we shall sometimes also consider a different topology on E_Δ taking as neighbourhoods for Δ the complements of compact sets in E, i.e., E_Δ is then the *one-point-compactification* of E.

Exercise 1.3. Consider the topology on E_Δ where the neighbourhoods of Δ are the complements of compact sets in E. Prove that E_Δ is Hausdorff if and only if E is locally compact.

Definition 1.4. We say that $(X_t)_{t\geq 0}$ is a *stochastic process with state space E and life time ζ* (on $(\Omega, \mathcal{M})$) if

   (i) $X_t : \Omega \to E_\Delta$ is an $\mathcal{M}/\mathcal{B}(E_\Delta)$-measurable map for all $t \in [0, \infty[$.

   (ii) $\zeta : \Omega \to [0, \infty]$ is $\mathcal{M}$-measurable.

   (iii) For all $\omega \in \Omega$, $X_t(\omega) \in E$ whenever $t < \zeta(\omega)$ and $X_t(\omega) = \Delta$ for all $t \geq \zeta(\omega)$.

A stochastic process $(X_t)_{t \geq 0}$ is said to be *measurable* if $(t, \omega) \mapsto X_t(\omega)$ is $\mathcal{B}([0, \infty[) \otimes \mathcal{M}/\mathcal{B}(E_\Delta)$-measurable. $(X_t)_{t \geq 0}$ is called $(\mathcal{M}_t)$-*adapted* for a filtration $(\mathcal{M}_t)$ on $(\Omega, \mathcal{M})$ if each X_t is $\mathcal{M}_t/\mathcal{B}(E_\Delta)$ measurable, $t \geq 0$. Below, for convenience we always set $X_\infty := \Delta$, unless otherwise specified.

Definition 1.5. A collection $\mathbf{M} := (\Omega, \mathcal{M}, (X_t)_{t \geq 0}, (P_z)_{z \in E_\Delta})$ is called a (time-homogeneous) *Markov process* (with state space E and life time ζ) if it has the following properties.

(M.1) There exists a filtration $(\mathcal{M}_t)$ on $(\Omega, \mathcal{M})$ such that $(X_t)_{t \geq 0}$ is an $(\mathcal{M}_t)$-adapted stochastic process with state space E and life time ζ.

(M.2) For each $t \geq 0$, there exists a *shift operator* $\theta_t : \Omega \to \Omega$ such that $X_s \circ \theta_t = X_{s+t}$ for all $s, t \geq 0$.

(M.3) P_z, $z \in E_\Delta$, are probability measures on $(\Omega, \mathcal{M})$ such that $z \mapsto P_z(\Gamma)$ is $\mathcal{B}(E_\Delta)^*$-measurable for each $\Gamma \in \mathcal{M}$ resp. $\mathcal{B}(E_\Delta)$-measurable if $\Gamma \in \sigma\{X_s \mid s \in [0, \infty[\}$ and $P_\Delta[X_0 = \Delta] = 1$.

(M.4) (*Markov property*) For all $A \in \mathcal{B}(E_\Delta)$ and any $t, s \geq 0$

$$P_z[X_{s+t} \in A \,|\, \mathcal{M}_s] = P_{X_s}[X_t \in A] \quad P_z\text{-a.s.}, \, z \in E_\Delta \ .$$

Remark. Clearly, if in 1.4 or 1.5, $\zeta \equiv \infty$ we can drop Δ. In this case 1.4, 1.5 have to be modified in the obvious way. We will do this below (e.g. in 1.22) without further notice.
Given $\mathbf{M}$ as in 1.5 and a positive measure μ on $(E_\Delta, \mathcal{B}(E_\Delta))$ we define a positive measure P_μ on $(\Omega, \mathcal{M})$ by

$$(1.4) \qquad P_\mu(\Gamma) := \int_{E_\Delta} P_z(\Gamma) \mu(dz) \ , \, \Gamma \in \mathcal{M} \ .$$

Note that if μ is equivalent to a positive measure ν on $(E_\Delta, \mathcal{B}(E_\Delta))$ then P_μ is equivalent to P_ν (i.e., they have the same null sets). If $\mu \in \mathcal{P}(E_\Delta)$ we denote as usual the expectation and conditional expectation w.r.t. some sub-σ-algebra $\mathcal{G}$ of $\mathcal{M}$ w.r.t. P_μ by E_μ, $E_\mu[\,\cdot\,|\mathcal{G}]$ respectively. Let $(\mathcal{M}_t)$ be a filtration on $(\Omega, \mathcal{M})$. An $\mathcal{M}$-measurable function $\tau : \Omega \to [0, \infty]$ is called an $(\mathcal{M}_t)$-*stopping time* if $\{\tau \leq t\} \in \mathcal{M}_t$ for all $t \geq 0$. For an $(\mathcal{M}_t)$-stopping time τ we define

$$(1.5) \qquad \mathcal{M}_\tau := \{\Gamma \in \mathcal{M} \mid \Gamma \cap \{\tau \leq t\} \in \mathcal{M}_t \text{ for all } t \geq 0\} \ .$$

Clearly, $\mathcal{M}_\tau$ is a sub-σ-field of $\mathcal{M}$.

Exercise 1.6. Let $\sigma, \tau, \sigma_n, n \in \mathbb{N}$, be $(\mathcal{M}_t)$-stopping times. Prove:

(i) $\sigma \wedge \tau$, $\sigma \vee \tau$ are $(\mathcal{M}_t)$-stopping times.

(ii) If $\sigma_n \leq \sigma_{n+1}$ for all $n \in \mathbb{N}$, then $\sup_{n \geq 1} \sigma_n$ is an $(\mathcal{M}_t)$-stopping time.

(iii) If $\sigma_n \geq \sigma_{n+1}$ for all $n \in \mathbb{N}$ and $(\mathcal{M}_t)$ is right continuous then $\inf_{n \geq 1} \sigma_n$ is an $(\mathcal{M}_t)$-stopping time.

(iv) If $(X_t)_{t\geq 0}$ is an $(\mathcal{M}_t)$-adapted process with state space E and life time ζ, then ζ is an $(\mathcal{M}_t)$-stopping time.

(Cf. e.g. [DM 78, IV.55].)

Exercise 1.7. Let σ, τ be $(\mathcal{M}_t)$-stopping times. Prove:

(i) σ is $\mathcal{M}_\sigma$-measurable.

(ii) If $\sigma \leq \tau$ then $\mathcal{M}_\sigma \subset \mathcal{M}_\tau$.

(iii) If $\Gamma \in \mathcal{M}_\sigma$ then $\Gamma \cap \{\sigma \leq \tau\}, \Gamma \cap \{\sigma < \tau\} \in \mathcal{M}_{\sigma \wedge \tau}$.

(iv) If $(X_t)_{t\geq 0}$ is an $(\mathcal{M}_t)$-adapted process such that $t \mapsto X_t(\omega)$ is right continuous for all $\omega \in \Omega$, then X_τ is $\mathcal{M}_\tau/\mathcal{B}(E_\Delta)$-measurable, where $X_\tau(\omega) := X_{\tau(\omega)}(\omega), \omega \in \Omega$.

(Cf. e.g. [DM 78, IV.56 and IV.64].)

Definition 1.8. Let $\mathbf{M} = (\Omega, \mathcal{M}, (X_t)_{t\geq 0}, (P_z)_{z\in E_\Delta})$ be a Markov process with state space E, life time ζ and corresponding filtration $(\mathcal{M}_t)$. $\mathbf{M}$ is called a *right process* (w.r.t. $(\mathcal{M}_t)$) if it has the following additional properties.

(M.5) (*Normal property*) $P_z[X_0 = z] = 1$ for all $z \in E_\Delta$.

(M.6) (*Right continuity*) For each $\omega \in \Omega$, $t \mapsto X_t(\omega)$ is right continuous on $[0, \infty[$.

(M.7) (*Strong Markov property*) $(\mathcal{M}_t)$ is right continuous and for every $(\mathcal{M}_t)$-stopping time σ and every $\mu \in \mathcal{P}(E_\Delta)$

$$P_\mu [X_{\sigma+t} \in A|\mathcal{M}_\sigma] = P_{X_\sigma}[X_t \in A] \qquad P_\mu\text{-a.s.}$$

for all $A \in \mathcal{B}(E_\Delta)$, $t \geq 0$.

Let $\mathbf{M}$ be a Markov process w.r.t. a filtration $(\mathcal{M}_t)$. We set for $t \in [0, \infty]$

$$(1.6) \qquad \mathcal{F}_t^0 \; := \; \sigma\{X_s \,|\, 0 \leq s \leq t\}$$

$$(1.7) \qquad \mathcal{F}_t \; := \; \bigcap_{\mu \in \mathcal{P}(E_\Delta)} (\mathcal{F}_t^0)^{P_\mu}.$$

$(\mathcal{F}_t)$ is called the *natural filtration of* $\mathbf{M}$.

Exercise 1.9. Let $\mathbf{M}$ be a Markov process w.r.t. $(\mathcal{M}_t)$. Prove:

(i) $\mathbf{M}$ has the Markov property w.r.t. $(\mathcal{F}_t^0)$.

(ii) If $\mathbf{M}$ has the Markov property w.r.t. $(\mathcal{F}_{t+}^0)$, then $((\mathcal{F}_t^0)^{P_\mu})$ is right continuous for every $\mu \in \mathcal{P}(E_\Delta)$. (Cf. [BG 68, I.(8.12)])

(iii) (*Blumenthal's 0-1-law*) Assume $\mathbf{M}$ satisfies (M.5) then

$$P_z(\Gamma) = 0 \text{ or } 1 \text{ for all } \Gamma \in \mathcal{F}_0 \text{ and } z \in E_\Delta \ .$$

(iv) Let $\Gamma \in \mathcal{F}_\infty$, then $z \mapsto P_z[\Gamma]$ is $\mathcal{B}(E_\Delta)^*$-measurable. (Cf. [BlHa 86, Proposition IV.4.5])

(v) If $\mathbf{M}$ satisfies (M.5), (M.6) and has the strong Markov property w.r.t. $(\mathcal{F}^0_{t+})$ then for any $(\mathcal{F}_t)$-stopping time σ, θ_σ is $\mathcal{F}_{\sigma+t}/\mathcal{F}_t$-measurable for all $t \in [0, \infty]$ and for every $\mathcal{F}_\infty$-measurable bounded function Y on Ω and every $\mu \in \mathcal{P}(E_\Delta)$

$$E_\mu[Y \circ \theta_\sigma \,|\, \mathcal{F}_\sigma] = E_{X_\sigma}[Y] \qquad P_\mu\text{-}a.s. \,.$$

(Cf. [BlHa 86, Proposition IV.6.8])

(vi) If $\mathbf{M}$ is a right process w.r.t. some filtration $(\mathcal{M}_t)$ then $\mathbf{M}$ is a right process w.r.t. $(\mathcal{F}_t)$.

Remark 1.10. In view of Exercise 1.9 when dealing with a right process $\mathbf{M}$ w.r.t. some filtration $(\mathcal{M}_t)$ in the sequel, we shall always assume that $(\mathcal{M}_t)$ is the natural filtration $(\mathcal{F}_t)$ and set $\mathcal{F} := \mathcal{F}_\infty$, unless otherwise stated.

For the rest of this section we fix a right process $\mathbf{M}$ with state space E and life time ζ. We use $E_z[\,\cdot\,]$, $E_z[\,\cdot\,|\mathcal{G}]$ for the expectation resp. conditional expectation given a sub-σ-algebra $\mathcal{G}$ of $\mathcal{F}$ w.r.t. P_z, $z \in E_\Delta$. Since $(X_t)_{t \geq 0}$ is measurable by (M.6),

$$(1.8) \qquad p_t f(z) := p_t(z, f) := E_z[f(X_t)] \,, \quad z \in E \,, \ t \geq 0 \,, \ f \in \mathcal{B}(E)^+ \,,$$

defines a sub-Markovian semigroup of kernels on $(E, \mathcal{B})$ (cf. Chap.II, Sect. 4; the semigroup property is an immediate consequence of the Markov property). Furthermore, we define

$$(1.9) \qquad \begin{aligned} R_\alpha f(z) &:= R_\alpha(z, f) := E_z\left[\int_0^\infty e^{-\alpha t} f(X_t) dt\right] \\ &\left(= \int_0^\infty e^{-\alpha t} p_t f(z) dt\right) \,, \alpha > 0 \,, \ z \in E \,, \ f \in \mathcal{B}(E)^+ \,. \end{aligned}$$

Exercise 1.11. Prove that if $\mathcal{B}(E) = \sigma(C(E))$ (i.e., the σ-algebra generated by all continuous functions on E), then $(p_t)_{t>0}$ in (1.8) is measurable (cf.II.5.1), and that hence $(R_\alpha)_{\alpha>0}$ in (1.9) is a sub-Markovian resolvent of kernels on $(E, \mathcal{B}(E))$.

We shall call $(p_t)_{t \geq 0}$ and $(R_\alpha)_{\alpha>0}$ the *transition semigroup* and *resolvent* of $\mathbf{M}$ respectively.

For $B \subset E_\Delta$ we define

$$(1.10) \qquad \sigma_B := \inf\{t > 0 | X_t \in B\} \qquad (\textit{first hitting time of } B)$$
$$(1.11) \qquad D_B := \inf\{t \geq 0 | X_t \in B\} \qquad (\textit{first entry time of } B\,).$$

It is known that if $B \in \mathcal{B}(E)$, then both σ_B and D_B are $(\mathcal{F}_t)$-stopping times (cf. e.g. [DM 78, IV.50]). But up to Sect.IV.4 below we shall only use this fact for open sets B.

Exercise 1.12. Let U be an open subset of E. Prove:

(i) $D_U = \sigma_U$.

(ii) σ_U is an $(\mathcal{F}_t)$-stopping time.

Note that for $B \in \mathcal{B}(E)$, $\{\sigma_B > 0\} \in \mathcal{F}_0$ and hence by 1.9(iii)

$$(1.12) \qquad\qquad P_z[\sigma_B > 0] = 0 \text{ or } 1 \ \text{ for all } z \in E_\Delta \ .$$

Definition 1.13. Let $\mathbf{M} = (\Omega, \mathcal{F}, (X_t)_{t \geq 0}, (P_z)_{z \in E_\Delta})$ be a right process with state space E and life time ζ.

(i) Let μ be a positive measure on $(E_\Delta, \mathcal{B}(E_\Delta))$. $\mathbf{M}$ is called μ-*tight* if there exists an increasing sequence $(K_n)_{n \in \mathbb{N}}$ of compact metrizable sets in E such that
$$P_\mu \left[\lim_{n \to \infty} \sigma_{E \setminus K_n} < \zeta \right] = 0 \ .$$

(ii) Let m be a σ-finite positive measure on $(E_\Delta, \mathcal{B}(E_\Delta))$. $\mathbf{M}$ is called an m-*special standard process* if for one (and hence all) measure(s) $\mu \in \mathcal{P}(E_\Delta)$ which are equivalent to m it has the following additional properties

    (M.8) *(left limits up to ζ)* $X_{t-} := \lim_{\substack{s \uparrow t \\ s < t}} X_s$ exists in E for all $t \in]0, \zeta[$ P_μ-a.s. .

    (M.9) *(quasi-left continuity up to ζ)* If τ, τ_n, $n \in \mathbb{N}$, are $(\mathcal{F}_t^{P_\mu})$-stopping times such that $\tau_n \uparrow \tau$ then $X_{\tau_n} \to X_\tau$ as $n \to \infty$ P_μ-a.s. on $\{\tau < \zeta\}$.

    (M.10) *(special)* If τ, τ_n, $n \in \mathbb{N}$, are as in (M.9) then X_τ is $\bigvee_{n \in \mathbb{N}} \mathcal{F}_{\tau_n}^{P_\mu}$-measurable.

(iii) $\mathbf{M}$ is called a *special standard process* if it is a μ-special standard process for all $\mu \in \mathcal{P}(E_\Delta)$.

(iv) $\mathbf{M}$ is called a *Hunt process* if (M.8) and (M.9) holds with ζ replaced by ∞ and E by E_Δ for all $\mu \in \mathcal{P}(E_\Delta)$.

Remark 1.14. (i) For "nice" spaces E (e.g. E Polish) (M.9) implies the almost sure-existence of left limits up to ζ (cf. [Sh 88, (47.10)]), and a Hunt process always has property (M.10).

(ii) The notion of an m-special standard process $\mathbf{M}$, in the case when E is a *metrizable Lusin* space (i.e., topologically isomorphic to a Borel subset of a complete separable metric space), was introduced in [GS 84]. In this case $\mathbf{M}$ is automatically m-tight as is implied immediately by the following general theorem.

Theorem 1.15. *Assume E is a metrizable Lusin space and $(X_t)_{t \geq 0}$ a stochastic process with state space E and life time ζ on $(\Omega, \mathcal{F})$. Assume that P is a complete probability measure on $(\Omega, \mathcal{F})$ such that $(X_t)_{t \geq 0}$ is cadlag P-a.s., i.e., P-a.s., $t \mapsto X_t$ is right continuous on $[0, \infty[$ and has left limits X_{t-} for all $t \in]0, \zeta[$.*

Then there exists an increasing sequence $(K_n)_{n \in \mathbb{N}}$ of compact subsets of E such that

$$P\left[\lim_{n \to \infty} \sigma_{E \setminus K_n} < \zeta\right] = 0 \ .$$

Remark 1.16. Theorem 1.15 will not be used before Section 6 below and generalizes to metrizable co-Souslin spaces (cf. [AMR 91]).

Before we prove 1.15 we need some preparations concerning the Skorohod topology which we shortly describe below following [Mai 72].

Let $\overline{E}$ be a complete separable metric space equipped with a bounded metric ρ. A function a from $[0, \infty[$ to $\overline{E}$ is called *cadlag* if $t \mapsto a(t)$ is right continuous on $[0, \infty[$ and has left limits $a(t-)$ for all $t \in]0, \infty[$. Let $\mathcal{D}$ be the set of all cadlag functions from $[0, \infty[$ to $\overline{E}$. Let Λ denote the set of all real-valued increasing functions λ on $[0, \infty[$ such that $\lambda(0) = 0$. For $\lambda \in \Lambda$ define

$$\|\lambda\| := \sup_{\substack{s,t \geq 0 \\ s \neq t}} \left|\log \frac{\lambda(t) - \lambda(s)}{t - s}\right| + \sup_{t \geq 0} |\lambda(t) - t| \ .$$

Note that if $\|\lambda\| < \infty$, then λ is continuous, strictly increasing and onto. In this case we have for its inverse λ^{-1} that $\|\lambda\| = \|\lambda^{-1}\|$. For $a, b \in \mathcal{D}$ define

$$(1.13) \qquad d(a, b) := \inf\{\|\lambda\| + \sup_{t \geq 0} e^{-t} \rho(a(t), b(\lambda(t))) \mid \lambda \in \Lambda\} \ .$$

For $t \geq 0$ we define maps π_t, π_{t-} from $\mathcal{D}$ to $\overline{E}$ by

$$\pi_t(a) := a(t) \ , \quad \pi_{t-}(a) := a(t-) \ , \quad a \in \mathcal{D} \ .$$

Let Π, Π_- denote the maps from $[0, \infty[\times \mathcal{D}$, $]0, \infty[\times \mathcal{D}$ respectively to $\overline{E}$ given by

$$\Pi(t, a) := \pi_t(a) \ , \quad \Pi_-(t, a) := \pi_{t-}(a) \ .$$

Proposition 1.17.   *(i) $(\mathcal{D}, d)$ is a complete separable metric space.*

(ii) If $\mathcal{B}(\mathcal{D})$ denotes the Borel σ-algebra of $(\mathcal{D}, d)$, then $\mathcal{B}(\mathcal{D}) = \sigma\{\pi_t \mid t \geq 0\}$.

Proof. [Mai 72]. $\qquad\qquad\qquad\qquad\qquad\qquad\qquad\qquad\qquad\qquad\qquad\qquad\quad \square$

Lemma 1.18. *Let $\Gamma \subset \mathcal{D}$, Γ compact, and $M > 0$. Then $\Pi([0, M] \times \Gamma)$ is a relatively compact subset of $\overline{E}$ and its closure is contained in $\Pi([0, M] \times \Gamma) \cup \Pi_-(]0, M] \times \Gamma)$.*

Proof. Let $a_n(t_n) \in \Pi([0, M] \times \Gamma)$, $n \in \mathbb{N}$. Selecting a subsequence if necessary we may assume that $t_n \to t$ and $a_n \to a$ in $(\mathcal{D}, d)$ as $n \to \infty$ for some $t \in [0, M]$, $a \in \Gamma$. By definition (1.13) there exist $\lambda_n \in \Lambda, n \in \mathbb{N}$, such that $\|\lambda_n\| \xrightarrow[n \to \infty]{} 0$ and

$$\lim_{n \to \infty} \sup_{t \geq 0} e^{-t} \rho(a_n(t), a(\lambda_n(t))) = 0 \ .$$

Hence $\lim\limits_{n\to\infty} \lambda_n(t_n) = t$ and again selecting a subsequence we may assume that either $\lambda_n(t_n) \downarrow t$ or $\lambda_n(t_n) \uparrow t$ as $n \to \infty$. Consequently,

$$\text{either} \quad \lim_{n\to\infty} a_n(t_n) = a(t) \in \overline{E} \quad \text{or} \quad \lim_{n\to\infty} a_n(t_n) = a(t-) \in \overline{E}$$

(since e.g. $\rho(a(t), a_n(t_n)) \le \rho(a(t), a(\lambda_n(t_n))) + \rho(a(\lambda_n(t_n)), a_n(t_n))$ for all $n \in \mathbb{N}$). $\qquad\square$

Proof of 1.15. Let $\overline{E}$ and $\mathcal{D}$ be as above with E a topological subspace of $\overline{E}$ and $E \in \mathcal{B}(\overline{E})$. Without loss of generality we may assume that for every $\omega \in \Omega$, $t \mapsto X_t(\omega)$ is right continuous on $[0, \infty[$ and has left limits $X_{t-}(\omega)$ for all $t \in]0, \zeta(\omega)[$. Define for $\omega \in \Omega$

$$Y_t(\omega) := \begin{cases} X_t(\omega) & \text{if } \zeta(\omega) = \infty \\ X_{\frac{t\zeta}{1+t}}(\omega) & \text{if } \zeta(\omega) < \infty \end{cases}$$

Then $Y := (Y_t)_{t\ge 0}$ defines a map from Ω to $\mathcal{D}$ such that $\pi_t \circ Y \ (= Y_t)$ is $\mathcal{F}$-measurable for all $t \ge 0$. Hence by 1.17(ii), Y is $\mathcal{F}/\mathcal{B}(\mathcal{D})$-measurable. Let $Q := P \circ Y^{-1}$ be the image measure of P under Y defined on $(\mathcal{D}, \mathcal{B}(\mathcal{D}))$. Let $\mathcal{D}_0 \subset \mathcal{D}$ denote the set of all $a \in \mathcal{D}$ such that $a(t), a(t-) \in E$ for all $t \ge 0, t > 0$ respectively, and set

$$S := \mathcal{D} \setminus \mathcal{D}_0 .$$

Then

$$S = \{a \in \mathcal{D} \,|\, (t, a) \in G \text{ for some } t \ge 0\}$$

where

$$G := \{(t, a) \in [0, \infty[\times \mathcal{D} \,|\, a(t) \in \overline{E} \setminus E \text{ or } a(t-) \in \overline{E} \setminus E\} .$$

Since both Π and Π_- are $\mathcal{B}([0, \infty[) \otimes \mathcal{B}(\mathcal{D})/\mathcal{B}(\overline{E})$-measurable and $\overline{E} \setminus E \in \mathcal{B}(\overline{E})$, we have that $G \in \mathcal{B}([0, \infty[) \otimes \mathcal{B}(\mathcal{D})$, hence $S \in \mathcal{B}(\mathcal{D})^Q$ (by [DM 78, III.13 and III.33]) and thus $\mathcal{D}_0 \in \mathcal{B}(\mathcal{D})^Q$. But obviously $Y(\Omega) \subset \mathcal{D}_0$, consequently $Q(\mathcal{D}_1) = 1$ for some $\mathcal{D}_1 \in \mathcal{B}(\mathcal{D})$ with $\mathcal{D}_1 \subset \mathcal{D}_0$. Since by 1.17(i), Q is inner regular on $(\mathcal{D}, \mathcal{B}(\mathcal{D}))$ (cf. e.g. [B 78,41.3]) we can find an increasing sequence $(\Gamma_n)_{n\in\mathbb{N}}$ of compact subsets of $\mathcal{D}_1$ such that

$$Q(\Gamma_n) \ge Q(\mathcal{D}_1) - \frac{1}{n} = 1 - \frac{1}{n} \text{ for all } n \in \mathbb{N} .$$

Fix $n \in \mathbb{N}$ and let K_n denote the closure of $\Pi([0, n] \times \Gamma_n)$ in $\overline{E}$, then K_n is a compact subset of E by 1.18. Furthermore, if $\omega \in Y^{-1}(\Gamma_n)$ then for all $t \in \left[0, \frac{n\zeta(\omega)}{1+n} \wedge n\right]$

$$X_t(\omega) = \begin{cases} Y_t(\omega) \in K_n & \text{if } \zeta(\omega) = \infty \\ Y_{\frac{t}{\zeta-t}}(\omega) \in K_n & \text{if } \zeta(\omega) < \infty , \end{cases}$$

which implies that

$$\sigma_{E\backslash K_n}(\omega) \geq \frac{n\zeta(\omega)}{1+n} \wedge n$$

where σ_A is the first hitting time of $A \subset E$ w.r.t. $(X_t)_{t\geq 0}$. Consequently,

$$P\left[\lim_{m\to\infty} \sigma_{E\backslash K_m} \geq \frac{n\zeta}{1+n} \wedge n\right] \geq P\left[\sigma_{E\backslash K_n} \geq \frac{n\zeta}{1+n} \wedge n\right]$$
$$\geq P\left[Y^{-1}(\Gamma_n)\right] = Q[\Gamma_n]$$
$$\geq 1 - \frac{1}{n}$$

and letting $n \to \infty$ we obtain the assertion. $\qquad\qquad\square$

During the preparation of this book P.A. Meyer suggested to us the following shorter alternative proof of 1.15 which is based on Choquet's Capacitability Theorem (cf. III.2.9) rather than the Skorohod topology. We include it below and would like to thank P.A. Meyer for it at this point. Also this proof generalizes to metrizable co-Souslin spaces.

Alternative proof of 1.15. We may assume that E is a Borel subset of a compact metric space $\overline{E}$. We only prove the assertion for the E-valued process $(Y_t)_{t\geq 0}$ which is defined as in the above proof. We set for $\omega \in \Omega$ and $t \in [0,\infty[$

$$\overline{Y_0^t}(\omega) := \text{closure of } \{Y_s(\omega) \,|\, s \in [0,t]\} \text{ in } \overline{E} \ .$$

Fix $n \in \mathbb{N}$ and define for $A \subset \overline{E}$

$$T_A^{(n)} := \inf\{0 \leq t \leq n \,|\, \overline{Y_0^t} \cap A \neq \emptyset\}$$

(with the convention that $\inf \emptyset := +\infty$). Note that by the right continuity of the sample paths one easily checks that for open sets $U \subset \overline{E}$

$$T_U^{(n)} = \inf\{0 < t \leq n \,|\, Y_t \in U\} =: \sigma_U^{(n)} \ .$$

Consequently, $\{T_U^{(n)} < \infty\} \in \mathcal{F}$. Define for $U \subset \overline{E}, \overline{E}$ open,

$$\Gamma_n(U) := P[T_U^{(n)} < \infty]$$

and for arbitrary $A \subset \overline{E}$

$$\Gamma_n(A) := \inf\{\Gamma_n(U) \,|\, A \subset U \subset \overline{E}, U \text{ open } \} \ .$$

Since Γ_n is strictly subadditive on the open sets it follows as in the proof of III.2.8 that Γ_n is a Choquet capacity. If $(K_l)_{l\in\mathbb{N}}$ is a decreasing sequence of compact subsets of $\overline{E}$ and $K := \bigcap K_l$ then $T_K^{(n)} < \infty$ if and only if $T_{K_l}^{(n)} < \infty$ for all $l \in \mathbb{N}$. Let $K \subset \overline{E}$, K compact. Then, since $K = \bigcap \overline{U}_l$ for some open sets $U_l \subset \overline{E}$ with $K \subset U_l$, $l \in \mathbb{N}$, it follows that $[T_k^{(n)} < \infty] \in \mathcal{F}$ and

$$\Gamma_n(K) = P[T_K^{(n)} < \infty] \ .$$

By Choquet's capacitability theorem (cf. III.2.9 and [DM 78, III.2.8]) it follows for every Borel subset $A \subset \overline{E}$ that $[T_A^{(n)} < \infty] \in \mathcal{F}$ and

$$\Gamma_n(A) = P[T_A^{(n)} < \infty] .$$

In particular, we have that

$$\Gamma_n(\overline{E}\backslash E) = P[T_{\overline{E}\backslash E}^{(n)} < \infty] = 0 ,$$

since $t \mapsto Y_t(\omega)$ is cadlag in E for all $\omega \in \Omega$. Consequently, we can find a decreasing sequence $(U_{n,l})_{l \in \mathbb{N}}$ of open subsets of $\overline{E}$ such that $\overline{E}\backslash E \subset U_{n,l}$ and

$$P[\sigma_{U_{n,l}}^{(n)} < \infty] = P[T_{U_{n,l}}^{(n)} < \infty] = \Gamma_n(U_{n,l}) < \frac{1}{l2^n} ,$$

$l \in \mathbb{N}$. Setting

$$U_l := \bigcap_{1 \leq n \leq l} U_{n,l} , \quad l \in \mathbb{N},$$

we easily obtain that

$$P[\lim_{l \to \infty} \sigma_{U_l} < \infty] = 0 .$$

Hence $K_l := E\backslash U_l$, $l \in \mathbb{N}$, form an increasing sequence of compact subsets of E having the desired property. $\qquad\square$

To conclude this section we state some basic facts on Ray resolvents (cf. [G 75]) which we shall use in Sect.3 below. The proofs are included in the Appendix. Assume for the rest of this section that E is a compact metric space.

Definition 1.19. A Markovian resolvent of kernels $(R_\alpha)_{\alpha>0}$ on $(E, \mathcal{B}(E))$ is called a *Ray resolvent* if

(R.1) $R_\alpha C(E) \subset C(E)$ for all $\alpha > 0$

(R.2) $\bigcup_{\alpha>0} C(E) \cap S^\alpha$ separates the points of E.

Here S^α denotes the set of all functions f which are *α-supermedian* for $(R_\beta)_{\beta>0}$, i.e., $f \in \mathcal{B}(E)^+$ and $\beta R_{\alpha+\beta} f \leq f$ for all $\beta > 0$.

Theorem 1.20. *Let $(R_\alpha)_{\alpha>0}$ be a Ray resolvent. Then there exists a unique semigroup $(p_t)_{t\geq 0}$ of Markovian kernels on $(E, \mathcal{B}(E))$ having the following properties:*

(i) $R_\alpha f = \int_0^\infty e^{-\alpha t} p_t f \, dt$ for all $\alpha > 0$, $f \in C(E)$.

(ii) $t \mapsto p_t f(z)$ is right continuous on $[0, \infty[$ for all $z \in E$, $f \in C(E)$.

If

$$(1.14) \qquad D := \{z \in E \mid \lim_{\alpha \to \infty} \alpha R_\alpha f(z) = f(z) \text{ for all } f \in C(E)\}$$

then in addition,

(iii) $D \in \mathcal{B}(E)$, $p_0(z, \cdot) = \epsilon_z$ *if and only if* $z \in D$ *and* $p_t(z, E \setminus D) = 0$ *for all*
$t \geq 0$, $z \in E$.

Here ϵ_z denotes the Dirac measure in $z \in E$.

Proof. See [G 75, Theorem (3.6)] and Appendix, Subsection 3 a).   □

Remark 1.21. *D in (1.14) is called the set of non-branching points.* Clearly,
1.20(iii) implies that the restriction of $(p_t)_{t\geq0}$ to $(D, \mathcal{B}(E) \cap D)$ is again a semi-group of Markovian kernels.

Theorem 1.22. *Let $(p_t)_{t\geq0}$ and D be as in 1.20. Then there exists a right process* $\mathbf{M} = (\Omega, \mathcal{F}, (X_t)_{t\geq0}, (P_z)_{z\in D})$ *with state space D and transition semigroup* $(p_t)_{t\geq0}$ *such that for all $\omega \in \Omega$, $t \mapsto X_t(\omega)$ has left limits in E for all $t \in]0, \infty[$.*

Proof. See [Sh 88, Theorem (9.13)] and Appendix, Subsection 3 b).   □

2 Association of right processes and Dirichlet forms

In this section we assume that $\mathcal{B}(E) = \sigma(C(E))$ and fix a σ-finite positive measure m on $(E, \mathcal{B}(E))$. We start with an important result for right processes $\mathbf{M}$ with state space E and introduce the following class $\mathcal{D}(\mathbf{M})$ of functions on E. $\mathcal{D}(\mathbf{M})$ shall denote all $f \in \mathcal{B}_b(E) \cap L^1(E; m)$ such that $t \mapsto f(X_t(\omega))$ is right continuous on $[0, \infty[$ for all $\omega \in \Omega_0$ for some $\Omega_0 \in \mathcal{F}$ with $P_z(\Omega_0) = 1$ for all $z \in E$. Note that, of course, $C_b(E) \cap L^1(E; m) \subset \mathcal{D}(\mathbf{M})$.

Lemma 2.1. *Let $\mathbf{M}$ be a right process with state space E and transition semigroup $(p_t)_{t>0}$ such that m is p_t-supermedian for all $t > 0$. Then*

$$(2.1) \qquad \textit{there exists } \rho \in \mathcal{D}(\mathbf{M}) \textit{ such that } \rho > 0 \textit{ m-a.e. .}$$

Proof. Since m is σ-finite, there exists $\varphi \in \mathcal{B}_b(E) \cap L^1(E; m)$ such that $\varphi(z) > 0$ for all $z \in E$. Let $(R_\alpha)_{\alpha>0}$ be the resolvent of $\mathbf{M}$ and define

$$(2.2) \qquad\qquad\qquad \rho := R_1\varphi .$$

Despite the fact that $\varphi(\Delta) = 0$, it follows from the normal property (M.5) and the right continuity (M.6) that $\rho(z) > 0$ for all $z \in E$. By Claim 1 in the proof of 5.14 below, if $\Omega_0 := \{\omega \in \Omega \,|\, t \mapsto R_1\varphi(X_t(\omega))$ is right-continuous on $[0, \infty[\}$, then $\Omega_0 \in \mathcal{F}$ and $P_z(\Omega_0) = 1$ for all $z \in E$, i.e., $\rho \in \mathcal{D}(\mathbf{M})$. (Note that for this part of the proof of 5.14 we only use that $\mathbf{M}$ is a right process (cf. 5.13) and that we can take $\mu = \epsilon_z$, $z \in E$.) Since m is p_t-supermedian for all $t > 0$, hence αR_α-supermedian for all $\alpha > 0$, we have that $\int \rho\, dm = \int R_1\varphi\, dm \leq \int \varphi\, dm < \infty$, and (2.1) follows.   □

Remark 2.2. Claim 1 in 5.14 which we used in the proof of 2.1 above is based on some results from advanced (though quite standard) probability theory (cf. 5.8 and 5.10 below) which we shall not prove in this book, but refer to [DM 78,82]. In order to keep this book up to Chap.IV, Sect.4 reasonably self-contained (as announced in the introduction) we shall prove directly that the process $\mathbf{M}$ constructed in the next section satisfies (2.1) without using 2.1 (cf. Remark 3.23 below). Hence all subsequent results of this section (which all rely on (2.1)) are applicable to $\mathbf{M}$ and we can ignore 2.1 up to Chap.IV, Sect. 4.

Proposition 2.3. *Let* $\mathbf{M}$ *be a right process with state space* E *satisfying (2.1). Then*

$$(2.3) \qquad\qquad \mathcal{D}(\mathbf{M}) \text{ is dense in } L^2(E;m) \ .$$

Proof. We prove (2.3) in two steps.

Step 1. Suppose first that $m(E) < \infty$. Then $C_b(E) \subset \mathcal{D}(\mathbf{M})$, hence we only need to show that $C_b(E)$ is dense in $L^2(E;m)$. To this end, let H be the set of all bounded functions which are limits of sequences in $C_b(E)$ in $L^2(E;m)$. Then H is a real vector space containing $C_b(E)$. Furthermore, if $f_n \in H$, $n \in \mathbb{N}$, such that $f_n \uparrow f \in \mathcal{B}_b(E)$ on E, then $f_n \xrightarrow[n\to\infty]{} f$ in $L^2(E;m)$ and hence there exist $\varphi_n \in C_b(E)$, $n \in \mathbb{N}$, such that $\varphi_n \xrightarrow[n\to\infty]{} f$ in $L^2(E;m)$, i.e., $f \in H$. Since $\mathcal{B}(E) = \sigma(C(E))$ a monotone class argument (cf. [Sh 88, A.0.6]) yields $H = \mathcal{B}_b(E)$ hence $C_b(E)$ is dense in $L^2(E;m)$.

Step 2. For general m we know by (2.1) that there exists $\rho \in \mathcal{D}(\mathbf{M})$, $\rho > 0$ m-a.e. . Let $f \in L^2(E;m)$ and $\rho_n := (n \cdot \rho) \wedge 1$, $n \in \mathbb{N}$. Then by Step 1 there exist $g_k \in C_b(E) \cap L^2(E;\rho_n^2 \cdot m)$ such that $g_k \xrightarrow[k\to\infty]{} f$ in $L^2(E;\rho_n^2 \cdot m)$, i.e., $g_k\rho_n \xrightarrow[k\to\infty]{} f\rho_n$ in $L^2(E;m)$. But clearly, $f\rho_n \uparrow f$ m-a.e. on E as $n \to \infty$ and (2.3) follows. $\square$

Exercise 2.4. Let $\mathbf{M}, \hat{\mathbf{M}}$ be two right processes with state space E. Let $(p_t)_{t>0}$, $(\hat{p}_t)_{t>0}$ and $(R_\alpha)_{\alpha>0}$, $(\hat{R}_\alpha)_{\alpha>0}$ be the corresponding transition semigroups resp. resolvents. Prove that $(p_t)_{t>0}$ is in *duality with* $(\hat{p}_t)_{t>0}$ w.r.t. m (i.e., p_t is in duality with $\hat{p}_t$ w.r.t. m for all $t > 0$) if and only if $(R_\alpha)_{\alpha>0}$ is *in duality with* $(\hat{R}_\alpha)_{\alpha>0}$ w.r.t. m , (i.e., R_α is in duality with $\hat{R}_\alpha$ w.r.t. m for all $\alpha > 0$).

Let us now explain the connection with Dirichlet forms. Consider a right process $\mathbf{M}$ with state space E and transition semigroup $(p_t)_{t>0}$ such that m is p_t-supermedian for all $t > 0$. Property (2.3) implies that $(p_t)_{t>0}$ satisfies II.(4.6), hence by II.4.3 there exists a corresponding strongly continuous contraction semigroup $(T_t)_{t>0}$ on $L^2(E;m)$ which is sub-Markovian. Suppose that $(e^{-t}T_t^\mathbb{C})_{t>0}$ is the restriction of a holomorphic contraction semigroup on some sector or (cf. I.2.21) equivalently $1 - L$ satisfies the sector condition (2.5) where L is the generator of $(T_t)_{t>0}$. Then by the results of Chap.I, Sections 2 and 4, there exists a corresponding Dirichlet form $(\mathcal{E}, D(\mathcal{E}))$. $\mathbf{M}$ is associated with $(\mathcal{E}, D(\mathcal{E}))$ in the following sense.

Definition 2.5. Let $(\mathcal{E}, D(\mathcal{E}))$ be a Dirichlet form on $L^2(E;m)$ and $(T_t)_{t>0}$, $(\hat{T}_t)_{t>0}$ the associated sub-Markovian strongly continuous contraction semigroups on $L^2(E;m)$.

(i) A right process $\mathbf{M}$ with state space E and transition semigroup $(p_t)_{t>0}$ is called *associated* (*coassociated* respectively) *with* $(\mathcal{E}, D(\mathcal{E}))$ if $p_t f$ is an m-version of $T_t f$ ($\hat{T}_t f$ respectively) for all $t > 0$, $f \in \mathcal{B}_b(E) \cap L^2(E; m)$. $\mathbf{M}$ is called *properly associated* (*properly coassociated* respectively) *with* $(\mathcal{E}, D(\mathcal{E}))$ if in addition, $p_t f$ is $\mathcal{E}$-quasi-continuous for all $t > 0$, $f \in \mathcal{B}_b(E) \cap L^2(E; m)$.

(ii) A pair $(\mathbf{M}, \hat{\mathbf{M}})$ is called *(properly) associated with* $(\mathcal{E}, \mathcal{D}(\mathcal{E}))$ if $\mathbf{M}$ is (properly) associated and $\hat{\mathbf{M}}$ is (properly) coassociated with $(\mathcal{E}, \mathcal{D}(\mathcal{E}))$.

Remark 2.6. (i) Note that by our notation convention (i.e., if $f \in \mathcal{B}_b(E)$ then f also denotes the associated m-class) p_t respects m-classes if $\mathbf{M}$ is associated with $(\mathcal{E}, D(\mathcal{E}))$ (cf. the Remark following II.4.1).

(ii) Note that if $\mathbf{M}$ is associated with $(\mathcal{E}, D(\mathcal{E}))$ then $p_t f$ is an m-version of $T_t f$ for all $t > 0$ and all $\mathcal{B}(E)$-measurable, m-square integrable functions f on E by II.4.2. A corresponding statement for "properly associated" is contained in 2.9 below.

(iii) If $(\mathbf{M}, \hat{\mathbf{M}})$ is associated with $(\mathcal{E}, D(\mathcal{E}))$ and if $(p_t)_{t>0}$, $(\hat{p}_t)_{t>0}$ denote the corresponding transition semigroups, then clearly $(p_t)_{t>0}$ is in duality with $(\hat{p}_t)_{t>0}$ w.r.t. m. Hence m is p_t- and $\hat{p}_t$-supermedian for all $t > 0$ (cf. Chap.II, Subsection 4a)).

Sometimes it is easier to check "(co)association" in terms of the resolvents which is possible by

Exercise 2.7. Let $\mathbf{M}$ be a right process with state space E, transition semigroup $(p_t)_{t>0}$ and resolvent $(R_\alpha)_{\alpha>0}$. Let $(T_t)_{t>0}$ be a strongly continuous contraction semigroup on $L^2(E; m)$ with associated resolvent $(G_\alpha)_{\alpha>0}$. Prove that $p_t f$ is an m-version of $T_t f$ for all $t > 0$, $f \in L^2(E; m)$, if and only if $R_\alpha f$ is an m-version of $G_\alpha f$ for all $\alpha > 0$, $f \in L^2(E; m)$.

The corresponding statement for "proper (co)association" is more difficult to prove.

Proposition 2.8. *Let* $(\mathcal{E}, D(\mathcal{E}))$ *be a Dirichlet form on* $L^2(E; m)$ *and* $\mathbf{M}$ *a right process with state space* E *and corresponding resolvents* $(G_\alpha)_{\alpha>0}$, $(R_\alpha)_{\alpha>0}$ *respectively. Then the following are equivalent:*

(i) $\mathbf{M}$ *is properly associated with* $(\mathcal{E}, D(\mathcal{E}))$.

(ii) $R_\alpha f$ *is an* $\mathcal{E}$*-quasi-continuous* m*-version of* $G_\alpha f$ *for all* $\alpha > 0$, $f \in \mathcal{B}_b(E) \cap$ $L^2(E; m)$.

Proof. Suppose (i) holds and let $(T_t)_{t>0}$, $(p_t)_{t>0}$ be the semigroups corresponding to $(\mathcal{E}, D(\mathcal{E}))$, $\mathbf{M}$ respectively. Let $\alpha > 0$, $f \in \mathcal{B}_b(E) \cap L^2(E; m)$. Since $T_s f \xrightarrow[s \downarrow 0]{} f$ in $L^2(E; m)$ we have by I.2.9(i) that

$$T_s G_\alpha f = G_\alpha T_s f \xrightarrow[s \downarrow 0]{} G_\alpha f \ \text{ in } \mathcal{D}(\mathcal{E}) \ (\text{w.r.t. } \tilde{\mathcal{E}}_1^{1/2}) \ .$$

By 2.7, $R_\alpha f$ is an m-version of $G_\alpha f$, hence by assumption for $s > 0$, $p_s(R_\alpha f)$ is an $\mathcal{E}$-quasi-continuous m-version of $T_s G_\alpha f$, and by III.3.5 a subsequence of $(p_s(R_\alpha f))_{s>0}$ converges $\mathcal{E}$-quasi-uniformly. Hence, if we can prove that it converges to $R_\alpha f$ $\mathcal{E}$-q.e., $R_\alpha f$ is $\mathcal{E}$-quasi-continuous. But this follows since (by (1.9))

$$p_s(R_\alpha f)(z) = e^{\alpha s} \int_s^\infty e^{-\alpha t} p_t f(z) dt \xrightarrow[s\downarrow 0]{} R_\alpha f(z)$$

for all $z \in E$.

Suppose (ii) holds and let $f \in \mathcal{B}_b(E) \cap L^2(E; m)$, $t > 0$. We know by I.2.21 and I.2.20 that $T_t f \in D(L) \subset D(\mathcal{E})$ where L is the generator of $(T_t)_{t>0}$. But by I.2.13(ii)

$$\lim_{\alpha \to \infty} \mathcal{E}_1(\alpha G_\alpha u - u, \alpha G_\alpha u - u) = 0$$

for all $u \in D(\mathcal{E})$. Hence, since $p_t f$ is an m-version of $T_t f$ by 2.7, it follows by assumption and III.3.5 that $(\alpha_n R_{\alpha_n}(p_t f))_{n\in\mathbb{N}}$ converges $\mathcal{E}$-quasi-uniformly for some sequence $(\alpha_n)_{n\in\mathbb{N}}$ in $]0, \infty[$ increasing to ∞. Furthermore,

$$\alpha R_\alpha(p_t f)(z) = \int_0^\infty e^{-s} E_z\left[f\left(X_{t+\frac{s}{\alpha}}\right)\right] ds , \quad z \in E , \ \alpha > 0 .$$

Consequently, if $f \in \mathcal{D}(\mathbf{M})$, then $\lim_{\alpha\to\infty} \alpha R_\alpha(p_t f)(z) = p_t f(z)$ for all $z \in E$. Hence $p_t f$ is $\mathcal{E}$-quasi-continuous in this case. Now let ρ be as in (2.1) and fix $n \in \mathbb{N}$. Set $\rho_n := n\rho \wedge 1$ and define

$$H := \{f \in \mathcal{B}_b(E) \,|\, p_t(f\rho_n) \text{ is } \mathcal{E}\text{-quasi-continuous}\} .$$

Then H is a real vector space, containing $C_b(E)$ which is closed under bounded monotone convergence. The latter can be seen as follows. If $f_k \in H$ with $0 \le f_k \uparrow f$ pointwise on E as $k \to \infty$ for some $f \in \mathcal{B}_b(E)$ then $p_t(f_k \rho_n) \uparrow p_t(f\rho_n)$ pointwise on E as $k \to \infty$. But each $p_t(f_k \rho_n)$ is an $\mathcal{E}$-quasi-continuous m-version of $T_t(f_k \rho_n)$ and $(T_t(f_k \rho_n))_{k\in\mathbb{N}}$ is bounded w.r.t. $\tilde{\mathcal{E}}_1^{1/2}$ by I.2.21 and I.2.20. Consequently, by I.2.12 and III.3.5 the Cesaro mean of a subsequence of $(p_t(f_k \rho_n))_{k\in\mathbb{N}}$ converges $\mathcal{E}$-quasi-uniformly to $p_t(f\rho_n)$ which is therefore $\mathcal{E}$-quasi-continuous. Now, since $\mathcal{B}(E) = \sigma(C(E))$, a monotone class argument yields that $H = \mathcal{B}_b(E)$ (cf. e.g. [Sh 88, A.0.6]). If $f \in \mathcal{B}_b^+(E) \cap L^2(E; m)$, then $p_t(f\rho_n) \uparrow p_t f$ pointwise on E as $n \to \infty$. Each $p_t(f\rho_n)$ is an $\mathcal{E}$-quasi-continuous m-version of $T_t(f\rho_n)$ and again $(T_t(f\rho_n))_{n\in\mathbb{N}}$ is bounded w.r.t. $\mathcal{E}_1^{1/2}$ by I.2.21 and I.2.20. The same arguments as above now show that $p_t f$ is $\mathcal{E}$-quasi-continuous. Since for $f \in \mathcal{B}_b(E) \cap L^2(E; m)$, $p_t f = p_t f^+ - p_t f^-$ m-a.e., (i) is proved. $\qquad\square$

Exercise 2.9. Consider the situation of 2.8. Suppose that $\mathbf{M}$ is properly associated with $(\mathcal{E}, D(\mathcal{E}))$. Prove that $p_t f$, $R_\alpha f$ are $\mathcal{E}$-quasi-continuous m-versions of $T_t f$, $G_\alpha f$ respectively, for all $\alpha, t > 0$, $f \in L^2(E; m)$.

3 Quasi-regularity and the construction of the process

In this section we identify a class of Dirichlet forms (called *quasi-regular*) for which an associated special standard process exists (cf. Theorem 3.5 below). In Section 5 we shall prove that these are the only Dirichlet forms for which this is the case. Therefore, Definition 3.1 below is fundamental for the theory of Dirichlet forms. In this section we assume again that $\mathcal{B}(E) = \sigma(C(E))$ and we fix a σ-finite positive measure m on $(E, \mathcal{B}(E))$. For a subset $Y \subset E$, $\mathcal{B}(Y)$ denotes the Borel-σ-algebra of Y equipped with the trace topology inherited from E. Recall that $\mathcal{B}(Y) = \mathcal{B}(E) \cap Y$ ($:=$ the trace σ-algebra of $\mathcal{B}(E)$ on Y).

Definition 3.1. A Dirichlet form $(\mathcal{E}, D(\mathcal{E}))$ on $L^2(E; m)$ is called *quasi-regular* if:

  (i) There exists an $\mathcal{E}$-nest $(E_k)_{k \in \mathbb{N}}$ consisting of compact sets.

  (ii) There exists an $\tilde{\mathcal{E}}_1^{1/2}$-dense subset of $D(\mathcal{E})$ whose elements have $\mathcal{E}$-quasi-continuous m-versions.

  (iii) There exist $u_n \in D(\mathcal{E})$, $n \in \mathbb{N}$, having $\mathcal{E}$-quasi-continuous m-versions $\tilde{u}_n$, $n \in \mathbb{N}$, and an $\mathcal{E}$-exceptional set $N \subset E$ such that $\{\tilde{u}_n | n \in \mathbb{N}\}$ separates the points of $E \setminus N$.

Remark 3.2. (i) We emphasize that the notion of quasi-regularity only depends on the symmetric part of $(\mathcal{E}, D(\mathcal{E}))$.

(ii) By III.2.11 property 3.1(i) is equivalent to the *tightness* of $\mathrm{Cap}_{h,g}$ (where h, g are as in III.2.11). $\mathrm{Cap}_{h,g}$ is called *tight* if there exist compact sets $K_n \subset E$, $n \in \mathbb{N}$, such that $\lim_{n \to \infty} \mathrm{Cap}_{h,g}(E \setminus K_n) = 0$. Hence 3.1(i) can be viewed as a substitute for local compactness. It is easy to see that any *regular* Dirichlet form on a locally compact separable metric state space E (in the sense of [F 80]) is quasi-regular. For details we refer to Subsection 4a) below.

(iii) If 3.1(i) and (iii) hold then by making E_k, $k \in \mathbb{N}$, smaller if necessary, we may assume that $N \subset \cap E_k^c$ and that $\tilde{u}_n \in C(\{E_k\})$ for all $n \in \mathbb{N}$ (cf. III.3.3). Hence for each $k \in \mathbb{N}$

$$\rho(x, y) := \sum_{n=1}^{\infty} \frac{1}{2^n} \left(|\tilde{u}_n(x) - \tilde{u}_n(y)| \wedge 1 \right) \; ; \; x, y \in E_k \, ,$$

defines a (separating) metric on E_k which by the compactness of E_k is compatible with the original topology on E_k inherited from E. Therefore, we can always assume all E_k in 3.1(i) to be metrizable. In particular, then the set $Y := \bigcup_{k \geq 1} E_k$ by [Sch 73, Corollary 2, p.102] is a *(topological) Lusin space* (i.e., the continuous one-to-one image of a Polish space). Since $\mathcal{B}(Y) = \mathcal{B}(E) \cap Y$ and since $E \setminus Y$ is $\mathcal{E}$-exceptional, hence $m(E \setminus Y) = 0$ by III.2.3, $L^2(E; m)$ can be identified with $L^2(Y; m)$ canonically. Therefore, when dealing with quasi-regular Dirichlet forms one could assume without loss of generality that E is a (topological) Lusin space.

(iv) Under the assumption that property 3.1(i) holds, property 3.1(iii) is equivalent to:

$$
(3.1) \qquad
\begin{array}{l}
\text{There exist } u_n \in D(\mathcal{E}),\ n \in \mathbb{N},\ \text{having } \mathcal{E}\text{-quasi-continuous} \\
m\text{-versions } \tilde{u}_n,\ n \in \mathbb{N},\ \text{and an } \mathcal{E}\text{-exceptional set } N \subset E \\
\text{such that } \sigma\{\tilde{u}_n | n \in \mathbb{N}\} \supset \mathcal{B}(E \setminus N)\ .
\end{array}
$$

Indeed, if we assume (3.1) to hold we can assume that $N \in \mathcal{B}(E)$. Since $\sigma\{\tilde{u}_n \,|\, n \in \mathbb{N}\} \supset \mathcal{B}(E \setminus N)$, it is easy to check that for all $z \in E \setminus N$ the set e_z defined as the intersection of sets of type $\{\tilde{u}_n \geq a\} \cap E \setminus N$ with $n \in \mathbb{N}$, $a \in \mathbb{Q}$, such that $\tilde{u}_n(z) \geq a$, and of type $\{\tilde{u}_n < a\} \cap E \setminus N$ with $n \in \mathbb{N}$, $a \in \mathbb{Q}$, such that $\tilde{u}_n(z) < a$, is the smallest set in $\mathcal{B}(E \setminus N)$ containing z. Hence $e_z = \{z\}$ and therefore $\{\tilde{u}_n \,|\, n \in \mathbb{N}\}$ separates the points of $E \setminus N$. Conversely, suppose that 3.1(iii) holds. We may assume that $\tilde{u}_n = 0$ on N for all $n \in \mathbb{N}$. By 3.1(i) and 3.2(iii) we may assume that $E \setminus N \in \mathcal{B}(Y)$ where Y is as in 3.2(iii). Consider the map $\Lambda : z \mapsto (\tilde{u}_n(z))_{n \in \mathbb{N}}$. Then Λ is one to one from $E \setminus N$ into $\mathbb{R}^{\mathbb{N}}$. If we equip $\Lambda(E \setminus N)$ with the σ-algebra $\mathcal{A}$ generated by all canonical projections π_n, $n \in \mathbb{N}$, from $\mathbb{R}^{\mathbb{N}}$ to $\mathbb{R}$, then, because $E \setminus N \in \mathcal{B}(Y)$, we can apply [P 67, Theorem 2.4, p.135] to conclude that $\mathcal{B}(E \setminus N) = \sigma\{\pi_n \circ \Lambda \,|\, n \in \mathbb{N}\}$. But $\tilde{u}_n = \pi_n \circ \Lambda$ on $E \setminus N$ for all $n \in \mathbb{N}$, and (3.1) is proved.

Now let us discuss some properties of quasi-regular Dirichlet forms.

Proposition 3.3. *Let $(\mathcal{E}, D(\mathcal{E}))$ be a quasi-regular Dirichlet form on $L^2(E; m)$. Then:*

(i) $D(\mathcal{E})$ is separable w.r.t. $\tilde{\mathcal{E}}_1^{1/2}$.

(ii) Each element $u \in D(\mathcal{E})$ has an $\mathcal{E}$-quasi-continuous m-version denoted by $\tilde{u}$.

(iii) If f is $\mathcal{E}$-quasi-continuous and $f \geq 0\ m - a.e.$ on an open subset U of E, then $f \geq 0\ \mathcal{E}$-q.e. on U. In particular, $\tilde{u}$ is $\mathcal{E}$-q.e. unique for all $u \in D(\mathcal{E})$.

Proof. (i): By 3.2(iv) we may assume that E is a Lusin space. Then there exists a countable set $\mathcal{D}_0$ which is dense in $L^2(E; m)$. By I.2.13(ii) we know that $\{G_1 f \,|\, f \in L^2(E; m)\}$ is dense in $D(\mathcal{E})$ w.r.t. $\tilde{\mathcal{E}}_1^{1/2}$. But for $f \in L^2(E; m)$ we can find $f_n \in \mathcal{D}_0$, $n \in \mathbb{N}$, such that $f_n \xrightarrow[n \to \infty]{} f$ in $L^2(E; m)$ hence $G_1 f_n \xrightarrow[n \to \infty]{} G_1 f$ in $D(\mathcal{E})$ by I.2.9(i). Consequently, $\{G_1 f \,|\, f \in \mathcal{D}_0\}$ is dense in $D(\mathcal{E})$ w.r.t. $\tilde{\mathcal{E}}_1^{1/2}$.
(ii) is a consequence of 3.1(ii) and III.3.5.
(iii): Let $(F_k)_{k \in \mathbb{N}}$ be an $\mathcal{E}$-nest such that $f \in C(\{F_k\})$. For $k \in \mathbb{N}$ set $F_k' := F_k \cap E_k$ where E_k is as in 3.1(i) with E_k metrizable (cf. 3.2 (iii)). Since each F_k' is second countable as a compact metrizable space, it is strongly Lindelöf. Hence by III.3.8, $\tilde{F}_k := supp[1_{F_k'} \cdot m]$, $k \in \mathbb{N}$, form a regular $\mathcal{E}$-nest. Now by III.3.9, assertion (iii) follows, since $f \in C(\{\tilde{F}_k\})$ and $E \setminus \bigcup_{k \geq 1} \tilde{F}_k$ is $\mathcal{E}$- exceptional. $\square$

The following is a bit technical, but will be very useful later.

Proposition 3.4. *Let $(\mathcal{E}, D(\mathcal{E}))$ be a quasi-regular Dirichlet form.*

(i) If $\mathcal{D}_1$ is a dense subset of $D(\mathcal{E})$, then there exists an $\mathcal{E}$-exceptional set $N \subset E$ and $\mathcal{E}$-quasi-continuous m-versions $\tilde{u}$, $u \in \mathcal{D}_1$, such that $\{\tilde{u} \mid u \in \mathcal{D}_1\}$ separates the points of $E \setminus N$.

(ii) There exists a countable subset $\mathcal{D}_0^+$ of $D(\mathcal{E})$ consisting of bounded 1-excessive functions such that $\mathcal{D}_0^+ - \mathcal{D}_0^+$ is dense in $\mathcal{D}(\mathcal{E})$, an $\mathcal{E}$-exceptional set $N \subset E$ and $\mathcal{E}$-quasi-continuous m-versions $\tilde{u}, u \in \mathcal{D}_0^+$, such that $\{\tilde{u} \mid u \in \mathcal{D}_0^+\}$ separates the points of $E \setminus N$.

Proof. (i): By quasi-regularity there exist $u_n \in D(\mathcal{E})$, $n \in \mathbb{N}$, and an $\mathcal{E}$-exceptional set N_0 such that $\{\tilde{u}_n \mid n \in \mathbb{N}\}$ separates the points of $E \setminus N_0$. By assumption and III.3.5 there exist $u_{n,k} \in \mathcal{D}_1$, $k \in \mathbb{N}$, such that $\tilde{u}_{n,k} \xrightarrow[k \to \infty]{} \tilde{u}_n$ $\mathcal{E}$-quasi-uniformly, hence pointwise outside some $\mathcal{E}$-exceptional set $N_n \supset N_0$. Consequently, $\{\tilde{u} \mid u \in \mathcal{D}_1\}$ separates the points of $E \setminus \bigcup_{n \geq 1} N_n$.

(ii): By quasi-regularity there exist $u_n \in D(\mathcal{E})$, $n \in \mathbb{N}$, such that $\{u_n \mid n \in \mathbb{N}\}$ is dense in $\mathcal{D}(\mathcal{E})$ and $\{\tilde{u}_n \mid n \in \mathbb{N}\}$ separates the points of $E \setminus N$ for some $\mathcal{E}$-exceptional set $N \subset E$. Let L be the generator of $(\mathcal{E}, D(\mathcal{E}))$, $\mathbb{Q}_+$ the positive rationals, and define

$$
\begin{aligned}
\mathcal{D}_1 &:= \left\{ ((1 - L)\alpha G_\alpha u_n)^+, \; ((1 - L)\alpha G_\alpha u_n)^- \mid n \in \mathbb{N}, \; \alpha \in \mathbb{Q}_+ \right\}, \\
\mathcal{D}_2^+ &:= G_1(\mathcal{D}_1).
\end{aligned}
$$

Then $\mathcal{D}_2^+$ is countable, consists of 1-excessive functions by III.1.3(i), and $\mathcal{D}_2^+ - \mathcal{D}_2^+$ is dense in $D(\mathcal{E})$ by I.2.9(i) and I.2.13(ii). By (i), $\mathcal{D}_2^+ - \mathcal{D}_2^+$ separates the points of E up to an $\mathcal{E}$-exceptional set, hence the same is true for $\mathcal{D}_2^+$. Hence $\mathcal{D}_0^+ := \{u \wedge n \mid u \in \mathcal{D}_2^+, \; n \in \mathbb{N}\}$ is by I.4.17 the desired set. $\qquad\square$

Now we can formulate the existence theorem.

Theorem 3.5. *Let $(\mathcal{E}, D(\mathcal{E}))$ be a quasi-regular Dirichlet form on $L^2(E; m)$. Then there exists a pair $(\mathbf{M}, \hat{\mathbf{M}})$ of m-tight special standard processes which is properly associated with $(\mathcal{E}, D(\mathcal{E}))$.*

Remark 3.6. (i) It will turn out that for m-a.e. $z \in E$, P_z-a.e. $\mathbf{M}$ and $\hat{\mathbf{M}}$ take values in the Lusin space Y introduced in 3.2(iii).

(ii) The fact that the converse of 3.5 also holds is a consequence of Theorem 5.1 below.

(iii) Since the notion of quasi-regularity depends on the symmetric part of $(\mathcal{E}, D(\mathcal{E}))$ only, it follows from 3.5 that there also exists an m-tight special standard process $\tilde{\mathbf{M}}$ associated with $(\tilde{\mathcal{E}}, D(\tilde{\mathcal{E}}))$.

The rest of this section is devoted to the proof of 3.5. It will be carried out via a series of lemmas and propositions and is structured in six steps. Clearly, it is enough to prove the existence of $\mathbf{M}$; the existence of $\hat{\mathbf{M}}$ then follows by considering the quasi-regular Dirichlet form $\mathcal{E}^*(u, v) := \mathcal{E}(v, u)$; $u, v \in D(\mathcal{E}^*) := D(\mathcal{E})$.

We fix a quasi-regular Dirichlet form $(\mathcal{E}, D(\mathcal{E}))$ on $L^2(E; m)$ with associated strongly continuous contraction resolvent $(G_\alpha)_{\alpha>0}$ and strongly continuous contraction semigroup $(T_t)_{t>0}$. The idea of the proof is to construct a set Y_2 with $E \setminus Y_2$ $\mathcal{E}$-exceptional and via a nice countable set $J_0 \subset D(\mathcal{E})$ of 1-excessive ($\mathcal{E}$-quasi-continuous) functions a compactification $\overline{E}$ of $Y_2 \cup \{\Delta\}$ and from $(G_\alpha)_{\alpha>0}$ a corresponding Ray-resolvent $(\overline{R}_\alpha)_{\alpha>0}$ on $\overline{E}$. Then, in particular using 3.1(i), we show that the corresponding right process $\overline{\mathbf{M}}$ on $\overline{E}$ can be restricted to $Y_2 \cup \{\Delta\}$ and that this restriction $\mathbf{M}$ is properly associated with $(\mathcal{E}, D(\mathcal{E}))$.
To start let us fix $(E_k)_{k\in\mathbb{N}}$ as in 3.1(i) with each E_k metrizable (cf. 3.2(iii)) and $\mathcal{D}_0^+$, N as in 3.4. Of course, we may assume that $N \subset E \setminus Y$ where we set

$$(3.2) \qquad Y := \bigcup_{k \geq 1} E_k .$$

Step 1. Construction of $\mathcal{E}$-quasi-continuous kernels $\tilde{R}_\alpha$ from G_α, $\alpha > 0$.

Proposition 3.7. *Let $\alpha > 0$. There exists a kernel $\tilde{R}_\alpha$ from $(E, \mathcal{B}(E))$ to $(Y, \mathcal{B}(Y))$ satisfying the following conditions:*

(i) $\tilde{R}_\alpha f$ is an $\mathcal{E}$-quasi-continuous m-version of $G_\alpha f$ for each $f \in L^2(Y; m)$.

(ii) $\alpha \tilde{R}_\alpha(z, Y) \leq 1$ for all $z \in E$.

The kernel is $\mathcal{E}$-q.e. unique in the sense that if K is a kernel satisfying (i), then $K(z, \cdot) = \tilde{R}_\alpha(z, \cdot)$ for $\mathcal{E}$-q.e. $z \in E$.

Proof. For each $f \in L^2(Y; m)$ $(\cong L^2(E; m))$ by 3.3(ii) we can choose an $\mathcal{E}$-quasi-continuous m-version $(G_\alpha f)\tilde{}$ of $G_\alpha f$. We denote the set of all $\mathcal{E}$-quasi-continuous functions on E by $qC(E)$. Then by 3.3(iii) and I.4.2(i), $\tilde{G}_\alpha : L^2(Y; m) \to qC(E)$ defined by $\tilde{G}_\alpha f := (G_\alpha f)\tilde{}$, $f \in L^2(Y; m)$, has the following properties:

(3.3) $\tilde{G}_\alpha(c_1 f + c_2 g) = c_1 \tilde{G}_\alpha f + c_2 \tilde{G}_\alpha g$ $\mathcal{E}$-q.e. for all $c_1, c_2 \in \mathbb{R}, f, g \in L^2(Y; m)$.

(3.4) If $f \in L^2(Y; m), f \geq 0$, then $\tilde{G}_\alpha f \geq 0$ $\mathcal{E}$-q.e. .

(3.5) If $f \in L^2(Y; m), 0 \leq f \leq 1$, then $0 \leq \alpha \tilde{G}_\alpha f \leq 1$ $\mathcal{E}$-q.e. .

Since Y is a (topological) Lusin space (cf. 3.2(iii)) we can identify $(Y, \mathcal{B}(Y))$ as a measurable space with a Borel subset of $([0, 1], \mathcal{B}([0, 1]))$ (cf. e.g. [DM 78, III.20]). Let $(A_k)_{k\in\mathbb{N}}$ be an increasing sequence in $\mathcal{B}(Y)$ such that $\bigcup_{k \geq 1} A_k = Y$ and $m(A_k) < \infty$ for all $k \in \mathbb{N}$. Let $\mathbb{Q}_0$ denote all rational numbers in $[0, 1]$. Set

$$(3.6) \qquad I^0_{k,s}(z) := \tilde{G}_\alpha \left(1_{A_k \cap [0, s]} \right)(z) , \quad z \in E , \ k \in \mathbb{N} , \ s \in \mathbb{Q}_0 .$$

Then by (3.4), (3.5), we can find an $\mathcal{E}$-exceptional set $N_1 \in \mathcal{B}(E)$ such that for all $z \in E \setminus N_1$ we have that

$$(3.7) \qquad 0 \leq I^0_{k,s}(z) \leq I^0_{k',s'}(z) \qquad \text{for all } k \leq k',\ s \leq s',$$

$$(3.8) \qquad 0 \leq \alpha I^0_{k,s}(z) \leq 1 \qquad \text{for all } k \in \mathbb{N},\, s \in \mathbb{Q}_0.$$

Now define for $t \in [0,1]$, $k \in \mathbb{N}$,

$$I_{k,t}(z) := \begin{cases} \lim_{s \downarrow t} I^0_{k,s}(z) & \text{if } z \in E \setminus N_1 \\ 0 & \text{if } z \in N_1 \ . \end{cases}$$

Then $t \mapsto I_{k,t}(z)$ is right continuous on $[0,1]$ for all $z \in E$ and $I_{k,t}(z)$ inherits the properties (3.7), (3.8). Hence for each $k \in \mathbb{N}$ and each $z \in E$ fixed, $t \mapsto I_{k,t}(z)$ generates a finite measure $K_k(z, \cdot)$ on $([0,1], \mathcal{B}([0,1]))$ such that

$$K_k(z, [0,t]) = I_{k,t}(z) \text{ for all } t \in [0,1] \ .$$

By the usual monotone class arguments we obtain by (3.3), (3.5) and III.3.5 that for all $k \in \mathbb{N}$ and any $f \in \mathcal{B}_b([0,1])$

$$(3.9) \qquad K_k f := \int f(y) K_k(\cdot, dy) = \tilde{G}_\alpha \left(f\, 1_{A_k} \right) \ \mathcal{E}\text{-q.e.} \ .$$

In particular, there exists an $\mathcal{E}$-exceptional set $N_2 \in \mathcal{B}(E)$ such that for all $z \in E \setminus N_2$

$$(3.10) \qquad K_{k+j}(z, A_k \cap [0,t]) = K_k(z, [0,t]) \quad \text{for all } k, j \in \mathbb{N},\ t \in \mathbb{Q}_0 \ .$$

$$(3.11) \qquad 0 \leq \alpha K_k(z, Y) \leq 1 \qquad \text{for all } k \in \mathbb{N} \ .$$

We now define for $A \in \mathcal{B}(Y)$

$$\tilde{R}_\alpha(z, A) := \begin{cases} \lim_{k \to \infty} K_k(z, A) & \text{if } z \in E \setminus N_2 \\ 0 & \text{if } z \in N_2 \ . \end{cases}$$

Then $\tilde{R}_\alpha$ is a kernel from $(E, \mathcal{B}(E))$ to $(Y, \mathcal{B}(Y))$, since $k \mapsto K_k(z, A)$ is increasing because of (3.10). Now (3.9), I.2.9(i), and III.3.5 imply that $\tilde{R}_\alpha$ satisfies condition (i). (ii) follows from (3.11), and thus the existence of $\tilde{R}_\alpha$ is proved. To show uniqueness, let K be another kernel from $(E, \mathcal{B}(E))$ to $(Y, \mathcal{B}(Y))$ satisfying (i). Then there exists an $\mathcal{E}$-exceptional set $N_3 \subset E$ such that for all $z \in E \setminus N_3$

$$K(z, A_k \cap [0,t]) = \tilde{R}_\alpha(z, A_k \cap [0,t]) \text{ for all } k \in \mathbb{N}\ ,\ t \in \mathbb{Q}_0 \ ,$$

and hence (by monotone class arguments) $K(z, \cdot) = \tilde{R}_\alpha(z, \cdot)$. $\qquad \square$

Remark 3.8. Note that having established (3.3)-(3.5) one can also deduce 3.7 from [AM 91e, Theorem 4.2].

Step 2. Construction of a "nice" set $J_0 \subset D(\mathcal{E})$ and an adapted "large enough" set $Y_2 \subset E$.

Let $\mathbb{Q}_+^*$ denote the strictly positive rational numbers. Let $\mathcal{U}_0$ be a countable family of open subsets of E such that

$$(3.12) \qquad\qquad E_k^c \in \mathcal{U}_0 \text{ for all } k \in \mathbb{N}.$$

$$(3.13) \qquad \begin{aligned} &\{U \cap E_k \,|\, U \in \mathcal{U}_0\} \text{ forms a base for the relative topology} \\ &\text{of } E_k \text{ for all } k \in \mathbb{N}. \end{aligned}$$

Define

$$(3.14) \qquad \mathcal{U} := \Big\{ U \subset E \,|\, U = \bigcup_{i=1}^n U_i \text{ for some } U_i \in \mathcal{U}_0 \Big\} \cup \{E\} \,.$$

Let $\varphi \in L^2(E; m)$ with $0 < \varphi \le 1$ and set $h := \tilde{R}_1\varphi$. Then h is an $\mathcal{E}$-quasi-continuous m-version of $G_1\varphi$ (cf. 3.7). For each $U \in \mathcal{U}$ let h_U be the corresponding 1-reduced function, and we fix an $\mathcal{E}$-quasi-continuous m-version of it (cf. 3.3(ii)) also denoted by h_U. For $\mathcal{D}_0^+$ we choose the $\mathcal{E}$-quasi-continuous m-versions specified in 3.4(ii) corresponding to N and we take $h_E \equiv h$ (cf. III.1.6(i)).

Let J_0 be the smallest family of bounded $\mathcal{E}$-quasi-continuous 1-excessive functions in $D(\mathcal{E})$, having the following properties:

$$(3.15) \qquad J_0 \supset \mathcal{D}_0^+ \cup \{h_U \,|\, U \in \mathcal{U}\} \,.$$
$$(3.16) \qquad \tilde{R}_\alpha u, \ u_U \in J_0 \text{ if } u \in J_0, \ \alpha \in \mathbb{Q}_+^*, \ U \in \mathcal{U} \,.$$
$$(3.17) \qquad u \wedge 1, \ u \wedge v, \ (u+1) \wedge v \in J_0 \text{ if } u, v \in J_0 \,.$$
$$(3.18) \qquad c_1 u + c_2 v \in J_0 \text{ if } u, v \in J_0, c_1, c_2 \in \mathbb{Q}_+^* \,.$$

By III.1.3(iv) and the following exercise (with F consisting of all $\mathcal{E}$-quasi-continuous m-versions of elements in $D(\mathcal{E})$) J_0 is countable.

Exercise 3.9. (cf. [F 80, Lemma 6.1.1]) Let F be a set, G a countable subset of F and S a countable collection of mappings from $F \times F$ into F. Then there exists a countable subset H of F containing G such that $s(H \times H) \subset H$ for all $s \in S$.

In the following lemma we construct a set $Y_1 \subset E$ on which J_0 and $\tilde{R}_\alpha, \alpha \in \mathbb{Q}_+^*$, have some nice properties that will be useful below.

Lemma 3.10. *There exists a regular $\mathcal{E}$-nest $(F_k)_{k \in \mathbb{N}}$ consisting of compact metrizable sets $F_k \subset E_k$, $k \in \mathbb{N}$, such that the following properties hold:*

(i) $J_0 \subset C(\{F_k\})$.

(ii) $\inf\{h(x) \,|\, x \in F_k\} > 0$ for all $k \in \mathbb{N}$,

and if $Y_1 := \bigcup\limits_{k \geq 1} F_k$, then for all $x \in Y_1$ and all $u \in J_0$

(iii) $0 \leq u_U(x) \leq u_V(x) \leq u(x)$ for all $U, V \in \mathcal{U}$ with $U \subset V$.

(iv) $u_U(x) = u(x)$ if $x \in U$ for all $U \in \mathcal{U}$.

(v) $\beta \tilde{R}_{1+\beta} u(x) \leq u(x)$ for all $\beta \in \mathbb{Q}_+^*$.

(vi) $\tilde{R}_\alpha u(x) - \tilde{R}_\beta u(x) = (\beta - \alpha)\tilde{R}_\alpha \tilde{R}_\beta u(x)$ for all $\alpha, \beta \in \mathbb{Q}_+^*$.

(vii) $\lim\limits_{\substack{\alpha \to \infty \\ \alpha \in \mathbb{Q}_+^*}} \alpha \tilde{R}_\alpha u(x) = u(x)$.

Proof. By I.2.13(ii), III.3.5 and the usual diagonal argument we can find an increasing sequence $(\alpha_n)_{n \in \mathbb{N}}$ in $\mathbb{Q}_+^*$ such that $\alpha_n \tilde{R}_{\alpha_n} u \xrightarrow[n \to \infty]{} u$ $\mathcal{E}$-q.e. for all $u \in J_0$. Noting that by 3.3(iii), assertions (iii)-(vi) hold for all $u \in J_0$ and all x outside some $\mathcal{E}$-exceptional set (since they hold m-a.e. by III.1.5(iv),(iii), I.(1.1) respectively) and using III.3.3 we can construct an $\mathcal{E}$-nest $(F_k')_{k \in \mathbb{N}}$ such that $F_k' \subset E_k$, $k \in \mathbb{N}$, (i), (iii)-(vi) and $\lim\limits_{n \to \infty} \alpha_n \tilde{R}_{\alpha_n} u(x) = u(x)$ hold for all $u \in J_0$, $x \in \bigcup_{k \geq 1} F_k'$. In order to show (vii) let $u \in J_0$ and $x \in \bigcup_{k \geq 1} F_k'$. Note that $\alpha \tilde{R}_\alpha u(x) \xrightarrow[\alpha \to \infty]{} u(x)$ is equivalent with $\alpha \tilde{R}_{\alpha+1} u(x) \xrightarrow[\alpha \to \infty]{} u(x)$. By (v) and (vi), $\alpha \mapsto \alpha \tilde{R}_{\alpha+1} u(x)$ is increasing on $\mathbb{Q}_+^*$ (cf. III.1.3.(vi)). Hence (vii) follows. Applying III.3.6 and III.3.8 consecutively we obtain a desired regular $\mathcal{E}$-nest $(F_k)_{k \in \mathbb{N}}$. $\square$

Note that $\tilde{R}_\alpha$, $\alpha \in \mathbb{Q}_+^*$, is not a kernel on $(Y_1, \mathcal{B}(Y_1))$, therefore we need

Lemma 3.11. *There exists $Y_2 \in \mathcal{B}(E)$, $Y_2 \subset Y_1$ such that $E \setminus Y_2$ is $\mathcal{E}$-exceptional and*

$$\tilde{R}_\alpha(x, Y \setminus Y_2) = 0 \quad \text{for all } x \in Y_2 \, , \, \alpha \in \mathbb{Q}_+^* \, .$$

Proof. Since $G_\alpha\left(1_{Y \setminus Y_1}\right) = 0$ m-a.e. and $\tilde{R}_\alpha\left(1_{Y \setminus Y_1}\right)$ is an $\mathcal{E}$-quasi-continuous m-version of $G_\alpha\left(1_{Y \setminus Y_1}\right)$, by 3.3(iii) there exists $Z_1 \in \mathcal{B}(Y)$, $Z_1 \subset Y_1$ such that $E \setminus Z_1$ is $\mathcal{E}$-exceptional and $\tilde{R}_\alpha(x, Y \setminus Y_1) = 0$ for all $x \in Z_1$, $\alpha \in \mathbb{Q}_+^*$. Repeating this argument we obtain a decreasing sequence $(Z_n)_{n \in \mathbb{N}}$ in $\mathcal{B}(Y)$ such that $E \setminus Z_{n+1}$ is $\mathcal{E}$-exceptional and $\tilde{R}_\alpha(x, Y \setminus Z_n) = 0$ for all $x \in Z_{n+1}$, $\alpha \in \mathbb{Q}_+^*$ and $n \in \mathbb{N}$. Hence $Y_2 := \bigcap\limits_{n \geq 1} Z_n$ is a desired set. $\square$

Step 3. The compactification $\overline{E}$ of $Y_\Delta := Y_2 \cup \{\Delta\}$ via $J := J_0 + \mathbb{Q}_+ 1_{Y_\Delta}$ and the construction of a Ray resolvent $(\overline{R}_\alpha)_{\alpha > 0}$ on $\overline{E}$ from $(R_\alpha)_{\alpha \in \mathbb{Q}_+^*}$.

From now on we consider each function in J_0 as a function on Y_2. Again we adjoin an isolated point Δ to Y_2 (Y_2 equipped with the trace topology) and define for $\alpha \in \mathbb{Q}_+^*$, $A \in \mathcal{B}\left(Y_2 \cup \{\Delta\}\right)$

$$(3.19) \quad R_\alpha(x, A) := \begin{cases} \tilde{R}_\alpha(x, A \cap Y_2) + (\frac{1}{\alpha} - \tilde{R}_\alpha(x, Y_2))1_A(\Delta) & \text{if } x \in Y_2 \\ \frac{1}{\alpha}1_A(\Delta) & \text{if } x = \Delta \end{cases}$$

Then by 3.10(vi), $(R_\alpha)_{\alpha \in \mathbb{Q}_+^*}$ is a Markovian resolvent of kernels on $Y_2 \cup \{\Delta\}$. From now on we regard every function $u \in D(\mathcal{E})$ as a function on $Y_2 \cup \{\Delta\}$ by (restricting it to Y_2 and) setting $u(\Delta) = 0$. Set

$$(3.20) \qquad J := \{u + c\, 1_{Y_2 \cup \{\Delta\}} \mid u \in J_0 \, , \, c \in \mathbb{Q}_+\} \, .$$

Then J is countable and $u|_{F_k \cap Y_2}$ (i.e., the restriction of u to $F_k \cap Y_2$) is continuous for all $u \in J$, $k \in \mathbb{N}$, where $(F_k)_{k \in \mathbb{N}}$ is as in 3.10. J inherits the following properties of J_0.

Lemma 3.12. $\quad$ (i) $u \wedge v \in J$ and $R_\alpha u \in J$ for all $\alpha \in \mathbb{Q}_+^*$, $u, v \in J$.

$\quad$ (ii) If $u \in J$ then $\beta R_{1+\beta} u(x) \leq u(x)$ for all $x \in Y_2 \cup \{\Delta\}$, $\beta \in \mathbb{Q}_+^*$ (i.e., u is 1-supermedian w.r.t. $(R_\alpha)_{\alpha \in \mathbb{Q}_+^*}$).

$\quad$ (iii) $\displaystyle \lim_{\substack{\alpha \to \infty \\ \alpha \in \mathbb{Q}_+^*}} \alpha R_\alpha u(x) = u(x)$ for all $u \in J$, $x \in Y_2 \cup \{\Delta\}$.

$\quad$ (iv) J separates the points of $Y_2 \cup \{\Delta\}$.

Proof. (i) follows from the identity $(f + c_1) \wedge (g + c_2) = (f + c_1 - c_2) \wedge g + c_2$ if $c_1 \geq c_2$, and (3.17) resp. (3.16). (ii) is a consequence of 3.10(v), and (iii) follows immediately from 3.10(vii).

(iv): J_0 separates the points of Y_2 because $\mathcal{D}_0^+$ does by 3.4(ii) and $\mathcal{D}_0^+ \subset J_0$ by (3.15). Since $Y_2 \subset Y_1$ it follows by 3.10(ii) that for each $x \in Y_2$, $h(x) > 0$, but $h(\Delta) = 0$ by our definition. $\qquad\qquad \square$

Let $J = \{u_n \mid n \in \mathbb{N}\}$ and set $g_n := \frac{2}{\pi} \arctan u_n$, $n \in \mathbb{N}$. Define

$$\rho(x, y) := \sum_{n \geq 1} \frac{1}{2^n} |g_n(x) - g_n(y)| \; ; \quad x, y \in Y_2 \cup \{\Delta\} \, .$$

Since J separates the points of $Y_2 \cup \{\Delta\}$, ρ is a metric on $Y_2 \cup \{\Delta\}$. Define

$$(3.21) \qquad \overline{E} := \text{ completion of } Y_2 \cup \{\Delta\} \text{ w.r.t. } \rho \, .$$

Below we shall use the prefix "$\rho-$" to denote concepts which are meant relative to the topology induced by ρ.

Lemma 3.13. $\quad$ (i) $(\overline{E}, \rho)$ is a compact metric space.

$\quad$ (ii) $\rho - \mathcal{B}(Y_2 \cup \{\Delta\}) = \mathcal{B}(Y_2 \cup \{\Delta\})$

$\quad$ (iii) $Y_2 \cup \{\Delta\} \in \rho - \mathcal{B}(\overline{E})$

$\quad$ (iv) If $F \subset F_k$ for some k (where F_k is as in 3.10), F a compact subset of Y_2, then F is ρ-compact and the ρ-topology induced on F coincides with the topology on F inherited from E.

Proof. (i): Since J separates the points (cf. 3.12(iv))

$$\Phi : x \mapsto (g_n(x))_{n \in \mathbb{N}}$$

defines an isometry from $(Y_2 \cup \{\Delta\}, \rho)$ to $[0,1]^{\mathbb{N}}$ with the product metric. Since $[0,1]^{\mathbb{N}}$ is compact, $(\overline{E}, \rho)$ is a compact metric space.

(ii): By definition

$$\rho - \mathcal{B}(Y_2 \cup \{\Delta\}) = \sigma\{g_n \mid n \in \mathbb{N}\} .$$

Since $Y_2 \cup \{\Delta\}$ is a topological Lusin space and J separates the points of $Y_2 \cup \{\Delta\}$, it follows as in 3.2(iv) (or cf. [Sch 73, Lemma 18, p.108]) that

$$\mathcal{B}(Y_2 \cup \{\Delta\}) = \sigma\{g_n \mid n \in \mathbb{N}\} .$$

(iii): By (ii), $(Y_2 \cup \{\Delta\}, \rho - \mathcal{B}(Y_2 \cup \{\Delta\}))$ is a standard Borel space, hence (iii) follows from [P 67, Theorem 2.4, p.135].

(iv): Since each g_n is continuous on F, the trace topology on F (inherited from Y_2 or E) is stronger than the ρ-topology on F. Since F is compact in Y_2, the two topologies on F coincide. $\square$

By definition each $u \in J$ is ρ-uniformly continuous on $Y_2 \cup \{\Delta\}$ and hence has a unique ρ-continuous extension to $\overline{E}$. Set

$$\overline{J} := \{u \in C(\overline{E}) \mid u|_{Y_2 \cup \{\Delta\}} \in J\}$$

(where $C(\overline{E})$ denotes the set of all ρ-continuous functions on $\overline{E}$).

Exercise 3.14. Prove that $\overline{J}$ is inf-stable (i.e., $u \wedge v \in \overline{J}$ if $u, v \in \overline{J}$), hence so is $\overline{J} - \overline{J}$. Show furthermore, that $1 \in \overline{J}$, that $\overline{J}$ separates the points of $\overline{E}$ and that hence $\overline{J} - \overline{J}$ is dense in $C(\overline{E})$ w.r.t. uniform norm. $\| \ \|_\infty$

Let $\overline{u} \in \overline{J}$ and $u := \overline{u}|_{Y_2 \cup \{\Delta\}}$. Let $\alpha \in \mathbb{Q}_+^*$. Since $R_\alpha u \in J$ we can define

$$(3.22) \qquad \overline{R_\alpha \overline{u}} := \rho\text{-uniformly continuous extension of } R_\alpha u \text{ to } \overline{E}$$

and extend it to $\overline{J} - \overline{J}$ by linearity. Clearly, $\overline{R}_\alpha$ is increasing and $\alpha \overline{R}_\alpha 1_{\overline{E}} = 1_{\overline{E}}$. Hence each $\overline{R}_\alpha$ extends to a continuous linear operator from $C(\overline{E})$ to $C(\overline{E})$ ($C(\overline{E})$ equipped with the uniform norm $\| \ \|_\infty$) which is *positive*, i.e., $\overline{R}_\alpha f \geq 0$ if $f \in C(\overline{E})$, $f \geq 0$. In particular, each $\alpha \overline{R}_\alpha$ is a contraction.

Exercise 3.15. Let T be a positive continuous linear operator on $C(\overline{E})$. Prove that there exists a kernel K on $(\overline{E}, \mathcal{B}(\overline{E}))$ such that for all $u \in C(\overline{E})$

$$Tu(x) = \int u(y) K(x, dy) , \quad x \in \overline{E} .$$

Below, we use the symbol $\overline{R}_\alpha$ also for the corresponding kernel on $(\overline{E}, \mathcal{B}(\overline{E}))$. It is easy to check that also $(\overline{R}_\alpha)_{\alpha \in \mathbb{Q}_+^*}$ satisfies the resolvent equation. Hence it follows that

$$\alpha \mapsto \overline{R}_\alpha , \quad \alpha \in \mathbb{Q}_+^* ,$$

is a locally uniformly continuous map into the space of bounded linear operators on $C(\overline{E})$ equipped with the usual operator norm. Consider its unique continuous extension $\alpha \mapsto \overline{R}_\alpha$, $\alpha \in [0, \infty[$. By construction and 3.12(ii) each $\overline{u} \in \overline{J}$ is

$\alpha \overline{R}_{1+\alpha}$-supermedian for all $\alpha > 0$, i.e., $\overline{J} \subset S^1$ (cf. 1.19(R.2)). Since $\overline{J}$ separates the points of $\overline{E}$ this means that the kernels $\overline{R}_\alpha$, $\alpha \geq 0$, form a Ray resolvent.

Step 4. The corresponding right process $\overline{M}$ on $\overline{E}$

Let D be the set of non-branching points corresponding to $(\overline{R}_\alpha)_{\alpha>0}$ (cf. (1.14)). Since each $\overline{u} \in \overline{J}$ is 1-supermedian for $(\overline{R}_\alpha)_{\alpha>0}$, by the resolvent equation $\alpha \mapsto \alpha\overline{R}_{1+\alpha}\overline{u}$ is increasing, hence by 3.12(iii), $\lim\limits_{\alpha\to\infty} \alpha\overline{R}_\alpha\overline{u} = \overline{u}$ on $Y_2 \cup \{\Delta\}$. Consequently, since each αR_α is a contraction on $(C(\overline{E}), \|\ \|_\infty)$,

$$(3.23) \qquad\qquad Y_2 \cup \{\Delta\} \subset D .$$

Let $(\overline{p}_t)_{t>0}$ be the semigroup of Markovian kernels on $(\overline{E}, \mathcal{B}(\overline{E}))$ corresponding to $(\overline{R}_\alpha)_{\alpha>0}$ as in 1.20 and $\overline{M} = \left(\overline{\Omega}, \mathcal{F}, (\mathcal{F}_t)_{t\geq0}, (\overline{X}_t)_{t\geq0}, (\overline{P}_x)_{x\in D}\right)$ a corresponding right process with state space D as in 1.22. Then we have, in particular, that

$$(3.24) \qquad \overline{R}_\alpha f(x) = \overline{E}_x\left[\int_0^\infty e^{-\alpha t} f(\overline{X}_t)\,dt\right] = \int_0^\infty e^{-\alpha t}\overline{p}_t f(x)\,dt$$
$$\text{for all } f \in C(\overline{E}), x \in D, \alpha > 0 ,$$

(where $\overline{E}_x$ denotes expectation w.r.t. $\overline{P}_x$) and

$$(3.25) \qquad \begin{array}{l} t \mapsto \overline{X}_t(\omega) \text{ is right } \rho\text{-continuous and has left } \rho\text{-limits in } \overline{E} \\ \text{for each } \omega \in \overline{\Omega} . \end{array}$$

Below for simplicity we set

$$(3.26) \qquad\qquad Y_\Delta := Y_2 \cup \{\Delta\}$$

and for $u \in J$ let $\overline{u} \in \overline{J}$ denote its unique continuous extension to $\overline{E}$. Recall that by our convention $\mathcal{F} = \mathcal{F}_\infty$ (cf. (1.2)).

Lemma 3.16. *Let $u \in J$. Then $\left(e^{-t}\overline{u}(\overline{X}_t), \mathcal{F}_t, \overline{P}_x\right)_{t\in[0,\infty]}$ is a supermartingale for all $x \in D$ where $e^{-\infty}\overline{u}(\overline{X}_\infty) := 0$.*

Proof. Observe that for $f \in \rho - \mathcal{B}^+(\overline{E})$

$$e^{-t}\overline{p}_t\overline{R}_1 f = \int_t^\infty e^{-s}\overline{p}_s f \, ds \leq \overline{R}_1 f$$

(cf. III.1.3(i)). Hence we have for all $\alpha > 0$, $t > 0$, by the resolvent equation that

$$\begin{aligned} e^{-t}\overline{p}_t\overline{R}_{\alpha+1}\overline{u} &= e^{-t}\overline{p}_t(\overline{R}_1(\overline{u} - \alpha\overline{R}_{\alpha+1}\overline{u})) \\ &\leq \overline{R}_1(\overline{u} - \alpha\overline{R}_{\alpha+1}\overline{u}) = \overline{R}_{\alpha+1}\overline{u} \end{aligned}$$

since $\overline{u}$ is $\alpha\overline{R}_{\alpha+1}$-supermedian. Multiplying by α and using that

$$\|\alpha\overline{R}_{\alpha+1}\overline{u}\|_\infty \leq (\alpha + 1)\|\overline{R}_{\alpha+1}\overline{u}\|_\infty \leq \|\overline{u}\|_\infty$$

we conclude by 1.20(iii) and Lebesgue's dominated convergence theorem that

$$(3.27) \qquad e^{-t}\overline{p}_t\overline{u} = \lim_{\alpha\to\infty} e^{-t}\overline{p}_t\alpha\overline{R}_{\alpha+1}\overline{u} \le \lim_{\alpha\to\infty} \alpha\overline{R}_{\alpha+1}\overline{u} = \overline{u} \text{ on } D .$$

Now let $s < t$, then by the Markov property of $\overline{M}$, and (3.27)

$$\begin{aligned}
\overline{E}_x\left[e^{-t}\overline{u}(\overline{X}_t)|\overline{\mathcal{F}}_s\right] &= \overline{E}_{\overline{X}_s}\left[e^{-t}\overline{u}(\overline{X}_{t-s})\right] \\
&= e^{-t}\overline{p}_{t-s}\overline{u}(\overline{X}_s) \\
&\le e^{-s}\overline{u}(\overline{X}_s)
\end{aligned}$$

$\overline{P}_x$-a.e., for each $x \in D$. $\qquad\qquad\qquad\qquad\qquad\qquad\qquad\qquad\square$

Step 5. $\overline{M}$ can be "restricted" to live on E.

Lemma 3.17. *Let $x \in Y_\Delta$, $\alpha > 0$, then*

$$\overline{E}_x\left[\int_0^\infty e^{-\alpha t}1_{\{\overline{X}_t\notin Y_\Delta\}}\,dt\right] = 0 .$$

Proof. It is enough to prove the assertion for $\alpha \in \mathbb{Q}_+^*$. Let $\overline{u} \in \overline{J}$. Then by (3.24), (3.22)

$$\overline{E}_x\left[\int_0^\infty e^{-\alpha t}\overline{u}(\overline{X}_t)\,dt\right] = \overline{R}_\alpha\overline{u}(x) = R_\alpha(\overline{u}|_{Y_\Delta})(x)$$

since $\rho - \mathcal{B}(\overline{E}) = \sigma(\overline{J})$, monotone class arguments imply that this equality holds for all bounded $\rho - \mathcal{B}(\overline{E})$-measurable functions $\overline{u}$. Taking $\overline{u} := 1_{\overline{E}\setminus Y_\Delta}$ the assertion follows. $\qquad\qquad\square$

For $A \subset \overline{E}$, let $\overline{\sigma}_A$ denote the first hitting time of A w.r.t. $\overline{M}$ (cf. (1.10)). The following is crucial:

Proposition 3.18. *Let F be a compact subset of Y_2 with $F \subset F_k$ for some $k \in \mathbb{N}$ (F_k as in 3.10). Let $A := \overline{E} \setminus F$ and $u \in J_0$. Then there exists an $\mathcal{E}$-exceptional set $N \subset E$ such that*

$$(3.28) \qquad \overline{E}_x\left[e^{-\overline{\sigma}_A}\overline{u}(\overline{X}_{\overline{\sigma}_A})\right] \le u_{E\setminus F}(x) \quad \text{for all } x \in (Y_2 \setminus N)\cup\{\Delta\} .$$

Proof. Since $F \subset F_k \subset E_k$ we can find an increasing sequence $(U_n)_{n\in\mathbb{N}}$ in $\mathcal{U}$ such that $\bigcup_{n\ge1} U_n = E \setminus F =: U$. By 3.10(iii) $(u_{U_n}(x))_{n\in\mathbb{N}}$ is increasing for all $x \in Y_\Delta$. Hence $(\overline{u_{U_n}})_{n\in\mathbb{N}}$ is increasing on $\overline{E}$. Let $f := \lim_{n\to\infty} \overline{u_{U_n}}$. By 3.10(iv) we obtain

$$(3.29) \qquad f(x) = u(x) \text{ for every } x \in (Y_2 \setminus F)\cup\{\Delta\} = A\cap Y_\Delta .$$

Furthermore, by III.2.13(ii) and III.3.5

$$(3.30) \qquad\qquad\qquad f = u_U \ \mathcal{E}\text{-q.e. on } Y_2 .$$

By 3.16, $\left(e^{-t}\overline{u_{U_n}}(\overline{X}_t), \overline{\mathcal{F}}_t, \overline{P}_x\right)_{t\in[0,\infty]}$ is a supermartingale for all $n \in \mathbb{N}$, hence $\left(e^{-t}f(\overline{X}_t), \overline{\mathcal{F}}_t, \overline{P}_x\right)_{t\in[0,\infty]}$ is a supermartingale for all $x \in D$.

Now fix $x \in Y_\Delta$. By 3.17 there exists a countable dense subset I_x of $]0, \infty[$ such that if $\Omega' := \bigcap_{s\in I_x} \{\overline{X}_s \in Y_\Delta\}$ then

$$(3.31) \qquad\qquad \overline{P}_x[\Omega'] = 1 \ .$$

Let

$$I_x := \{t_i \,|\, i \in \mathbb{N}\}$$

and define for $n \in \mathbb{N}$

$$\sigma_n := \min\{t_i \,|\, 1 \le i \le n \text{ and } \overline{X}_{t_i} \in A\}$$

(where as usual we set $\min \emptyset = +\infty$). Hence by the optional sampling theorem (cf. e.g. [B 78, 58.5]) and (3.29)-(3.31)

$$
\begin{aligned}
\overline{E}_x\left[e^{-\sigma_n}\overline{u}(\overline{X}_{\sigma_n})\right] &= \overline{E}_x\left[e^{-\sigma_n}u(\overline{X}_{\sigma_n})1_{\Omega'}\right] \\
&= \overline{E}_x\left[e^{-\sigma_n}f(\overline{X}_{\sigma_n})\right] \\
&\le \overline{E}_x[f(\overline{X}_0)] = f(x) = u_U(x)
\end{aligned}
$$

for $\mathcal{E}$-q.e. $x \in Y_2$. The inequality also holds for $x = \Delta$ because $f(\Delta) = 0 = u_U(\Delta)$. Since A is ρ-open, $(\sigma_n)_{n\in\mathbb{N}}$ decreases to $\overline{\sigma}_A$, and hence (3.28) follows. $\square$

We need the following elementary fact.

Lemma 3.19. *Let* $(\Omega, \mathcal{M}_t, Y_t, P)_{t\in[0,\infty]}$ *be a right continuous supermartingale such that* $Y_\infty \equiv 0$ *and* $Y_t \ge 0$ *for all* $t \in [0, \infty[$. *Then for every* $t \in [0, \infty]$

$$\Omega_t := \{\inf_{s\le t} Y_s = 0\} \cap \{Y_t > 0\} \in \mathcal{M}_t$$

and $P[\Omega_t] = 0$.

Proof. Let

$$\tau_n := \inf\{s \ge 0 \,|\, Y_s < \frac{1}{n}\}, \ n \in \mathbb{N} \ .$$

Then τ_n is an $(\mathcal{M}_t)$-stopping time for each n and $Y_{\tau_n} \le \frac{1}{n}$. By the optional sampling theorem it follows that for all $t > 0$

$$E\left[Y_t 1_{\{\tau_n \le t\}}\right] = E\left[Y_{t\vee\tau_n}1_{\{\tau_n\le t\}}\right] \le E\left[Y_{\tau_n}1_{\{\tau_n\le t\}}\right] \le \frac{1}{n} \ ,$$

(where $E[\,\cdot\,]$ denotes the expectation w.r.t. P) hence

$$E\left[Y_t \, 1_{\bigcap_{n\ge 1}\{\tau_n\le t\}}\right] = 0 \ .$$

Since

$$\Omega_t = \{Y_t > 0\} \cap \bigcap_{n \geq 1} \{\tau_n \leq t\} \in \mathcal{M}_t ,$$

the assertion follows. $\square$

We define

$$(3.32) \qquad \zeta := \overline{\sigma}_{\overline{E} \setminus Y_2} \ (= \inf\{t > 0 \,|\, \overline{X}_t \notin Y_2\}) .$$

Lemma 3.20. *There exist* $\Omega \in \overline{\mathcal{F}}$ *and* $S \in \mathcal{B}(Y_2)$ *with* $E \setminus S$ $\mathcal{E}$*-exceptional having the following properties.*

(i) *ζ is an $(\overline{\mathcal{F}}_t)$-stopping time, $\zeta > 0$ on $\{\overline{X}_0 = x\} \cap \Omega$ for all $x \in Y_2$, and $\zeta = 0$ on $\{\overline{X}_0 = \Delta\} \cap \Omega$.*

(ii) *$\overline{P}_x[\Omega] = 1$ for all $x \in S_\Delta := S \cup \{\Delta\}$.*

(iii) *If $\omega \in \Omega$ then $t \mapsto \overline{X}_t(\omega)$ is right-continuous on $[0, \zeta(\omega)[$, has left limits $\overline{X}_{t-}(\omega)$, $\rho - \overline{X}_{t-}(\omega)$ in the original topology resp. ρ-topology, and $\overline{X}_{t-}(\omega) = \rho - \overline{X}_{t-}(\omega)$ for all $t \in]0, \zeta(\omega)[$.*

(iv) *If $\omega \in \Omega$ then $\overline{X}_t(\omega)$, $\overline{X}_{t-}(\omega) \in S$ for all $t \in [0, \zeta(\omega)[$, $t \in]0, \zeta(\omega)[$ respectively.*

(v) *$R_1 u(x) = \overline{E}_x \left[\int_0^\zeta e^{-s} \overline{u}(\overline{X}_s) ds \right]$ for all $x \in S_\Delta$, $u \in J_0$.*

(vi) *$\overline{\theta}_t(\Omega) \subset \Omega$ for all $t \in [0, \infty[$ (where $\overline{\theta}_t$ is the shift operator corresponding to $\overline{\mathbf{M}}$).*

(vii) *There exist compact sets $K_n \subset F_n \cap Y_2$ with $K_n \subset K_{n+1}$, $n \in \mathbb{N}$, such that $\lim_{n \to \infty} \overline{\sigma}_{\overline{E} \setminus K_n}(\omega) = \zeta(\omega)$ for each $\omega \in \Omega$.*

Proof. Let Ω_0 denote the set of all $\omega \in \overline{\Omega}$ such that for all $u \in J_0$ and all $t \in \mathbb{Q}_+^*$, and hence all $t > 0$,

$$\inf_{s \leq t} e^{-s} \overline{u}(\overline{X}_s)(\omega) > 0 \ \text{or} \ e^{-t} \overline{u}(\overline{X}_t)(\omega) = 0 .$$

By 3.16, 3.19, $\Omega_0 \in \overline{\mathcal{F}}$ and $\overline{P}_x[\Omega_0] = 1$ for all $x \in D$. Let $(K_n)_{n \in \mathbb{N}}$ be an $\mathcal{E}$- nest with $K_n \subset F_n \cap Y_2$, $n \in \mathbb{N}$, and define

$$\sigma_n := \overline{\sigma}_{\overline{E} \setminus K_n} , \quad n \in \mathbb{N} ; \qquad \sigma := \lim_{n \to \infty} \sigma_n \ (= \sup_{n \in \mathbb{N}} \sigma_n) .$$

Note that by 3.18, III.2.12 and III.3.5 we can find $Z_1 \in \mathcal{B}(Y_2)$ such that $E \setminus Z_1$ is $\mathcal{E}$-exceptional and

$$(3.33) \quad \overline{E}_x \left[e^{-\sigma} \overline{u}(\rho - \lim_{n \to \infty} \overline{X}_{\sigma_n}) \right] = 0 \ \text{for all} \ x \in Z_1 \cup \{\Delta\} \ \text{and all} \ u \in J_0 .$$

Defining Ω_1 to be the set of all $\omega \in \Omega_0$ such that

$$e^{-\sigma(\omega)} \overline{u}(\rho - \lim_{n \to \infty} \overline{X}_{\sigma_n}(\omega)) = 0 \ \text{for all} \ u \in J_0 ,$$

we have by (3.33) that $\overline{P}_x[\Omega_1] = 1$ for all $x \in Z_1 \cup \{\Delta\}$. Since $\Omega_1 \subset \Omega_0$ we thus obtain that

$$(3.34) \qquad e^{-t}\overline{u}(\overline{X}_t(\omega)) = 0 \quad \text{for all } u \in J_0 \ , \ \omega \in \Omega_1 \ , \ t \geq \sigma(\omega) \ .$$

In particular, $\overline{\theta}_t(\Omega_1) \subset \Omega_1$ for all $t \geq 0$. Taking $u = h$ (cf. 3.10(ii)) it follows that $\zeta \leq \sigma$ on Ω_1 and that $\sigma > 0$ on $\{\overline{X}_0 = x\} \cap \Omega_1$ for all $x \in Y_2$, since $\overline{h} = h > 0$ on Y_2. But $\sigma_n \leq \zeta$ for all $n \in \mathbb{N}$ hence $\sigma \leq \zeta$; consequently,

$$(3.35) \qquad \zeta = \sigma \text{ on } \Omega_1 \text{ and } \zeta > 0 \text{ on } \{\overline{X}_0 = x\} \cap \Omega_1 \text{ for all } x \in Y_2 \ .$$

Furthermore, if $\omega \in \Omega_1$ and $t < \zeta(\omega) = \sigma(\omega)$ then $t < \sigma_n(\omega)$ for some $n \in \mathbb{N}$. Hence $\overline{X}_s(\omega) \in K_n$ for all $s \in [0, t + \delta]$ for some $\delta > 0$. Since the ρ-topology and the original topology coincide on K_n by 3.13(iv), we obtain that $s \mapsto \overline{X}_s(\omega)$ is right continuous on $[0, \zeta(\omega)[$ and has left limits on $]0, \zeta(\omega)[$ which coincide with the left ρ-limits and belong to $\bigcup_{n \geq 1} K_n \subset Y_2$. Furthermore by (3.22), (3.24) and (3.34), (3.35)

$$(3.36) \quad R_1 u(x) = \overline{E}_x\left[\int_0^\zeta e^{-s}\overline{u}(\overline{X}_s)ds\right] \quad \text{for all } u \in J_0 \ , \ x \in Z_1 \cup \{\Delta\} \ .$$

Repeating the above argument (with Z_1 taking the role of Y_2 in the next step except keeping ζ fixed) we can find decreasing sequences $(Z_n)_{n \in \mathbb{N}}$ in $\mathcal{B}(Y_2)$ and $(\Omega_n)_{n \in \mathbb{N}}$ in $\overline{\mathcal{F}}$ having the following properties for all $n \in \mathbb{N}$

(3.37) $E \setminus Z_n$ is $\mathcal{E}$-exceptional .

(3.38) $\overline{P}_x[\Omega_n] = 1$ for all $x \in Z_n \cup \{\Delta\}$.

(3.39) $\overline{\theta}_t(\Omega_n) \subset \Omega_n$.

(3.40) $\overline{X}_t(\omega), \overline{X}_{t-}(\omega) \in Z_{n-1}$ for all $\omega \in \Omega_n$ and all $t < \zeta(\omega)$.

(3.41) $R_1 u(x) = \overline{E}_x\left[\int_0^\zeta e^{-s}\overline{u}(\overline{X}_s)ds\right]$ for all $u \in J_0$, and all $x \in Z_n \cup \{\Delta\}$.

Defining $S := \bigcap_{n \geq 1} Z_n$, $\Omega := \bigcap_{n \geq 1} \Omega_n$ we obtain sets satisfying assertions (ii)-(vi). (vii) holds because of (3.35). The first part of (i) follows from (3.35). To show the second suppose $\omega \in \Omega$ with $X_0(\omega) = \Delta$. If $\zeta(\omega) > 0$ then $X_t(\omega) \in Y_2$ for all $t < \zeta(\omega)$. We have seen that $t \mapsto \overline{X}_t(\omega)$ is right continuous on $[0, \zeta(\omega)[$ w.r.t. the original topology on Y_2. Hence $\lim_{t \to 0} \overline{X}_t(\omega) = \overline{X}_0(\omega) = \Delta$, but this contradicts the fact that Δ is isolated from Y_2. $\qquad \square$

Step 6. The restriction M of $\overline{M}$ is special standard and associated with $(\mathcal{E}, D(\mathcal{E}))$.

Lemma 3.21. *Let $(\tau_n)_{n \geq 1}$ be an increasing sequence of $(\overline{\mathcal{F}}_t)$- stopping times with limit τ. Set*

$$V(\omega) := \begin{cases} \rho - \lim\limits_{n \to \infty} \overline{X}_{\tau_n}(\omega) \ (\in \overline{E}) & \text{if } \tau(\omega) < \infty \\ \Delta & \text{if } \tau(\omega) = \infty \ . \end{cases}$$

Then $\overline{P}_x[V \neq \overline{X}_\tau, V \in Y_2] = 0$ for all $x \in S_\Delta$. (Here as before we set $\overline{X}_\infty \equiv \Delta$.)

Proof. Fix $x \in S_\Delta$ and let us first assume that τ is bounded. Let $f, g \in C(\overline{E})$, $\alpha > 0$. Then by (3.24) and the strong Markov property of $\overline{\mathbf{M}}$

$$
\begin{aligned}
\overline{E}_x \left[g(V)\overline{R}_\alpha f(\overline{X}_\tau) \right] &= \overline{E}_x \left[g(V)\overline{E}_{\overline{X}_\tau} \left[\int_0^\infty e^{-\alpha t} f(\overline{X}_t) dt \right] \right] \\
&= \overline{E}_x \left[g(V) e^{\alpha \tau} \int_\tau^\infty e^{-\alpha t} f(\overline{X}_t) dt \right] \\
&= \lim_{n \to \infty} \overline{E}_x \left[g(\overline{X}_{\tau_n}) e^{\alpha \tau_n} \int_{\tau_n}^\infty e^{-\alpha t} f(\overline{X}_t) dt \right] \\
&= \lim_{n \to \infty} \overline{E}_x \left[g(\overline{X}_{\tau_n}) \overline{R}_\alpha f(\overline{X}_{\tau_n}) \right] \\
&= \overline{E}_x \left[g(V)\overline{R}_\alpha f(V) \right] \ .
\end{aligned}
$$

By monotone class arguments it follows that

$$
\overline{E}_x \left[g(V)\alpha\overline{R}_\alpha f(\overline{X}_\tau) \right] = \overline{E}_x \left[g(V)\alpha\overline{R}_\alpha f(V) \right]
$$

for all $\rho - \mathcal{B}(\overline{E})$ measurable bounded functions g. Hence, since $Y_2 \in \rho - \mathcal{B}(\overline{E})$ (cf. 3.13(iii)), we may replace g by $1_{Y_2} \cdot g$, and then we can let α tend to infinity to obtain that for all $f, g \in C(\overline{E})$

$$
\overline{E}_x \left[g(V)f(\overline{X}_\tau)1_{\{V \in Y_2\}} \right] = \overline{E}_x \left[g(V)f(V)1_{\{V \in Y_2\}} \right] \ .
$$

Using monotone class arguments again we obtain that

$$
\overline{E}_x \left[h(V, \overline{X}_\tau)1_{\{V \in Y_2\}} \right] = \overline{E}_x \left[h(V, V)1_{\{V \in Y_2\}} \right]
$$

for all $\mathcal{B}(\overline{E} \times \overline{E})$-measurable bounded functions h. Choosing $h = $ indicator function of the diagonal in $\overline{E} \times \overline{E}$, we obtain the assertion for bounded τ. The general case now follows, since

$$
\begin{aligned}
& \overline{P}_x \left[V \neq \overline{X}_\tau, \ V \in Y_2 \right] \\
&= \overline{P}_x \left[V \neq \overline{X}_\tau, \ V \in Y_2, \tau < \infty \right] \\
&= \sum_{n=1}^\infty \overline{P}_x \left[V \neq \overline{X}_\tau, \ V \in Y_2, \ n - 1 \leq \tau < n \right] \\
&= \sum_{n=1}^\infty \overline{P}_x \left[\rho - \lim_{k \to \infty} \overline{X}_{\tau_k \wedge n} \neq \overline{X}_{\tau \wedge n}, \ \rho - \lim_{k \to \infty} \overline{X}_{\tau_k \wedge n} \in Y_2, \ n - 1 \leq \tau < n \right]
\end{aligned}
$$

which is equal to zero by what we proved for bounded τ. $\square$

Now define $\mathbf{M}_S := (\Omega, \mathcal{F}, (\mathcal{F}_t)_{t \geq 0}, (X_t)_{t \geq 0}, (P_x)_{x \in S_\Delta})$ by $\mathcal{F} := \overline{\mathcal{F}} \cap \Omega$, $\mathcal{F}_t := \overline{\mathcal{F}}_t \cap \Omega$, $P_x := \overline{P}_{x | \overline{\mathcal{F}} \cap \Omega}$, $x \in S_\Delta$, and for $\omega \in \Omega$

$$
(3.42) \qquad X_t(\omega) := \begin{cases} \overline{X}_t(\omega) & \text{if } t \in [0, \zeta(\omega)[\\ \Delta & \text{if } t \in [\zeta(\omega), \infty[\ , \end{cases}
$$

where Ω, S_Δ are as in 3.20 and ζ as in (3.32).

Proposition 3.22. $\mathbf{M}_S$ *is a special standard process with state space S and life time ζ such that*

(i) $E_x\left[\int_0^\infty e^{-\alpha s} f(X_s)ds\right] = \tilde{R}_\alpha f(x)$ *for all $f \in \mathcal{B}_b(Y_2)$ and all $x \in S$, $\alpha \in \mathbb{Q}_+^*$ (where again we set $f(\Delta) = 0$).*

(ii) $t \mapsto u(X_t(\omega))$ *is right continuous on $[0, \infty[$ and has left limits on $]0, \zeta(\omega)[$ for all $\omega \in \Omega$ and all $u \in J_0$.*

Proof. It is straightforward to check that $\mathbf{M}_S$ is a right process with state space S and life time ζ. Indeed $\mathbf{M}_S$ satisfies (M.5) by 3.20(i), (M.6) by 3.20(iii),(iv) and (M.6) by 1.7(iii) and 1.9(v). We leave the details of this part as an exercise to the reader. (M.8), i.e., the existence of left limits up to ζ, is satisfied by 3.20(iii), (iv). In order to prove (M.9), i.e., the quasi-left continuity up to ζ, let $(\tau_n)_{n \in \mathbb{N}}$ be an increasing sequence of $(\mathcal{F}_t)$-stopping times with limit τ. If $\omega \in \{\tau < \zeta\}$ then by 3.20(iii), (iv)

$$\rho - \lim_{n \to \infty} \overline{X}_{\tau_n}(\omega) = \lim_{n \to \infty} X_{\tau_n}(\omega) \in S \subset Y_2 .$$

Hence by 3.21

$$P_x\left[\lim_{n \to \infty} X_{\tau_n} \neq X_\tau , \tau < \zeta\right] = 0 \ \text{ for all } x \in S_\Delta$$

and (M.9) follows. To prove (M.10) let τ, τ_n, $n \in \mathbb{N}$, be as above and $\mu \in P(S_\Delta)$. We have to show that X_τ is $\bigvee_{n \geq 1} \mathcal{F}_{\tau_n}^{P_\mu}$-measurable. It follows by the definition of Ω, (3.34), and (3.35) that for h as in 3.10(ii)

$$E_x\left[e^{-\tau}\overline{h}(\overline{X}_\tau)1_{\{\tau=\zeta<\infty\}}\right] = 0 \ \text{ for all } x \in S_\Delta .$$

Hence since $h(x) > 0$ for all $x \in Y_2$,

$$P_x\left[\tau = \zeta < \infty , \overline{X}_\tau \in S\right] = 0 \ \text{ for all } x \in S_\Delta .$$

Since on the other hand by 3.21

$$P_x\left[\tau = \zeta < \infty , \rho - \lim_{n \to \infty} \overline{X}_{\tau_n} \in S , \overline{X}_\tau \notin S\right] = 0 ,$$

we conclude that

$$(3.43) \qquad P_x\left[\tau = \zeta < \infty , \rho - \lim_{n \to \infty} \overline{X}_{\tau_n} \in S\right] = 0 \ \text{ for all } x \in S_\Delta .$$

By 3.20(iii) and 3.21 we know that

$$(3.44) \qquad \rho - \lim_{n \to \infty} X_{\tau_n} = X_\tau \in S \ P_\mu - a.s. \text{ on } \{\tau < \zeta\}$$

and by (3.42)

$$(3.45) \qquad \rho - \lim_{n \to \infty} X_{\tau_n} = \Delta \text{ on } \{\tau > \zeta\} .$$

Combining (3.43)-(3.45) we deduce that for P_μ-a.e. $\omega \in \Omega$

$$X_\tau(\omega) = \begin{cases} \rho - \lim_{n\to\infty} X_{\tau_n}(\omega) & \text{if } \omega \in \{\rho - \lim_{n\to\infty} X_{\tau_n} \in S \ , \ \tau < \infty\} \\ \Delta & \text{if } \omega \in \{\rho - \lim_{n\to\infty} X_{\tau_n} \notin S \ , \ \tau < \infty\} \cup \{\tau = \infty\} \ . \end{cases}$$

Since τ and $\rho - \lim_{n\to\infty} X_{\tau_n}$ are $\bigvee_{n\geq 1} \mathcal{F}^{P_\mu}_{\tau_n}$-measurable, so is X_τ.

Assertion (ii) is obvious by (3.42) and 3.20(iii) since $\overline{u}$ is ρ-continuous on $\overline{E}$ for each $u \in J_0$. Hence it remains to prove (i). Define for $f \in \mathcal{B}_b(E)$, $\alpha \in \mathbb{Q}^*_+$,

$$V_\alpha f(x) := E_x\left[\int_0^\infty e^{-\alpha s} f(X_s)ds\right] \ , \ x \in S \ .$$

By the Markov property of $\mathbf{M}_S$ it follows that for all $\alpha, \beta \in \mathbb{Q}^*_+$, $x \in S$,

$$(3.46) \qquad\qquad V_\alpha f(x) - V_\beta f(x) = (\beta - \alpha)V_\alpha V_\beta f(x)$$

for all $f \in \mathcal{B}_b(E)$ (cf. 1.11). Fix $x \in S$ and note that (i) holds if $\alpha = 1$, $f \in J_0$ by (3.19), 3.20(v), (3.34) and (3.35). Consequently, by (3.46) and 3.10(vi) for all $\alpha \in \mathbb{Q}^*_+ \setminus \{1\}$, $f \in J_0$,

$$(3.47) \qquad \begin{aligned} V_\alpha f(x) + (\alpha - 1)V_1 V_\alpha f(x) &= V_1 f(x) \\ &= \tilde{R}_1 f(x) \\ &= \tilde{R}_\alpha f(x) + (\alpha - 1)\tilde{R}_1 \tilde{R}_\alpha f(x) \\ &= \tilde{R}_\alpha f(x) + (\alpha - 1)V_1 \tilde{R}_\alpha f(x) \end{aligned}$$

since $\tilde{R}_\alpha f \in J_0$. But if $g \in \mathcal{B}_b(Y_2)$ such that

$$g(x) + (\alpha - 1)V_1 g(x) = 0 \ \text{ for all } x \in S$$

then by (3.46)

$$g = (1 - \alpha)V_1 g = (\alpha - 1)\left[V_\alpha g + (\alpha - 1)V_\alpha \frac{g}{1-\alpha}\right] = 0 \text{ on } S \ .$$

By (3.47) it follows that

$$g := V_\alpha f - \tilde{R}_\alpha f = 0 \text{ on } S,$$

i.e., (i) holds for all $\alpha \in \mathbb{Q}^*_+$, $f \in J_0$. Since for h as in 3.10(ii) $Y_2 = \{h > 0\}$, hence $\sup_{n\in\mathbb{N}}(nh \wedge 1) = 1_{Y_2}$, (i) holds for $f = 1_{Y_2}$. Since J_0 is inf-stable, a monotone class argument (cf. e.g. [Sh 88, A.0.6]) now shows that (i) holds for all $f \in \mathcal{B}_b(Y_2)$. $\square$

It is now easy to complete the proof of 3.5 by extending $\mathbf{M}_S$ to a process $\mathbf{M} = (\Omega', \mathcal{F}', (X'_t)_{t\geq 0}, (P'_z)_{z\in E_\Delta})$ with state space E in such a way that each $z \in N := E \setminus S$ is a trap for $\mathbf{M}$ (cf. (3.49) below). More precisely, adjoin N to Ω as an extra set and define

$$
\begin{aligned}
\Omega' &:= \Omega \cup N \\
\mathcal{F}' &:= \{\Gamma \cup \Gamma_0 \,|\, \Gamma \in \mathcal{F},\ \Gamma_0 \in \mathcal{B}(N)\} \\
\zeta'(\omega) &:= \begin{cases} \zeta(\omega) & \text{if } \omega \in \Omega \\ +\infty & \text{if } \omega \in N \end{cases} \\
X_t'(\omega) &:= \begin{cases} X_t(\omega) & \text{if } \omega \in \Omega, & t \in [0,\infty[\\ z & \text{if } \omega = z \in N, & t \in [0,\infty[\end{cases} \\
P_z'(\Gamma) &:= \begin{cases} P_z(\Gamma \cap \Omega) & \text{if } z \in S_\Delta, \Gamma \in \hat{\mathcal{F}} \\ \epsilon_z(\Gamma) & \text{if } z \in E \setminus S_\Delta, \Gamma \in \hat{\mathcal{F}}, \end{cases}
\end{aligned}
$$

(3.48)

(and $\mathcal{F}_t'$, $t \geq 0$, as in (1.6), (1.7)). Then it is trivial to check that $\mathbf{M}$ is a special standard process with state space E and life time ζ' such that

(3.49) $$ P_z'\,[X_t' = z \text{ for all } t \geq 0] = 1 \quad \text{for all } z \in E \setminus S\,. $$

By 3.22(i), 3.7(i) and 3.11 we can apply 2.8 to conclude that $\mathbf{M}$ is properly associated with $(\mathcal{E}, D(\mathcal{E}))$. The m-tightness follows by 3.20(vii).

Remark 3.23. (i) In later chapters we shall refer to the extension procedure described by (3.48) as *trivial extension*.
(ii) By 3.22(ii) with $h = u$, h as in 3.10(ii), we see that $\mathbf{M}$ satisfies condition (2.1) with $\rho = h$. Note that indeed $h = \hat{R}_1\varphi \in L^1(E;m)$, as easily follows from 3.22(i), and 2.6(iii) (which is applicable by 3.5).

4   Examples of quasi-regular Dirichlet forms

a)   Regular Dirichlet forms

Assume E is a locally compact separable metric space and m a positive Radon measure on E. Let $(\mathcal{E}, D(\mathcal{E}))$ be a *regular* Dirichlet form on $L^2(E;m)$, i.e., $C_0(E) \cap D(\mathcal{E})$ is dense both in $D(\mathcal{E})$ w.r.t. $\tilde{\mathcal{E}}_1^{1/2}$ and in $C_0(E)$ w.r.t. the uniform norm $\|\ \|_\infty$. Then $(\mathcal{E}, D(\mathcal{E}))$ is quasi-regular. Indeed, 3.1(ii), (iii) are obvious. To show 3.1(i) let $(K_n)_{n\in\mathbb{N}}$ be a sequence of compact subsets of E such that K_{n+1} contains K_n in its interior for all $n \in \mathbb{N}$ and $E = \bigcup_{n\geq 1} K_n$. Clearly,

$$ C_0(E) \cap D(\mathcal{E}) \subset \bigcup_{n\geq 1} D(\mathcal{E})_{K_n}\,, $$

hence 3.1(i) holds by the regularity of $(\mathcal{E}, D(\mathcal{E}))$. This implies that 3.5 applies here to ensure the existence of a pair $(M, \hat{M})$ of (m-tight) special standard processes associated with $(\mathcal{E}, D(\mathcal{E}))$. We note that as in [Si 74], [F 80], [Ca-Me 75], [Le 82, 83] we shall prove that they are indeed Hunt processes (cf. V.2.12(ii), V.2.13 below).
In particular, all this applies to those examples $(\mathcal{E}, D(\mathcal{E}))$ discussed in Chap.II,

Sections 1,2, where E was an open subset U of $\mathbb{R}^d$ and $C_0^\infty(U)$ was dense in $D(\mathcal{E})$ w.r.t $\tilde{\mathcal{E}}_1^{1/2}$. In the case discussed in Chap.II, Subsection 1b) one just obtains *classical Brownian motion* on U with *absorbing boundary* .

b)   Dirichlet forms on infinite dimensional state space

Let E be a separable real Banach space with dual E' and μ a positive finite measure on its Borel σ-algebra $\mathcal{B}(E)$ with $\operatorname{supp}\mu = E$. Let K_0 be a (finite or) countable set of μ-admissible elements (cf. II.3.1) such that

$$(4.1) \qquad \begin{array}{c} \text{there exists a constant } c > 0 \text{ such that} \\ \sum_{k \in K_0} {}_{E'}\langle l, k \rangle_E^2 \leq c^2 \|l\|_{E'}^2, \ \text{ for all } l \in E' . \end{array}$$

(Note that (4.1) is, of course, always fulfilled if K_0 is finite.) In particular, condition II.(3.6) is satisfied, hence we know by II.3.5 that the closure of

$$(4.2) \qquad \mathcal{E}(u,v) := \sum_{k \in K_0} \int \frac{\partial u}{\partial k} \frac{\partial v}{\partial k} \, d\mu \ ; \ \ u, v \in \mathcal{F}C_b^\infty \ ,$$

denoted by $(\mathcal{E}, D(\mathcal{E}))$ exists and is a Dirichlet form on $L^2(E; \mu)$. To show that $(\mathcal{E}, D(\mathcal{E}))$ is quasi-regular note that 3.1(ii) holds by definition. By the Hahn-Banach theorem (cf. e.g. [Ch 69b]) $\{\sin l \,|\, l \in E'\}$ separates the points of E, i.e.,

$$(E \times E) \setminus d = \bigcup_{l \in E'} (\sin l, \sin l)^{-1} ((\mathbb{R} \times \mathbb{R}) \setminus d')$$

where d, d' denotes the diagonal in $E \times E$, $\mathbb{R} \times \mathbb{R}$ respectively. Since $E \times E$ is strongly Lindelöf as a separable metric space the above open cover of $(E \times E) \setminus d$ has a countable subcover, i.e., there exist $l_n \in E'$, $n \in \mathbb{N}$, such that $\{\sin l_n \,|\, n \in \mathbb{N}\}$ separates the points of E. Hence 3.1(iii) holds. In order to show 3.1(i) recall that by 3.2(ii) it is sufficient to prove that $\operatorname{Cap} := \operatorname{Cap}_{1,1}$ (i.e., $h = g \equiv 1$, which belongs to $\mathcal{F}C_b^\infty \subset D(\mathcal{E})$) is tight. This will be done below using the same arguments as in the proof of [RS 91a, Proposition 3.1]. We need the following lemma.

Lemma 4.1.   *(i) Let $\varphi \in C^1(\mathbb{R})$ with bounded derivative φ'. Then $\varphi \circ u \in D(\mathcal{E})$ whenever $u \in D(\mathcal{E})$ and for every $k \in K_0$*

$$\frac{\partial}{\partial k}\varphi(u) = \varphi'(u)\frac{\partial u}{\partial k} \ ,$$

where $\frac{\partial}{\partial k}$ also denotes the closure of $\frac{\partial}{\partial k} : \mathcal{F}C_b^\infty \to L^2(E; \mu)$ on $L^2(E; \mu)$.

(ii) Let $u, v \in D(\mathcal{E})$ then for all $k \in K_0$

$$\frac{\partial}{\partial k}(u \vee v) = 1_{\{u>v\}}\frac{\partial u}{\partial k} + 1_{\{u<v\}}\frac{\partial v}{\partial k} + \frac{1}{2}1_{\{u=v\}}\left(\frac{\partial u}{\partial k} + \frac{\partial v}{\partial k}\right)$$

and

$$\frac{\partial}{\partial k}(u \wedge v) = 1_{\{u<v\}}\frac{\partial u}{\partial k} + 1_{\{u>v\}}\frac{\partial v}{\partial k} + \frac{1}{2}1_{\{u=v\}}\left(\frac{\partial u}{\partial k} + \frac{\partial v}{\partial k}\right) .$$

(iii) For all $u, v \in D(\mathcal{E})$

$$\sum_{k \in K_0}\left(\frac{\partial}{\partial k}(u \vee v)\right)^2 \leq \left(\sum_{k \in K_0}\left(\frac{\partial u}{\partial k}\right)^2\right) \vee \left(\sum_{k \in K_0}\left(\frac{\partial v}{\partial k}\right)^2\right)$$

and

$$\sum_{k \in K_0}\left(\frac{\partial}{\partial k}(u \wedge v)\right)^2 \leq \left(\sum_{k \in K_0}\left(\frac{\partial u}{\partial k}\right)^2\right) \vee \left(\sum_{k \in K_0}\left(\frac{\partial v}{\partial k}\right)^2\right) .$$

Proof. (i): The fact that $\varphi \circ u \in D(\mathcal{E})$ is a consequence of I.4.14(ii) since

$$|\varphi(u)(x) - \varphi(u)(y)| \leq \|\varphi'\|_\infty |u(x) - u(y)| \text{ for all } x, y \in E$$

(though it is essentially proved again in what follows now). So, let $u \in D(\mathcal{E})$ and $u_n \in \mathcal{F}C_b^\infty$, $n \in \mathbb{N}$, such that $u_n \underset{n\to\infty}{\longrightarrow} u$ in $D(\mathcal{E})$. Then $\varphi(u_n) \underset{n\to\infty}{\longrightarrow} \varphi(u)$ in $L^2(E; \mu)$ and the closedness of $(\mathcal{E}, D(\mathcal{E}))$ implies that $\frac{\partial}{\partial k}\varphi(u_n) \underset{n\to\infty}{\longrightarrow} \frac{\partial}{\partial k}\varphi(u)$ in $L^2(E; \mu)$ for all $k \in K_0$, since clearly,

$$\frac{\partial\varphi(u_n)}{\partial k} = \varphi'(u_n)\frac{\partial u_n}{\partial k} \underset{n\to\infty}{\longrightarrow} \varphi'(u)\frac{\partial u}{\partial k} \text{ in } L^2(E; \mu).$$

Now the assertion follows.

(ii): Let $(\delta_n)_{n\in\mathbb{N}}$ be a Dirac sequence (i.e., $\delta_n \in C_0^\infty(\mathbb{R})$, $\delta_n \geq 0$, $\int \delta_n ds = 1$, $\delta_n(s) = \delta_n(-s)$, $s \in \mathbb{R}$, and $\operatorname{supp}\delta_n \subset \,]-\frac{1}{n}, \frac{1}{n}[$ for all $n \in \mathbb{N}$). Set $\varphi_n(t) := \int |t-s|\delta_n(s)\,ds$, $t \in \mathbb{R}, n \in \mathbb{N}$. Then $\varphi_n \in C^\infty(\mathbb{R})$, $\|\varphi_n'\|_\infty \leq 1$, $\varphi_n'(t) \underset{n\to\infty}{\longrightarrow} \operatorname{sign}(t)$ for all $t \in \mathbb{R}$ and $\varphi_n \underset{n\to\infty}{\longrightarrow} |\cdot|$ locally uniformly. Let $u \in D(\mathcal{E})$, u bounded, then $(\|\varphi_n(u)\|_\infty)_{n\in\mathbb{N}}$ is bounded, hence $\varphi_n(u) \underset{n\to\infty}{\longrightarrow} |u|$ in $L^2(E; \mu)$. The closedness of $(\mathcal{E}, D(\mathcal{E}))$ now implies that $\frac{\partial|u|}{\partial k} = \lim_{n\to\infty}\frac{\partial\varphi_n(u)}{\partial k}$ in $L^2(E; \mu)$ for all $k \in K_0$, since by (i) we have that

$$\frac{\partial\varphi_n(u)}{\partial k} = \varphi_n'(u)\frac{\partial u}{\partial k} \underset{n\to\infty}{\longrightarrow} \operatorname{sign}(u)\frac{\partial u}{\partial k} \text{ in } L^2(E; \mu) .$$

Consequently,

$$\frac{\partial|u|}{\partial k} = \operatorname{sign}(u)\frac{\partial u}{\partial k} \text{ for all } k \in K_0 .$$

An application of I.4.17 yields that this is true for all $u \in D(\mathcal{E})$. Using the identities $u \vee v = \frac{1}{2}(u + v) + \frac{1}{2}|u - v|$ and $u \wedge v = \frac{1}{2}(u + v) - \frac{1}{2}|u - v|$ we obtain (ii).

(iii) immediately follows from (ii). $\square$

Proposition 4.2. Cap *is tight.*

Proof. Since E is separable we may choose a fixed countable dense set $\{y_m \mid m \in \mathbb{N}\}$ in E. By the Hahn-Banach theorem we can find $l_m \in E'$ so that $\|l_m\|_{E'} = 1$ and $l_m(y_m) = \|y_m\|_E$. Note that hence

$$\|z\|_E = \sup_{m \in \mathbb{N}} l_m(z) \; ;$$

(indeed, $l_m(z) \leq |l_m(z)| \leq \|z\|_E$ and for some subsequence $(m_n)_{n \in \mathbb{N}}$,

$$\|z\|_E = \lim_{n \to \infty} \|y_{m_n}\|_E = \lim_{n \to \infty} l_{m_n}(y_{m_n}) = \lim_{n \to \infty} l_{m_n}(z) \leq \sup_{m \in \mathbb{N}} l_m(z)) \;.$$

Let $\varphi \in C_b^\infty(\mathbb{R})$, φ increasing with $\varphi(t) = t$ for all $t \in [-1,1]$, $\|\varphi'\|_\infty \leq 1$, and for $m \in \mathbb{N}$ define $v_m : E \to \mathbb{R}$ by

$$v_m(z) = \varphi(\|z - y_m\|_E) \;, \quad z \in E \;.$$

Suppose we can show that

(4.3) $w_n := \inf_{m \leq n} v_m, n \in \mathbb{N}$, converges $\mathcal{E}$-quasi-uniformly to zero on E .

Then (by III.2.11) for every $k \in \mathbb{N}$ there exists a closed set F_k such that $\mathrm{Cap}(F_k^c) < \frac{1}{k}$ and $w_n \xrightarrow[n \to \infty]{} 0$ uniformly on F_k. Hence for all $0 < \epsilon < 1$ there exists $n \in \mathbb{N}$ such that $w_n < \epsilon$ on F_k, i.e., by the definition of w_n,

$$F_k \subset \bigcup_{m=1}^{n} B(y_m, \epsilon)$$

where $B(y, \epsilon) := \{z \in E \mid \|z - y\|_E < \epsilon\}$. Consequently, F_k is totally bounded, hence compact, and Cap is tight. To show (4.3) we first fix $m \in \mathbb{N}$ and note that if for $n \in \mathbb{N}$

$$u_n(z) := \sup_{j \leq n} \varphi(l_j(z - y_m)) \;, \; z \in E \;,$$

then each u_n is a continuous version of an element in $D(\mathcal{E})$, $u_n \uparrow v_m$ on E as $n \to \infty$ hence in $L^2(E; \mu)$, and by 4.1(iii) (and induction)

$$\sum_{k \in K_0} \left(\frac{\partial u_n}{\partial k} \right)^2 \leq \sup_{j \leq n} \left(\sum_{k \in K_0} \varphi'(l_j(z - y_m))^2 {}_{E'}\langle l_j, k \rangle_E^2 \right) \;.$$

Since $\|\varphi'\|_\infty \leq 1$ and $\|l_j\|_{E'} = 1$ for all $j \in \mathbb{N}$, it follows by (4.1) that this is bounded by c^2. Applying I.2.12 we conclude that $v_m \in D(\mathcal{E})$ and

(4.4) $$\sum_{k \in K_0} \left(\frac{\partial v_m}{\partial k} \right)^2 \leq c^2 \text{ for all } m \in \mathbb{N} \;.$$

Since $\{y_m \mid m \in \mathbb{N}\}$ is dense in E, $w_n \downarrow 0$ on E as $n \to \infty$ hence in $L^2(E; \mu)$. But by 4.1(iii) (and induction) and (4.4)

$$\sum_{k\in K_0}\left(\frac{\partial w_n}{\partial k}\right)^2 \le \inf_{m\le n}\sum_{k\in K_0}\left(\frac{\partial v_m}{\partial k}\right)^2 \le c^2 \ \text{ for all } n\in\mathbb{N}\ .$$

Applying I.2.12 and III.3.5 we obtain that a subsequence of the Cesaro mean $\overline{w}_n = \frac{1}{n}\sum_{j=1}^{n} w_{n_j}$, $n\in\mathbb{N}$, of some subsequence $(w_{n_j})_{j\in\mathbb{N}}$ of $(w_n)_{n\in\mathbb{N}}$ converges to zero (in $D(\mathcal{E})$ and) $\mathcal{E}$-quasi-uniformly. But since $(w_n)_{n\in\mathbb{N}}$ is decreasing, $w_{n_n} \le \frac{1}{n}\sum_{j=1}^{n} w_{n_j}$ and thus $w_n \underset{n\to\infty}{\longrightarrow} 0$ $\mathcal{E}$-quasi-uniformly, i.e., (4.3) is shown and the proof is complete. $\qquad\square$

Remark. For an analogous result to 4.2 when E is replaced by the (non-metrizable) dual of a nuclear space we refer to [AR 89b, Sect.3.(c)].

By 3.5 we hence know that there exists a (μ-tight) special standard process associated with $(\mathcal{E}, D(\mathcal{E}))$ which is in fact a diffusion with infinite life time (cf. V.1.12(ii) and V.2.29(ii) below). All this remains true for the more general symmetric Dirichlet forms $(\mathcal{E}_A, D(\mathcal{E}_A))$ on $L^2(E;\mu)$ introduced in II.3.6 and hence also for those (not necessarily symmetric) Dirichlet forms studied in Chap.II, Subsection 3e). We leave the details to the reader by posing the following

Exercise 4.3. Let K_0 be as above and $A := (a_{k,k'})_{k,k'\in K_0}$ as in II.3.6. Let

$$\mathcal{E}_A(u,v) = \int \sum_{k,k'\in K_0} a_{k,k'}\frac{\partial u}{\partial k}\frac{\partial v}{\partial k'}\,d\mu\ ;\ \ u,v\in\mathcal{F}C_b^\infty\ ,$$

and assume that $(\mathcal{E}_A, \mathcal{F}C_b^\infty)$ is closable on $L^2(E;\mu)$ with closure $(\mathcal{E}_A, D(\mathcal{E}_A))$ (cf. II.3.6). Denote the corresponding capacity $\mathrm{Cap}_{h,g}$ with $h = g \equiv 1$ by Cap_A. Prove that Cap_A is tight and that hence $(\mathcal{E}_A, D(\mathcal{E}_A))$ is quasi-regular. Realize that, in particular, all this holds in the "coordinate free" case in II.3.9 where in addition, a separable real Hilbert space H is given which is continuously and densely embedded into H, and where K_0 is any orthonormal basis of H ($\subset E$) consisting of μ-admissible elements in E (cf. [RS 91a, Proposition 3.1]).

c)   Perturbation of Dirichlet forms

Let $(\mathcal{E}, D(\mathcal{E}))$ be a symmetric regular Dirichlet form on $L^2(E;m)$ where E is a locally compact separable metric space and m a Radon measure on E with $\mathrm{supp}[m] = E$. In this subsection we shall employ the knowledge of smooth measures and additive functionals developed in [F 80]. Set $\mathrm{Cap} := \mathrm{Cap}_{1,1}$ (see III.2.5(ii)). All "quasi-notions" are meant (in this subsection) with respect to Cap.

A positive measure μ on $(E, \mathcal{B}(E))$ is said to be *smooth* ($\mu\in S$ in notation) if $\mathrm{Cap}(A) = 0$ implies $\mu(A) = 0$ for all $A\in\mathcal{B}(E)$ and there exists an increasing sequence of compact sets $(F_n)_{n\in\mathbb{N}}$ such that

(4.5)

    (i) $\mu(F_n) < \infty$ for all $n \in \mathbb{N}$

    (ii) $\mu(E \setminus \cup F_n) = 0$

    (iii) $\lim_{n \to \infty} \mathrm{Cap}(K \setminus F_n) = 0$ for every compact set $K \subset E$.

Let $\mathbf{M} = (\Omega, \mathcal{F}, (X_t)_{t \geq 0}(P_z)_{z \in E_\Delta})$ be a Hunt process associated with $(\mathcal{E}, D(\mathcal{E}))$ with life time ζ and shift operator $(\theta_t)_{t \geq 0}$. Recall that we always assume that $(\mathcal{F}_t)_{t \geq 0}$ is the (universally completed) natural filtration of $(X_t)_{t \geq 0}$ and $\mathcal{F} := \mathcal{F}_\infty$ (cf. 1.10). A family $(A_t)_{t \geq 0}$ of positive functions on Ω is said to be a *positive continuous additive functional* (abbreviated *PCAF*) of $\mathbf{M}$ if:

    (i) $A_t(\cdot)$ is $\mathcal{F}_t$-measurable for all $t \geq 0$.

(4.6)

    (ii) There exists a *defining set* $\Lambda \in \mathcal{F}$ and an *exceptional set* $N \subset E$ with $\mathrm{Cap}(N) = 0$ such that $P_z(\Lambda) = 1$ for all $z \in E \setminus N$, $\theta_t\Lambda \subset \Lambda$ for all $t > 0$, and for each $\omega \in \Lambda$, $A.(\omega)$ is a positive continuous function satisfying: $A_0(\omega) = 0$, $A_t(\omega) < \infty$ for $t < \zeta(\omega)$, $A_t(\omega) = A_\zeta(\omega)$ for $t \geq \zeta(\omega)$, and $A_{t+s}(\omega) = A_t(\omega) + A_s(\theta_t\omega)$ for $s, t \geq 0$.

Two PCAF's (A_t) and (B_t) are said to be *equivalent* if they have a common defining set Λ and a common exceptional set N such that $A_t(\omega) = B_t(\omega)$ for all $\omega \in \Lambda$ and $t \geq 0$.

By [F 80, Theorem 5.1.3], for each smooth measures μ, there exists an (up to equivalence) unique PCAF (A_t^μ) such that

$$(4.7) \qquad \lim_{t \downarrow 0} E_{h \cdot m} \left[\frac{1}{t} \int_0^t f(X_s) \, dA_s^\mu \right] = \langle f \cdot \mu, h \rangle := \int_E h(z)(f \cdot \mu)(dz)$$

for any γ-excessive function h ($\gamma > 0$) and all $f \in \mathcal{B}^+(E)$. The formula (4.7), which is called *Revuz formula*, specifies a one to one correspondence between the family of all smooth measures and the family of all equivalent classes of PCAF's.

Let $\mu \in S$ correspond to (A_t^μ). We introduce the following notations (provided the corresponding integrals make sense)

$$(4.8) \qquad\qquad p_t^\mu f(z) := E_z \left[e^{-A_t^\mu} f(X_t) \right] \quad \text{for } t > 0$$

and

$$(4.9) \qquad\qquad R_\alpha^\mu f(z) := E_z \left[\int_0^\infty e^{-\alpha t - A_t^\mu} f(X_t) \, dt \right] \quad \text{for } \alpha > 0 \;.$$

For a closed set $F \subset E$ and $\alpha > 0$, we write also

$$(4.10) \qquad\qquad R_\alpha^{\mu, F} f(z) := E_z \left[\int_0^{\sigma_{F^c}} e^{-\alpha t - A_t^\mu} f(X_t) \, dt \right] \;, \quad z \in E \;.$$

(σ_{F^c} is the first hitting time of $F^c := E \setminus F$, see (1.10).)

We now consider $(\mathcal{E}^\mu, D(\mathcal{E}^\mu))$, the perturbation of $(\mathcal{E}, D(\mathcal{E}))$ by a smooth measure μ, which is defined by

$$(4.11) \qquad D(\mathcal{E}^\mu) := \{u \in D(\mathcal{E}) \,|\, \tilde{u} \ \mu\text{-square integrable}\}$$

(where $\tilde{u}$ denotes a quasi-continuous m-version of $u \in D(\mathcal{E})$) and

$$(4.12) \qquad \mathcal{E}^\mu(f,g) := \mathcal{E}(f,g) + \langle f,g \rangle_\mu \ ; \ f,g \in D(\mathcal{E}^\mu) \,,$$

where $\langle f,g \rangle_\mu := \int_E \tilde{f} \, \tilde{g} \, d\mu$. For $\alpha > 0$ we have

$$(4.13) \qquad \mathcal{E}^\mu_\alpha(f,g) = \mathcal{E}_\alpha(f,g) + \langle f,g \rangle_\mu \,.$$

Theorem 4.4. *Let μ be a smooth measure. Then:*

(i) *$(\mathcal{E}^\mu, D(\mathcal{E}^\mu))$ is a symmetric Dirichlet form on $L^2(E;m)$, and $(p^\mu_t)_{t>0}$ resp. $(R^\mu_\alpha)_{\alpha>0}$ is the strongly continuous sub-Markovian semigroup resp. sub-Markovian resolvent associated with $(\mathcal{E}^\mu, D(\mathcal{E}^\mu))$.*

(ii) *For any closed set $F \subset E$, $\alpha > 0$, and $f \in L^2(E;m)$, we have $R^{\mu,F}_\alpha f \in D(\mathcal{E}^\mu)_F$ and*

$$(4.14) \qquad \mathcal{E}^\mu_\alpha(R^{\mu,F}_\alpha f, u) = (f,u) \ \text{ for every } u \in D(\mathcal{E}^\mu)_F$$

where $D(\mathcal{E}^\mu)_F$ is specified by III.(1.1). Moreover, $R^{\mu,F}_\alpha f$ is quasi-continuous.

For the proof of Theorem 4.4 we refer to [AM 91d, 3.1 and 5.2]. See also [AM 91a].

Lemma 4.5. *Let μ be a smooth measure and $(F_n)_{n \in \mathbb{N}}$ an increasing sequence of closed sets. Then the following assertions are equivalent:*

(i) *$\displaystyle\lim_{n\to\infty} \mathrm{Cap}(K \setminus F_n) = 0$ for all compact sets $K \subset E$.*

(ii) *$P_z[\displaystyle\lim_{n\to\infty} \sigma_{F^c_n} \geq \zeta] = 1$ for q.e. $z \in E$.*

(iii) *$(F_n)_{n \in \mathbb{N}}$ is an $\mathcal{E}$-nest.*

(iv) *$(F_n)_{n \in \mathbb{N}}$ is an $\mathcal{E}^\mu$-nest.*

Proof. For the proof of $(i) \Longleftrightarrow (ii)$ see [F 80, Lemma 5.1.6]. The equivalence between (ii) and (iii) is obtained from $(ii) \Longleftrightarrow (iv)$ by setting $\mu \equiv 0$. Thus, we only need to prove $(ii) \Longleftrightarrow (iv)$. Let $(F_n)_{n \in \mathbb{N}}$ be an increasing sequence of closed sets. We set

$$\begin{aligned}
\sigma \ &:= \ \lim_{n\to\infty} \sigma_{F^c_n}, \\
U_n f \ &:= \ R^{\mu,F_n}_1 f = E_\cdot\left[\int_0^{\sigma_{F^c_n}} e^{-t-A^\mu_t} f(X_t)\,dt\right], \\
U_\infty f \ &:= \ E_\cdot\left[\int_0^{\sigma} e^{-t-A^\mu_t} f(X_t)\,dt\right], \\
U f \ &:= \ E_\cdot\left[\int_0^{\infty} e^{-t-A^\mu_t} f(X_t)\,dt\right], \\
H_n f \ &:= \ U f - U_n f, \ 1 \leq n \leq \infty,
\end{aligned}$$

provided the right hand sides make sense. Let $f \in L^2(E; m)$; then the following facts can be derived from 4.4(ii)

$$(4.15) \qquad U_n f \in D(\mathcal{E}^\mu)_{F_n} \,,$$

$$(4.16) \qquad \mathcal{E}_1^\mu(H_n f, g) = 0 \ \text{ for all } g \in D(\mathcal{E}^\mu)_{F_n},$$

$$(4.17) \qquad U_n f \to U_\infty f \ \text{q.e. and in } \mathcal{E}_1^\mu\text{-norm},$$

$$(4.18) \qquad H_n f \to U_\infty f \ \text{q.e. and in } \mathcal{E}_1^\mu\text{-norm} \,.$$

Suppose that (ii) holds. Then $U_\infty f = U f$ q.e. and (iv) follows from (4.16) and (4.17). Conversely, suppose that (iv) is true. Then from (4.16) and (4.18) we see that $H_\infty f = 0$ m-a.e. and hence $H_\infty f = 0$ q.e. because one can show that $H_\infty f$ is quasi-continuous. In particular, if we choose $f \in L^2(E; m)$ to be strictly positive everywhere, then since $H_\infty f = 0$ q.e., (ii) follows and the proof is complete. $\qquad \square$

Theorem 4.6. *Let μ be a smooth measure, then $(\mathcal{E}^\mu, D(\mathcal{E}^\mu))$ is a quasi-regular Dirichlet form.*

Proof. We have to verify that $(\mathcal{E}^\mu, D(\mathcal{E}^\mu))$ satisfies 3.1(i)-(iii). Let $(F_n)_{n \in \mathbb{N}}$ be an increasing sequence of compact sets satisfying condition (4.5). Then by 4.5 $(F_n)_{n \in \mathbb{N}}$ is an $\mathcal{E}^\mu$-nest. Hence $(\mathcal{E}^\mu, D(\mathcal{E}^\mu))$ satisfies 3.1(i). Also by 4.5 every quasi-continuous function is $\mathcal{E}^\mu$-quasi-continuous. Thus $(\mathcal{E}^\mu, D(\mathcal{E}^\mu))$ satisfies 3.1(ii) because every element of $D(\mathcal{E})$ admits a quasi-continuous modification. To verify the condition 3.1(iii), we choose $f_i \in C_0(E)$, $i \in \mathbb{N}$, such that $\mathcal{B}(E) = \sigma\{f_i \mid i \in \mathbb{N}\}$. Let N be an exceptional set of (A_t^μ) (see (4.6)(ii)). We set

$$f_{i,j}(z) := \begin{cases} E_z \left[j \int_0^\infty e^{-jt - A_t^\mu} f_i(X_t) \, dt \right] \,, & z \in X \setminus N \\ f_i(z) \,, & z \in N \,. \end{cases}$$

Then $f_{ij}(z) \to f_i(z)$ as $j \to \infty$ for all $z \in E$. . Taking $F = E$ in 4.4(ii) we see that $B_0 := \{f_{ij} \mid i, j \in \mathbb{N}\}$ consists of $\mathcal{E}^\mu$-quasi-continuous m-versions of elements in $D(\mathcal{E}^\mu)$. Obviously, we have $\sigma\{f_{ij} \mid f_{ij} \in B_0\} = \mathcal{B}(E)$ and the proof is thus complete. $\qquad \square$

We shall say that a Borel measure μ on E is *nowhere Radon* if $\mu(G) = \infty$ for all non-empty open subsets G of X.

Theorem 4.7. *Suppose that each single-point set of E is a set of zero capacity. Then, for each smooth measure ν with $\mathrm{supp}[\nu] = E$, there exists a smooth measure μ such that μ is equivalent to ν and μ is nowhere Radon.*

Proof. Put $B = \{z \in E \mid \nu \text{ is finite on a neighbourhood of } z\}$. If $B = \emptyset$, then ν itself is a desired nowhere Radon smooth measure. We assume now $B \neq \emptyset$. Let $\{x_j \mid j \in \mathbb{N}\}$ be a countable dense subset of B. We choose for each j a decreasing sequence of open sets $(G_{j,k})_{k \in \mathbb{N}}$ such that

$$\bigcap_{k \geq 1} G_{j,k} = \{x_j\}$$

and

$$\mathrm{Cap}(G_{j,k}) + \nu(G_{j,k}) \leq 2^{-k} \ .$$

Such a sequence exists because we have $\mathrm{Cap}(\{x_j\}) = 0$ which implies $\nu(\{x_j\}) = 0$, and ν is finite on a neighbourhood of x_j. We now define for each $j \in \mathbb{N}$

$$f_j(z) := \begin{cases} \dfrac{k}{\nu(G_{j,k})} & \text{if} \qquad z \in G_{j,k} - G_{j,k+1} \ , \ k \in \mathbb{N}, \\ 1 & \text{otherwise.} \end{cases}$$

Then f_j is bounded outside $G_{j,k}$ for each k. In particular, we can choose a positive number c_j such that

$$c_j \sup_{z \in E \backslash G_{j,j}} f_j(z) \leq 2^{-j} \ .$$

We define

$$f(z) := \sum_{j \geq 1} c_j f_j(z) \ , \quad z \in E \ ,$$

and

$$\mu := f \cdot \nu \ .$$

We claim that μ is a smooth measure with the required properties. Indeed, since $0 < f < \infty$ on $E \backslash N$ with $N := \bigcap_{n \geq 1} \bigcup_{j > n} G_{j,j}$, it is evident that μ is equivalent to ν because $\nu(N) = 0$. Consequently, μ charges no set of zero capacity. Let $(E_n)_{n \in \mathbb{N}}$ be an increasing sequence of compact sets satisfying (4.5) with respect to ν. We set for each $n \in \mathbb{N}$

$$G_n := \left(\bigcup_{1 \leq j \leq n} G_{j,2n} \right) \cup \left(\bigcup_{j > n} G_{j,j} \right)$$

and

$$F_n := E_n \backslash G_n \ .$$

Then $(F_n)_{n \in \mathbb{N}}$ is an increasing sequence of compact sets and it is easy to check that (4.5) holds for μ and $(F_n)_{n \in \mathbb{N}}$. Thus μ is a smooth measure. Let G be a non-empty open set of E. If $G \backslash B \neq \emptyset$, then $\mu(G) \geq c\nu(G) = \infty$ by the definition of B. Here $c := \sum_{j \geq 1} c_j > 0$. Suppose now $G \cap B \neq \emptyset$. Then $x_j \in G$ for some $j \in \mathbb{N}$, and for k large enough we have $G_{j,k} \subset G$. In this case

$$\mu(G) \geq c_j \int_{G_{j,k}} \frac{k}{\nu(G_{j,k})} \nu(dz) \geq c_j k \ .$$

Letting $k \to \infty$ we obtain $\mu(G) = \infty$. Thus μ is as desired. $\qquad \square$

We remark that the reference measure m is always a smooth measure with $\mathrm{supp}[m] = E$. Hence there always exists a nowhere Radon smooth measure provided each single-point set is a set of zero capacity. The latter property is satisfied, for example, if $(\mathcal{E}, D(\mathcal{E}))$ is the classical Dirichlet form associated with the Laplace operator on $\mathbb{R}^d$ with $d \geq 2$ (cf. Chap.II, Subsection 1b)). In this case a nowhere Radon smooth measure μ can be constructed more specifically by $\mu = V \cdot m$ with V being specified in II.(2.29) (cf. [AM 88, 1.3]).

Obviously, when μ is a nowhere Radon smooth measure, then the perturbation $(\mathcal{E}^\mu, D(\mathcal{E}^\mu))$ of $(\mathcal{E}, D(\mathcal{E}))$ by μ is a Dirichlet form with domain containing no continuous functions except the zero function. Nevertheless, since $(\mathcal{E}^\mu, D(\mathcal{E}^\mu))$ is quasi-regular, by 3.5 there always exists a special standard process $\mathbf{M}^\mu$ properly associated with $(\mathcal{E}^\mu, D(\mathcal{E}^\mu))$. In Chap. V, Subsection 2d) we shall prove that $\mathbf{M}^\mu$ is again a Hunt process.

5   Necessity of quasi-regularity and some probabilistic potential theory

In this section let E be a Hausdorff topological space such that $\mathcal{B}(E) = \sigma(C(E))$. In the same way as before we adjoin an isolated point Δ to E and set $E_\Delta := E \cup \{\Delta\}$. We fix a σ-finite positive measure m on $(E, \mathcal{B}(E))$ and extend m to $(E_\Delta, \mathcal{B}(E_\Delta))$ by setting $m(\{\Delta\}) = 0$. As before each function f on E is extended to E_Δ by $f(\Delta) := 0$. Let $(\mathcal{E}, D(\mathcal{E}))$ be a Dirichlet form with semigroups $(T_t)_{t>0}$, $(\hat{T}_t)_{t>0}$ and resolvents $(G_\alpha)_{\alpha>0}$, $(\hat{G}_\alpha)_{\alpha>0}$ on $L^2(E; m)$. This section is devoted to proving the following converse to 3.5.

Theorem 5.1. *Suppose that there exists an m-special standard process* $\mathbf{M} = (\Omega, \mathcal{F}, (X_t)_{t\geq0}, (P_z)_{z\in E_\Delta})$ *with state space E and life time ζ which is m-tight and associated with $(\mathcal{E}, D(\mathcal{E}))$. Then $(\mathcal{E}, D(\mathcal{E}))$ is quasi-regular, i.e., satisfies 3.1(i)-(iii). Moreover, $\mathbf{M}$ is properly associated with $(\mathcal{E}, D(\mathcal{E}))$.*

The proof of 5.1 is given in Subsections a)–c) below. Combining 5.1 with 3.5 we immediately obtain

Corollary 5.2. *Suppose there exists an m-tight m-special standard process $\mathbf{M}$ with state space E associated with $(\mathcal{E}, D(\mathcal{E}))$. Then there exists an m-tight special standard process $\hat{\mathbf{M}}$ coassociated with $(\mathcal{E}, D(\mathcal{E}))$ which is in duality with $\mathbf{M}$ w.r.t. m (i.e., its transition function is in duality with that of $\mathbf{M}$ w.r.t. m).*

Remark 5.3. (i) Recall that in 5.1, 5.2 (as well as in (ii) below) "m-tight" can be dropped if E is a metrizable Lusin space by 1.14(ii).

(ii) Consider the situation of 5.2. Clearly, by 5.1 $\hat{\mathbf{M}}$ is properly coassociated with $(\mathcal{E}, D(\mathcal{E}))$ and by 3.6(iii) we know that there also exists an m-tight special standard process $\tilde{\mathbf{M}}$ with state space E properly associated with the symmetric part $(\tilde{\mathcal{E}}, D(\mathcal{E}))$ of $(\mathcal{E}, D(\mathcal{E}))$. Further important consequences of 5.1 are discussed in Sect.6 below.

(iii) Note that by the correspondence of $(T_t)_{t>0}$ and $(G_\alpha)_{\alpha>0}$, in the situation of 5.1 we have by 2.7 that for all $f \in \mathcal{B}^+(E)$, f m-square integrable, and all $\alpha > 0$,

$$(5.1) \qquad R_\alpha f := E_\cdot\left[\int_0^\infty e^{-\alpha t} f(X_t)dt\right] = E_\cdot\left[\int_0^\zeta e^{-\alpha t} f(X_t)dt\right]$$

(cf.(1.9)) is an m-version of $G_\alpha f$.

a)   Proof of property 3.1(i)

Property 3.1(i) will be an immediate consequence of the following fundamental relationship between hitting times and capacities of open sets which will be generalized to Borel sets in Subsection c) (cf. 5.28 below). It can be considered as the first result connecting the analytic potential theory of $(\mathcal{E}, D(\mathcal{E}))$ (cf. Chapter III) and the probabilistic potential theory of **M**.

Theorem 5.4. *Let $U \subset E$, U open. For $\varphi \in \mathcal{B}_b(E)$, $\varphi > 0$ everywhere on E and $\int \varphi \, dm = 1$, set $\mu := \varphi \cdot m$, $g := \hat{G}_1 \varphi$ and $h := R_1 \varphi$. Then*

$$(5.2) \qquad \mathrm{Cap}_{h,g}(U) = E_\mu \left[\int_{\sigma_U}^{\zeta} e^{-s} \varphi(X_s) \, ds \right]$$

where σ_U is the first hitting time of U (cf.(1.10)).

Before we prove 5.4 we show that it implies property 3.1(i). Since **M** is m-tight there exists an increasing sequence $(K_n)_{n \in \mathbb{N}}$ of compact metrizable sets in E such that

$$P_z \left[\lim_{n \to \infty} \sigma_{E \setminus K_n} \geq \zeta \right] = 1 \quad \text{for} \quad m - a.e. \ z \in E \ .$$

Then

$$(5.3) \qquad E_z \left[\int_{\sigma_{E \setminus K_n}}^{\zeta} e^{-s} \varphi(X_s) \, ds \right] \downarrow 0 \quad \text{as } n \to \infty \text{ for} \quad m - a.e. \ z \in E \ ,$$

hence

$$\lim_{n \to \infty} \mathrm{Cap}_{h,g}(E \setminus K_n) = 0$$

by (5.2), and 3.1(i) is proved.

In order to prove 5.4 we need some preparations.

Remark 5.5. Below we only use that **M** is a right process, hence 5.4 holds in this more general case.

Define for $U \subset E$, U open, and $f \in \mathcal{B}^+(E)$

$$(5.4) \qquad R_1^{U^c} f(z) := E_z \left[\int_0^{\sigma_U} e^{-t} f(X_t) dt \right], \ z \in E \ .$$

(i.e., the *resolvent restricted to U^c*, cf. [F 80, Sect.4.4]) where $U^c := E \setminus U$.

Lemma 5.6. *Let $U \subset E$, U open, and $f \in \mathcal{B}^+(E) \cap L^2(E; m)$. Then $R_1^{U^c} f \in D(\mathcal{E})$ and*

$$(5.5) \qquad \mathcal{E}_1(R_1^{U^c} f, v) = (f, v) \quad \text{for all } v \in D(\mathcal{E})_{U^c}$$

(where $D(\mathcal{E})_{U^c}$ is defined as in III.(1.1)).

Proof. Let $z \in E$. Define

$$(5.6) \quad V_n f(z) := E_z \left[\int_0^\infty \exp(-t - n \int_0^t 1_U(X_s)\, ds)\, f(X_t)\, dt \right] \, , \quad n \in \mathbb{N} \, .$$

Since U is open and $\mathbf{M}$ has right continuous sample paths, $\sigma_U < t$ if and only if $\int_0^t 1_U(X_s)\, ds > 0$. Hence $V_n f(z) \downarrow R_1^{U^c} f(z)$ as $n \to \infty$. By the Markov property and Fubini's theorem we obtain that

$$(5.7) \qquad\qquad V_n f(z) = R_1 f(z) - n\, R_1(1_U V_n f)(z) \, .$$

Indeed, by Fubini's theorem

$$
\begin{aligned}
& n\, R_1(1_U V_n f)(z) \\
&= n \int_0^\infty E_z \left[e^{-t} 1_U(X_t) E_{X_t} \left[\int_0^\infty \exp(-s - n \int_0^s 1_U(X_{s'})\, ds')\, f(X_s)\, ds \right] \right] dt \\
&= n \int_0^\infty E_z \left[e^{-t} 1_U(X_t) \int_0^\infty \exp(-s - n \int_0^s 1_U(X_{s'+t})\, ds')\, f(X_{s+t})\, ds \right] dt \\
&= n\, E_z \left[\int_0^\infty 1_U(X_t) \int_t^\infty \exp(-s - n \int_t^s 1_U(X_{s'})\, ds')\, f(X_s)\, ds\, dt \right] \\
&= E_z \left[\int_0^\infty e^{-s} f(X_s) \int_0^s n\, 1_U(X_t) \exp(-n \int_t^s 1_U(X_{s'})\, ds')\, dt\, ds \right] \\
&= E_z \left[\int_0^\infty e^{-s} f(X_s)(1 - \exp(-n \int_0^s 1_U(X_{s'})\, ds'))\, ds \right] \\
&= R_1 f(z) - V_n f(z) \, .
\end{aligned}
$$

Since $G_1 f \in D(\mathcal{E})$, (5.7) implies that $V_n f \in D(\mathcal{E})$ and furthermore,

$$(5.8) \qquad \mathcal{E}_1(V_n f, v) = (f, v) - n(1_U V_n f, v) \quad \text{for all } v \in D(\mathcal{E}) \, .$$

In particular, since $V_n f \leq G_1 f$,

$$\mathcal{E}_1(V_n f, V_n f) = (f, V_n f) - n(1_U V_n f, V_n f) \leq (f, V_n f) \leq (f, G_1 f) \quad \text{for all } n \in \mathbb{N} \, .$$

Hence by I.2.12 $V_n f \to R_1^{U^c} f$ weakly in $(D(\mathcal{E}), \tilde{\mathcal{E}}_1)$, as $n \to \infty$. Clearly, $R_1^{U^c} f = 0$ on U, thus $R_1^{U^c} f \in \mathcal{D}(\mathcal{E})_{U^c}$. Since for all $v \in D(\mathcal{E})$, $\mathcal{E}_1(\cdot, v)$ is a continuous linear functional on $D(\mathcal{E})$ (with norm $\tilde{\mathcal{E}}_1^{1/2}$), (5.8) now implies (5.5). $\qquad\square$

Define for $U \subset E$, U open, and $f \in \mathcal{B}^+(E)$

$$(5.9) \qquad\qquad H^U f(z) := E_z \left[e^{-\sigma_U} f(X_{\sigma_U}) \right] \, , \quad z \in E \, .$$

Then by the strong Markov property for all $z \in E$

$$
\begin{aligned}
(5.10) \qquad H^U R_1 f(z) \;&=\; E_z \left[e^{-\sigma_U} E_{X_{\sigma_U}} \left[\int_0^\infty e^{-s} f(X_s)\, ds \right] \right] \\
&=\; E_z \left[e^{-\sigma_U} \int_0^\infty e^{-s} f(X_{s+\sigma_U})\, ds \right] \\
&=\; E_z \left[\int_{\sigma_U}^\infty e^{-s} f(X_s)\, ds \right] \, .
\end{aligned}
$$

Therefore,

$$(5.11) \qquad\qquad R_1 f = R_1^{U^c} f + H^U R_1 f$$

which is well-known as *Dynkin's formula*. It enables us to prove the following fundamental relationships between hitting probabilities and reduction of functions (cf. Chap.III). This will be generalized in V.1.6 below.

Proposition 5.7. *For all $U \subset E$, U open, and all $f \in \mathcal{B}^+(E) \cap L^2(E; m)$, $H^U R_1 f$ is a version of $(G_1 f)_U$ (cf. III.1.5).*

Proof. By (5.5) and (5.11) we have that

$$\mathcal{E}_1(H^U R_1 f, v) = 0 \quad \text{for all } v \in D(\mathcal{E})_{U^c} .$$

Since by (5.10) obviously $H^U R_1 f = R_1 f$ on U, the assertion follows by III.1.4 and III.1.6(i). $\qquad\qquad\square$

Proof of 5.4. By III.1.5(ii), 5.7 and (5.10) we obtain that

$$\begin{aligned}
\mathrm{Cap}_{h,g}(U) &= \mathcal{E}_1(h_U, \hat{g}_U) = \mathcal{E}_1(h_U, g) = (h_U, \varphi) \\
&= (H^U h, \varphi) = E_\mu \left[\int_{\sigma_U}^\infty e^{-s} \varphi(X_s)\, ds \right] .
\end{aligned}$$

$$\square$$

b)   Proof of property 3.1(ii)

First we recall some facts from the theory of stochastic processes which we need to prove Proposition 5.14 below. We refer to [DM 78,82] for more details.

Let $(\Omega, \mathcal{M}, P)$ be a probability space and $(\mathcal{M}_t)_{t \geq 0}$ a filtration on $(\Omega, \mathcal{M})$ satisfying the usual conditions (cf. 1.2). A subset $\Lambda \subset [0, \infty[\times \Omega$ is called *evanescent* if

$$\pi(\Lambda) := \{\omega \in \Omega \mid (t, \omega) \in \Lambda \text{ for some } t \in [0, \infty[\}$$

is a P-null set. As usual we consider a stochastic process $(Z_t)_{t \geq 0}$ as a map $(t, \omega) \mapsto Z_t(\omega)$ on $[0, \infty[\times \Omega$ denoted by Z. Two processes Z, Z' (on some state space) are called *indistinguishable* if $\{Z' \neq Z\}$ is evanescent. The *optional σ-algebra* $\mathcal{O}$ on $[0, \infty[\times \Omega$ is defined as the σ-algebra generated by all real-valued $(\mathcal{M}_t)$-adapted processes with cadlag trajectories. The *predictable σ-algebra* $\mathcal{P}$ on $[0, \infty[\times \Omega$ is defined as the σ-algebra generated by all real valued $(\mathcal{M}_t)$-adapted processes having left continuous trajectories. A stochastic process is called *optional* resp. *predictable* if it is $\mathcal{O}$-measurable resp. $\mathcal{P}$- measurable. An $(\mathcal{M}_t)$-stopping time τ is called a *predictable time* if its graph $[[\tau]] := \{(\tau(\omega), \omega) \mid \omega \in \Omega$ with $\tau(\omega) < \infty\}$ is an element of $\mathcal{P}$. Let $\mathcal{T}$ be the set of all $(\mathcal{M}_t)$-stopping times. For positive functions σ, τ on Ω we define

$$[[\sigma, \tau[[:= \{(t, \omega) \mid t \in [\sigma(\omega), \tau(\omega)[, \, \omega \in \Omega\}$$

and $[[\sigma, \tau]], \,]]\sigma, \tau[[, \,]]\sigma, \tau]]$ analogously.

Proposition 5.8.   (i) $\mathcal{O} = \sigma\{[[\sigma, \infty[[\mid \sigma \in \mathcal{T}\} = \sigma\{[[0, \sigma[[\mid \sigma \in \mathcal{T}\}.$

(ii) $\quad\begin{aligned}\mathcal{P} &= \sigma\{]]\sigma,\infty[[\,|\,\sigma\in\mathcal{T}\}\cup\{\{0\}\times B\,|\,B\in\mathcal{M}_0\}\\ &= \sigma\{]]0,\sigma[[\,|\,\sigma\in\mathcal{T}\}\cup\{\{0\}\times B\,|\,B\in\mathcal{M}_0\}\,.\end{aligned}$

(iii) $\mathcal{P}\subset\mathcal{O}\subset\mathcal{B}([0,\infty[)\otimes\mathcal{M}$.

(iv) *Any real valued $(\mathcal{M}_t)$-adapted process having right continuous trajectories is optional.*

Proof. [DM 78, IV.64-67]. $\qquad\qquad\qquad\qquad\qquad\qquad\qquad\qquad\square$

Proposition 5.9. *Let τ be a predictable time. Then there exists an increasing sequence $(\tau_n)_{n\in\mathbb{N}}$ of stopping times such that $\lim\limits_{n\to\infty}\tau_n=\tau$ and $\tau_n<\tau$ on $\{\tau>0\}$ for all $n\in\mathbb{N}$.*

Proof. [DM 78,IV.71]. $\qquad\qquad\qquad\qquad\qquad\qquad\qquad\qquad\qquad\quad\square$

A real-valued $(\mathcal{M}_t)$-martingale $(Z_t)_{t\geq0}$ is called a *strong martingale* if
(i) Z is *progressively measurable* (i.e., Z restricted to $[[0,t]]$ is $\mathcal{B}([0,t])\otimes\mathcal{M}_t$-measurable for all $t\geq0$);
(ii) $E[|Z_\tau|]<\infty$ for every bounded $\tau\in\mathcal{T}$;
(iii) $E[Z_\tau\,|\,\mathcal{M}_\sigma]=Z_\sigma$ for all bounded $\sigma,\tau\in\mathcal{T}$ with $\sigma\leq\tau$.

Proposition 5.10. *Let $(Z_t)_{t\geq0}$ be a strong $(\mathcal{M}_t)$-martingale. If (Z_t) is optional, then it is indistinguishable from a process, which is cadlag $P-a.s.$.*

Proof. [DM 82, VI.1.3 and VI.3]. $\qquad\qquad\qquad\qquad\qquad\qquad\qquad\square$

For $\sigma\in\mathcal{T}$ let $\mathcal{P}_\sigma$ denote the trace σ-algebra of $\mathcal{P}$ on $[[0,\sigma[[$.

Lemma 5.11. *Let $\sigma\in\mathcal{T}$ and let $(Z_t)_{t\geq0}$ be a real-valued $(\mathcal{M}_t)$-adapted process having left continuous trajectories on $]0,\sigma[$. Then Z restricted to $[[0,\sigma[[$ is $\mathcal{P}_\sigma$-measurable.*

Proof. [Sh 88, (A.6.8)]. $\qquad\qquad\qquad\qquad\qquad\qquad\qquad\qquad\qquad\square$

Proposition 5.12. *Let $\sigma\in\mathcal{T}$ and let $(Y_t)_{t\geq0}$, $(Z_t)_{t\geq0}$ be two bounded real-valued processes such that their restrictions to $[[0,\sigma[[$ are $\mathcal{P}_\sigma$-measurable. If for every predictable time τ*

$$E\left[Y_\tau 1_{\{0<\tau<\sigma\}}\right]=E\left[Z_\tau 1_{\{0<\tau<\sigma\}}\right]$$

then $Y\,1_{]]0,\sigma[[}$ and $Z\,1_{]]0,\sigma[[}$ are indistinguishable.

Proof. [Sh 88, (A.6.16), (A.6.11)]. $\qquad\qquad\qquad\qquad\qquad\qquad\qquad\square$

Now we turn to the proof of 3.1(ii).

Remark 5.13. In the sequel we shall not use the m-tightness of **M**. Furthermore, to prove Claim 1 in the proof of 5.14 below we only need **M** to be a right process.

Proposition 5.14. *Let $\tilde{C}$ denote the set of all $f \in \mathcal{B}_b(E)$ such that $t \mapsto f(X_t)$ is right continuous on $[0, \infty[$ and $t \mapsto f(X_{t-})$ is left continuous on $]0, \zeta[$ P_m-a.s.. Then $R_\alpha f \in \tilde{C}$ for all $f \in \mathcal{B}_b(E)$, $\alpha > 0$.*

Proof. Let $(\mathcal{F}_t)_{t\geq 0}$ be the natural filtration of $\mathbf{M}$. Then $(\mathcal{F}_t)$ is right continuous. For simplifying notations in this proof we also denote $\mathcal{F}_t^{P_\mu}$ (with μ as in 5.4) by $\mathcal{F}_t$ and $\mathcal{F}^{P_\mu}$ by $\mathcal{F}$. Thus the filtration $(\mathcal{F}_t)$ satisfies the usual conditions on $(\Omega, \mathcal{F}, P_\mu)$ (cf. 1.1, 1.2).
Let $f \in \mathcal{B}_b(E)$. $\alpha > 0$. Consider the following bounded process

$$Z_t := e^{-\alpha t} R_\alpha f(X_t) + \int_0^t e^{-\alpha s} f(X_s)\, ds \ , \ t \geq 0 \ .$$

Claim 1. $(R_\alpha f(X_t))_{t\geq 0}$ is cadlag $P_\mu - a.s.$
By the strong Markov property we have that for any stopping time τ

$$\begin{aligned}
Z_\tau &= E_\mu\left[\int_\tau^\infty e^{-\alpha s} f(X_s)ds \,|\, \mathcal{F}_\tau\right] + \int_0^\tau e^{-\alpha s} f(X_s)ds \\
&= E_\mu\left[\int_0^\infty e^{-\alpha s} f(X_s)ds \,|\, \mathcal{F}_\tau\right] \ P_\mu\text{-a.s.},
\end{aligned}$$

hence $(Z_t)_{t\geq 0}$ is a strong $(\mathcal{F}_t)$-martingale. If $g \in C_b(E)$, it follows by 5.8(iv) that $(g(X_t))_{t\geq 0}$ is optional (w.r.t. $(\mathcal{F}_t)$). Since $\mathcal{B}(E) = \sigma(C_b(E))$, monotone class theorems (cf. [Sh 88, A.0.6]) imply that this then holds for all $g \in \mathcal{B}_b(E)$. Consequently, $(R_\alpha f(X_t))_{t\geq 0}$ and hence $(Z_t)_{t\geq 0}$ is optional. Applying 5.10 we obtain that $(Z_t)_{t\geq 0}$ and hence $(R_\alpha f(X_t))_{t\geq 0}$ is P_μ-indistinguishable from a process which is cadlag $P_\mu - a.s..$

Claim 2. $R_\alpha f(X_{t-}) = R_\alpha f(X_t)_- \ P_\mu$-a.s. on $\{0 < t < \zeta\}$.
Let τ be a predictable time. By 5.9 there exists an increasing sequence of stopping times $(\tau_n)_{n\in\mathbb{N}}$ such that $\lim_{n\to\infty} \tau_n = \tau$ and $\tau_n < \tau$ on $\{\tau > 0\}$. Thus

$$(5.12) \qquad Z_{\tau-} = \lim_{n\to\infty} Z_{\tau_n} = \lim_{n\to\infty} E_\mu[Z_\tau \,|\, \mathcal{F}_{\tau_n}] = E_\mu[Z_\tau \,|\, \bigvee_{n\in\mathbb{N}} \mathcal{F}_{\tau_n}] = Z_\tau$$

P_μ-a.s. on $\{0 < \tau < \zeta\}$, where we used the martingale convergence theorem and (M.10) (i.e., the fact that $\mathbf{M}$ is "special") in the last equality. On the other hand by (M.9) (i.e., "quasi-left continuity")

$$(5.13) \qquad X_{\tau-} = \lim_{n\to\infty} X_{\tau_n} = X_\tau \ P_\mu\text{-a.s. on } \{0 < \tau < \zeta\} \ .$$

(5.12) and (5.13) imply that

$$R_\alpha f(X_{\tau-}) = R_\alpha f(X_\tau) = R_\alpha f(X_\tau)_- \ P_\mu\text{-a.s. on } \{0 < \tau < \zeta\} \ .$$

Since τ is an arbitrary predictable time this implies Claim 2 by 5.12 (since ζ is a stopping time by 1.6(iv)), if we can show that both $(R_\alpha f(X_{t-}))_{t\geq 0}$ and $(R_\alpha f(X_t)_-)_{t\geq 0}$ restricted to $[[0, \zeta[[$ are $\mathcal{P}_\zeta$-measurable. However, by 5.11 this is clear for the latter since it is left continuous on $]0, \zeta[$ and $(\mathcal{F}_t)$-adapted, and for the first by the same argument if $R_\alpha f$ is replaced by $g \in C_b(E)$. Since $\mathcal{B}(E) = \sigma(C_b(E))$ monotone class theorems imply that also $(R_\alpha f(X_{t-}))_{t\geq 0}$ restricted to $[[0, \zeta[[$ is P_ζ-measurable. $\qquad\square$

Since $\{G_\alpha f \,|\, f \in L^2(E;m), \alpha > 0\}$ is dense in $D(\mathcal{E})$ w.r.t. $\tilde{\mathcal{E}}_1^{1/2}$ (by I.2.13(ii)) property 3.1(ii) is now an immediate consequence of 5.14, 5.3(iii) and the following proposition.

Proposition 5.15. *Let $f \in \tilde{C}$ then there exists an $\mathcal{E}$-quasi-continuous function $\tilde{f}$ on E such that $f = \tilde{f} \, m$-a.e.*

In order to prove 5.15 we need to extend the study of the potential theory of **M** and its connections with the potential theory of $(\mathcal{E}, D(\mathcal{E}))$ started in Subsection a).

For $B \subset E$ we define the *first touching time* τ_B of B by

$$(5.14) \qquad\qquad \tau_B := \inf\{0 \leq t < \zeta \,|\, \overline{X_0^t} \cap B \neq \emptyset\} \wedge \zeta$$

where $\overline{X_0^t}(\omega)$ is the closure of $\{X_s(\omega) \,|\, s \in [0,t]\}$ in E (cf. the alternative proof of 1.15). We again use the convention that $\inf \emptyset := \infty$. Note that by (M.6) and (M.8) one has that

$$\tau_B = \inf\{0 \leq t < \zeta \,|\, X_t \in B \text{ or } X_{t-} \in B\} \wedge \zeta \qquad P_\mu\text{-a.s.} \ .$$

Hence by [DM 78,IV.50] τ_B is an $(\mathcal{F}_t^{P_\mu})$-stopping time if $B \in \mathcal{B}(E)$.

Exercise 5.16. Prove the following:

   (i) If $A \subset B$ then $\tau_A \geq \tau_B$.

   (ii) $\tau_B \leq \sigma_B \wedge \zeta$ for all $B \subset E$.

   (iii) If $U \subset E$, U open, then $\tau_U = \sigma_U \wedge \zeta$.

Definition 5.17. A subset N of E is called **M**-*exceptional* if there exists $\tilde{N} \in \mathcal{B}(E)$ such that $N \subset \tilde{N}$ and $P_m[\tau_{\tilde{N}} < \zeta] = 0$.

Lemma 5.18. *If $N \in \mathcal{B}(E)$ and N is **M**-exceptional then $m(N) = 0$.*

Proof. By (M.5) (i.e., **M** is "normal") we know that $\tau_N = 0 \ P_z$-a.s. for all $z \in N$ and that $\zeta > 0 \ P_z - a.s$, for all $z \in E$, hence $m(N) = 0$. $\qquad\square$

We introduce the following (semi-)norm $\| \ \|$ on $\mathcal{B}_b(E)$ which is a modified version of a norm introduced by Y. LeJan in [Le 83] and is the key tool in the analysis below. We fix φ and μ as in 5.4 and define

$$(5.15) \qquad \|f\| := E_\mu\left[\sup_{t\geq 0}[(\int_t^\infty e^{-s}\varphi(X_s)\,ds)(|f(X_t)| \vee |f(X_{t-})|)]\right] \ .$$

Exercise 5.19. Show that for all $B \in \mathcal{B}(E)$

$$\|1_B\| = E_\mu \left[\int_{\tau_B}^\infty e^{-s} \varphi(X_s) \, ds \right] \ .$$

As in 5.4 we set $h := R_1 \varphi$ and $g := \hat{G}_1 \varphi$.

Lemma 5.20.   *(i)* $\|1_U\| = \mathrm{Cap}_{h,g}(U)$ *for all* $U \subset E$, *U open.*

   (ii) $\|1_B\| \leq \mathrm{Cap}_{h,g}(B)$ *for all* $B \in \mathcal{B}(E)$.

   (iii) Let $f \in \mathcal{B}_b(E)$, *then* $\|f\| = 0$ *if and only if* $f = 0$ *except on an* **M**-*exceptional set.*

Proof. (i) is immediate by 5.19, 5.16(iii), 5.4, and (ii) is then obvious. To prove (iii) assume first that $\{f \neq 0\}$ is **M**-exceptional then by 5.19, $\|1_{\{f \neq 0\}}\| = 0$. But then

$$\|f\| = \|f \, 1_{\{f \neq 0\}}\| \leq \sup_{z \in E} |f(z)| \|1_{\{f \neq 0\}}\| = 0 \ .$$

Suppose now that $\|f\| = 0$. Set $B_n := \{|f| > \frac{1}{n}\}$, $n \in \mathbb{N}$, hence $1_{B_n} \leq n|f|$ and by 5.19

$$E_\mu \left[\int_{\tau_{B_n}}^\infty e^{-s} \varphi(X_s) ds \right] = \|1_{B_n}\| \leq n\|f\| = 0$$

which implies that $P_\mu[\tau_{B_n} < \zeta] = 0$ for all $n \in \mathbb{N}$, since $\varphi > 0$ on E. But since

$$\bigcup_{n \in \mathbb{N}} B_n = \{|f| > 0\} =: B,$$

we have that $\{\tau_B < \zeta\} = \bigcup_{n \in \mathbb{N}} \{\tau_{B_n} < \zeta\}$, consequently

$$P_\mu[\tau_B < \zeta] \leq \sum_{n=1}^\infty P_\mu[\tau_{B_n} < \zeta] = 0$$

and hence $\{|f| > 0\}$ is **M**-exceptional. $\qquad\square$

Remark 5.21. Using the m-tightness of **M** we shall prove in Subsection c) below that we actually have an equality in 5.20(ii) and as a consequence we shall obtain a stronger version of 5.15, namely that each $f \in \tilde{C}$ is already $\mathcal{E}$-quasi-continuous (cf. 5.28, 5.29 below).

Proposition 5.22. *Let* $f_n \in \tilde{C}$, $n \in \mathbb{N}$. *If* $f_n(z) \downarrow 0$ *as* $n \to \infty$ *for all* $z \in E \setminus N$ *for some* **M**-*exceptional set* $N \subset E$, *then* $\|f_n\| \downarrow 0$ *as* $n \to \infty$.

Proof. Let $\omega \in \Omega$ such that $t \mapsto f_n(X_t(\omega))$ is right continuous on $[0, \infty[$ and $t \mapsto f_n(X_{t-}(\omega))$ is left continuous on $]0, \zeta(\omega)[$ for all $n \in \mathbb{N}$ and such that $\tau_N(\omega) = \zeta(\omega)$. For $k \in \mathbb{N}$ and $T_k := \zeta - \frac{1}{k}$ we have that

$$\sup_{0 \leq t \leq T_k(\omega)} [(\int_t^\infty e^{-s} \varphi(X_s) \, ds) \, (|f_n(X_t)| \vee |f_n(X_{t-})|)] \, (\omega) \downarrow 0 \text{ as } n \to \infty \ ,$$

since $t \mapsto |f_n(X_t)| \vee |f_n(X_{t-})|(\omega)$ is upper semicontinuous on $[0, T_k]$. Consequently,

$$\limsup_{n \to \infty} \|f_n\| \leq E_\mu[e^{-T_k} - e^{-\zeta}] \sup_{z \in E} |f_1(z)| \sup_{z \in E} \varphi(z) \, ,$$

and letting $k \to \infty$ we obtain the assertion. $\square$

Define $\overline{C_b(E)}$ to be the closure of $C_b(E)$ in $\tilde{C}$ w.r.t. $\| \ \|$.

Lemma 5.23. $\overline{C_b(E)} = \tilde{C}$.

Proof. Since $\mathcal{B}(E) = \sigma(C_b(E)) = \sigma(\tilde{C})$ and because of 5.22 we can apply Daniell's theorem (cf. [DM 78, III.35]) and obtain that every positive bounded linear functional on the (semi-)normed vector lattice $(\tilde{C}, \| \ \|)$ is represented by some finite measure on $(E, \mathcal{B}(E))$. Since $|f| \leq |g|$ implies $\|f\| \leq \|g\|$, we conclude that every bounded linear functional l on $(\tilde{C}, \| \ \|)$ has a decomposition $l = l^+ - l^-$ where l^+, l^- are positive bounded linear functionals. (Set for $f \geq 0, l^+(f) := \sup_{0 \leq f' \leq f} l(f')$ and extend by linearity, then set $l^-(f) := l^+(f) - l(f)$.) Consequently, every bounded linear functional on $(\tilde{C}, \| \ \|)$ admits an integral representation in terms of some finite signed measure on $\mathcal{B}(E)$. Since any such measure is uniquely determined on $C_b(E)$ the assertion now follows by (a version of) the Hahn-Banach theorem (cf. e.g. [Ch 69b]). $\square$

We can now prove the following result which by virtue of 5.18 is a refinement of 5.15.

Proposition 5.24. *Let $f \in \tilde{C}$, then there exists an $\mathcal{E}$-quasi-continuous function $\tilde{f}$ on E such that $f = \tilde{f}$ outside some $\mathbf{M}$-exceptional set.*

Proof. By 5.23 there exist $f_n \in C_b(E)$, $n \in \mathbb{N}$, such that $\lim_{n \to \infty} \|f - f_n\| = 0$. We may assume that $\|f_{n+1} - f_n\| < 2^{-2n}$ for all $n \in \mathbb{N}$ and that $(f_n)_{n \in \mathbb{N}}$ is uniformly bounded (since f is bounded). Let

$$A_n := \{|f_{n+1} - f_n| > 2^{-n}\} \, , \quad n \in \mathbb{N} \, ,$$

and

$$B_N := \bigcup_{n \geq N} A_n \, , \quad N \in \mathbb{N} \, .$$

By 5.20(i) it follows that for all $N \in \mathbb{N}$

$$\begin{aligned}
\mathrm{Cap}_{h,g}(B_N) &\leq \sum_{n \geq N} \mathrm{Cap}_{h,g}(A_n) = \sum_{n \geq N} \|I_{A_n}\| \\
&\leq \sum_{n \geq N} \|2^n |f_{n+1} - f_n|\| \leq 2^{-N+1} \, .
\end{aligned}$$

Clearly, $(f_n)_{n \in \mathbb{N}}$ converges uniformly on the closed set $E \setminus B_N$ for all $N \in \mathbb{N}$. Set

$$\tilde{f}(z) := \begin{cases} \lim_{n \to \infty} f_n(z) & \text{for} \quad z \in \bigcup_{N \in \mathbb{N}}(E \setminus B_N) \\ f(z) & \text{else} \, , \end{cases}$$

then $\tilde{f}$ is $\mathcal{E}$-quasi-continuous and $\lim_{n\to\infty} \|f_n - \tilde{f}\| = 0$. The latter follows by 5.20(ii) since $\|f_n - \tilde{f}\|$ is up to a constant dominated by

$$E_\mu[\sup_{0\le t}(1_{B_N^c}|\tilde{f} - f_n|)(X_t) \vee (1_{B_N^c}|\tilde{f} - f_n|)(X_{t-})] + \|1_{B_N}\| \text{ for all } n, N \in \mathbb{N}.$$

Hence $\|f - \tilde{f}\| \le \lim_{n\to\infty} \|f - f_n\| + \lim_{n\to\infty} \|f_n - \tilde{f}\| = 0$. Thus the assertion follows by 5.20(iii). $\square$

c)   Proof of property 3.1(iii)

In this section we assume that $\mathbf{M} = (\Omega, \mathcal{F}, (X_t)_{t\ge 0}, (P_z)_{z\in E_\Delta})$ is an m-tight m-special standard process associated with $(\mathcal{E}, D(\mathcal{E}))$. Let φ, μ, h, g be as in 5.4. Define for $F \subset E$, F closed,

$$(5.16) \qquad \tilde{h}_F(z) := E_z\left[\int_{\tau_F}^\infty e^{-s}\varphi(X_s)\,ds\right]\ ,\ z \in E\ ,$$

and set

$$(5.17) \qquad R_1^{F^c}\varphi := h - \tilde{h}_F = E.\left[\int_0^{\tau_F} e^{-s}\varphi(X_s)\,ds\right]\ .$$

Note that by 5.16(iii) this is consistent with our notation introduced in Subsection a) (cf.(5.4)).

Proposition 5.25.   *(i) h is $\mathcal{E}$-quasi-continuous.*

(ii) Let $F \subset E$ be a closed set. Then $\tilde{h}_F$ is an $\mathcal{E}$-quasi-continuous m-version of an element in $D(\mathcal{E})$. Assume that F is compact if E is not metrizable, then $F = \{\tilde{h}_F = h\}$.

Before we prove 5.25 we show that it easily implies 3.1(iii):
Since $\mathbf{M}$ is m-tight there exists an increasing sequence $(K_k)_{k\in\mathbb{N}}$ of compact metrizable subsets of E such that

$$(5.18) \qquad P_m\left[\lim_{k\to\infty} \sigma_{K_k^c} < \zeta\right] = 0\ .$$

Since each K_k is totally bounded as a compact metrizable space we can find compact sets F_n, $n \in \mathbb{N}$, separating the points of $Y := \bigcup_{k=1}^\infty K_k$. Hence by 5.25, $\{\tilde{h}_{F_n} = h\}$, $n \in \mathbb{N}$, separate the points of Y and each $h - \tilde{h}_{F_n}$ is an $\mathcal{E}$-quasi-continuous m-version of an element in $D(\mathcal{E})$. Thus 3.1(iii) is proved.

To prove 5.25 we need some preparations which are also of independent interest. We fix a closed set $F \subset E$ and for $k \in \mathbb{N}$ we fix a metric ρ_k compatible with the trace topology on K_k induced by E. Set for $n, k \in \mathbb{N}$

$$B_{n,k} := \{z \in K_k \mid \rho_k(z, \bar{F}) < \frac{1}{n}\} \cup K_k^c\ .$$

Obviously, each $B_{n,k}$ is an open subset of E containing F; in particular, $\tau_{B_{n,k}} = \sigma_{B_{n,k}} \wedge \zeta$. Furthermore, for fixed k, $B_{n,k}$ is decreasing in n, hence $\left(\tau_{B_{n,k}}\right)_{n\in\mathbb{N}}$ is increasing.

Lemma 5.26. *Let $k \in \mathbb{N}$, then*

$$\lim_{n \to \infty} \tau_{B_{n,k}} = \tau_F \wedge \tau_{K_k^c} \ .$$

In particular, $\tau_F \wedge \tau_{K_k^c}$ is an $(\mathcal{F}_t)$-stopping time.

Proof. Since $K_k^c \cup F \subset B_{n,k}$, obviously

$$\tau_k := \lim_{n \to \infty} \tau_{B_{n,k}} \leq \tau_F \wedge \tau_{K_k^c} \ .$$

Suppose $\tau_k < \tau_{K_k^c}$, then for all $n \in \mathbb{N}$ there exists $t_n \in \left[\tau_{B_{n,k}}, (\tau_k + \frac{1}{n}) \wedge \tau_{K_k^c} \right[$ such that

$$X_{t_n} \in B_{n,k} \cap K_k.$$

Then $t_n \xrightarrow[n \to \infty]{} \tau_k$, hence $\overline{X_0^{\tau_k}} \cap F \cap K_k \neq \emptyset$. Consequently, $\tau_k \geq \tau_F$. $\square$

Lemma 5.27. *Let $k \in \mathbb{N}$ and set*

$$h_{n,k} := E. \left[\int_{\tau_{B_{n,k}}}^{\infty} e^{-s} \varphi(X_s) ds \right] \ , \quad n \in \mathbb{N},$$

and

$$h_k := E. \left[\int_{\tau_F \wedge \tau_{K_k^c}}^{\infty} e^{-s} \varphi(X_s) ds \right] \ .$$

Then:

(i) $\tilde{h}_F(z) \leq h_k(z) \leq h_{n,k}(z)$ for all $z \in E$,

(ii) $h_{n,k}(z) \downarrow h_k(z)$ as $n \to \infty$ for all $k \in \mathbb{N}$, $z \in E$,

(iii) $h_k(z) - \tilde{h}_F(z) \leq E_z \left[\int_{\tau_{K_k^c}}^{\infty} e^{-s} \varphi(X_s) \, ds \right]$ for all $z \in E$ and the latter decreases to zero for m-a.e. $z \in E$.

Proof. Since $F \subset B_{n,k}$, (i) is obvious and (ii) follows by 5.26. We have for $z \in E$

$$h_k(z) - \tilde{h}_F(z) = E_z \left[\int_{\tau_F \wedge \tau_{K_k^c}}^{\tau_F} e^{-s} \varphi(X_s) \, ds \right]$$

$$\leq E_z \left[\int_{\tau_{K_k^c}}^{\infty} e^{-s} \varphi(X_s) \, ds \right] \ .$$

Hence (iii) follows by (5.18). $\square$

Now we extend 5.20(i) to all Borel subsets of E.

Theorem 5.28. *Let $B \in \mathcal{B}(E)$. Then*

$$\mathrm{Cap}_{h,g}(B) = \|1_B\| \ \left(= E_\mu \left[\int_{\tau_B}^{\infty} e^{-s} \varphi(X_s) \, ds \right] \right) \ .$$

Proof. By 5.20(ii)

$$\mathrm{Cap}_{h,g}(B) \geq \|1_B\| \; .$$

To show the dual inequality we first consider a closed subset F of E. By 5.19 and the preceding lemma we then have that

$$\|1_F\| = (\tilde{h}_F, \varphi) = \lim_{k\to\infty}(h_k,\varphi) = \lim_{k\to\infty}\lim_{n\to\infty}(h_{n,k},\varphi) \; .$$

But by 5.4 and 5.16(iii) for all $n, k \in \mathbb{N}$

$$(h_{n,k},\varphi) = \mathrm{Cap}_{h,g}(B_{n,k}) \geq \mathrm{Cap}_{h,g}(F) \; ,$$

hence

$$(5.19) \qquad\qquad \mathrm{Cap}_{h,g}(F) \leq \|1_F\| \; .$$

Let now $B \in \mathcal{B}(E)$. For $k \in \mathbb{N}$, $\mathrm{Cap}_{h,g}$ restricted to K_k is a finite Choquet capacity (cf. III.2.8, III.2.9), hence there exist compact sets $F_n \subset B \cap K_k$, $n \in \mathbb{N}$, such that

$$\mathrm{Cap}_{h,g}(B \cap K_k) = \lim_{n\to\infty}\mathrm{Cap}_{h,g}(F_n) \; .$$

Hence by (5.19)

$$
\begin{aligned}
\mathrm{Cap}_{h,g}(B) \;&\leq\; \mathrm{Cap}_{h,g}(B \cap K_k) + \mathrm{Cap}_{h,g}(K_k^c)\\
&\leq\; \lim_{n\to\infty}\|1_{F_n}\| + \mathrm{Cap}_{h,g}(K_k^c)\\
&\leq\; \|1_B\| + \mathrm{Cap}_{h,g}(K_k^c) \; .
\end{aligned}
$$

But $\lim_{k\to\infty}\mathrm{Cap}_{h,g}(K_k^c) = 0$ by 5.4 and (5.18), thus the proof is complete. $\qquad\square$

As a consequence we get the following important result, connecting the potential theory of $\mathbf{M}$ and $(\mathcal{E}, D(\mathcal{E}))$.

Theorem 5.29.   *(i) A subset N of E is $\mathbf{M}$-exceptional if and only if it is $\mathcal{E}$-exceptional.*

  (ii) If $f \in \tilde{C}$ (where $\tilde{C}$ is as in 5.14) then f is $\mathcal{E}$-quasi-continuous. In particular, $R_\alpha f$ is $\mathcal{E}$-quasi-continuous for all $f \in \mathcal{B}_b(E)$, $\alpha > 0$, and $\mathbf{M}$ is properly associated with $(\mathcal{E}, D(\mathcal{E}))$.

Proof. (i): By definition we may assume that $N \in \mathcal{B}(E)$. Then the assertion is an easy consequence of 5.28.

(ii): The first part follows by (i) and 5.24, the second by 5.14 and 2.8. $\qquad\square$

We are now prepared to prove Proposition 5.25.

Proof of 5.25. (i): By 5.29(ii), $h = R_1\varphi$ is $\mathcal{E}$-quasi-continuous.

(ii): Let U be an open subset of E and define $V_n\varphi : E \to \mathbb{R}_+$ as in (5.6). Then from (5.7) and 5.29(ii) it follows that $V_n\varphi$ is an $\mathcal{E}$-quasi-continuous m-version of an element in $D(\mathcal{E})$. In the proof of 5.6 we showed that $V_n\varphi \xrightarrow[n\to\infty]{} R_1^{U^c}\varphi$ weakly in the Hilbert space $(D(\mathcal{E}), \tilde{\mathcal{E}}_1)$ and pointwise. Hence by I.2.12 and III.3.5, $R_1^{U^c}\varphi$

is $\mathcal{E}$-quasi-continuous. Consequently, by (5.11) and 5.7, $\tilde{h}_U := h - R_1^{U^c}\varphi$ is also an $\mathcal{E}$-quasi-continuous m-version of $h_U \in D(\mathcal{E})$; in particular, $\tilde{h}_{K_k^c}$ is $\mathcal{E}$-quasi-continuous for all $k \in \mathbb{N}$. Since

$$(5.20) \qquad \tilde{h}_{K_k^c}(z) = E_z\left[\int_{\sigma_{K_k^c}}^{\infty} e^{-s}\varphi(X_s)ds\right] \quad \text{for all } z \in E ,$$

$\tilde{h}_{K_k^c}(z) \downarrow 0$ as $k \to \infty$ for m-a.e. $z \in E$. But by Subsection a), $(K_k)_{k\in\mathbb{N}}$ is an $\mathcal{E}$-nest, hence $h_{K_k^c} \to 0$ in $D(\mathcal{E})$ as $k \to \infty$ by III.2.12. Then by III.3.5 a subsequence of $\left(\tilde{h}_{K_k^c}\right)_{k\in\mathbb{N}}$ converges to 0 $\mathcal{E}$-quasi-uniformly. But since $\left(\tilde{h}_{K_k^c}\right)_{k\in\mathbb{N}}$ is decreasing this implies that $\tilde{h}_{K_k^c} \underset{k\to\infty}{\longrightarrow} 0$ $\mathcal{E}$-quasi-uniformly.

Let now F be a closed subset of E. By 5.27(iii), and 5.16(iii) we now see that

$$(5.21) \qquad h_k \downarrow \tilde{h}_F \quad \mathcal{E}\text{-quasi-uniformly as } k \to \infty .$$

But we also know by 5.27(ii) that $h_{n,k} \downarrow h_k$ as $n \to \infty$ on E. Since $h_{n,k} = \tilde{h}_{B_{n,k}}$ is an $\mathcal{E}$-quasi-continuous m-version of $h_{B_{n,k}} \in D(\mathcal{E})$, by I.2.12 and III.3.5 we see that h_k is an $\mathcal{E}$-quasi-continuous m-version of an element in $D(\mathcal{E})$, and hence that $\tilde{h}_F$ is an $\mathcal{E}$-quasi-continuous m-version of an element in $D(\mathcal{E})$. Since $\tau_F = 0$ $P_z - a.s.$ for every $z \in F$ we have that $\tilde{h}_F(z) = h(z)$ for all $z \in F$, hence $F \subset \{h = \tilde{h}_F\}$. Suppose that $z \in E\backslash F$. Then by assumption on F there is a neighbourhood U of z such that its closure $\overline{U}$ is contained in $E\backslash F$. Let $\omega \in \Omega$ with $X_0(\omega) = z$. By the right continuity of the sample paths there is $s(\omega) > 0$ such that $X_t(\omega) \in U$ for all $t \in [0, s(\omega)]$ and hence $\overline{X_0^{s(\omega)}} \subset \overline{U}$. Hence $\tau_F(\omega) \geq s(\omega) > 0$. Since $X_0 = z$ $P_z - a.s.$, we thus obtain that $\tau_F > 0$ $P_z - a.s.$ Consequently,

$$h(z) - \tilde{h}_F(z) = E_z\left[\int_0^{\tau_F} e^{-s}\varphi(X_s)\,ds\right] > 0$$

since $\varphi(y) > 0$ for all $y \in E$. $\qquad\qquad\square$

We conclude this section with a result related to 5.25 which will be very useful later.

Proposition 5.30. *Let* $\mathbf{M} = (\Omega, \mathcal{F}, (X_t)_{t\geq0}, (P_z)_{z\in E_\Delta})$ *be a right process with life time* ζ *on* E *such that* $\mathbf{M}$ *is properly associated with a Dirichlet form* $(\mathcal{E}, \mathcal{D}(\mathcal{E}))$ *on* $L^2(E; m)$.

(i) Let $(F_k)_{k\in\mathbb{N}}$ *be an* $\mathcal{E}$*-nest, then*

$$P_z[\lim_{k\to\infty} \sigma_{E\backslash F_k} < \zeta] = 0 \quad \text{for } \mathcal{E}\text{-q.e. } z \in E$$

In particular, if $N \in \mathcal{B}(E)$ *is* $\mathcal{E}$*-exceptional, then*

$$P_z[\sigma_N < \infty] = P_z[\sigma_N < \zeta] = 0 \quad \text{for } \mathcal{E}\text{-q.e. } z \in E .$$

(ii) Assume that **M** *is a special standard process. Let* $u : E \to \mathbb{R}$ *be* $\mathcal{E}$*-quasi-continuous, then for* $\mathcal{E}$*-q.e.* $z \in E$ *one has that* $u(X_{t-}) = u(X_t)_-$ *for all* $t \in]0, \zeta[$ P_z*-a.s., and*

$$P_z[t \mapsto u(X_t) \text{ is right continuous on } [0, \infty[\text{ and}$$
$$t \mapsto u(X_{t-}) \text{ is left continuous on }]0, \zeta[\,] = 1 \; .$$

Proof. (ii) is an immediate consequence of (i), so we only prove (i). Let φ, μ, h, g be as in 5.4, and set

$$h_k(z) := E_z \left[\int_{\sigma_{E \setminus F_k}}^{\zeta} e^{-s} \varphi(X_s) \, ds \right] \, , \quad k \in \mathbb{N} \; .$$

By 5.5, 5.4 and III.2.11(i) we conclude that $h_k \downarrow 0$ m-a.e. as $k \to \infty$. But **M** is properly associated with $(\mathcal{E}, \mathcal{D}(\mathcal{E}))$, hence if $(R_\alpha)_{\alpha > 0}$ is the resolvent of **M**, it follows from 2.8 that $R_\alpha f$ is $\mathcal{E}$-quasi-continuous for all $\alpha > 0$ and $f \in \mathcal{B}_b(E) \cap L^2(E; m)$. Now exactly the same arguments as in the proof of 5.25 imply that $h_k \downarrow 0$ $\mathcal{E}$-q.e. and hence (i) follows. $\qquad \Box$

6   One-to-one correspondences

a)   Discussion of the correspondence and its consequences

Assume as before that E is a Hausdorff topological space such that $\mathcal{B}(E) = \sigma(C(E))$ and m is a σ-finite positive measure on $(E, \mathcal{B}(E))$.

Definition 6.1. Let $\mathbf{M} = (\Omega, \mathcal{F}, (X_t)_{t \geq 0}, (P_z)_{z \in E_\Delta})$ be a right process with state space E and life time ζ. A set $S \in \mathcal{B}(E)$ is said to be **M**-*invariant* if there exists $\Omega_{E \setminus S} \in \mathcal{F}$ such that

$$(6.1) \qquad \Omega_{E \setminus S} \supset \{\overline{X_0^t} \cap (E \setminus S) \neq \emptyset \text{ for some } 0 \leq t < \zeta\}$$

and $P_z[\Omega_{E \setminus S}] = 0$ for all $z \in S$. Here $\overline{X_0^t}$ is defined as in (5.14).

Remark 6.2. (i) Note that if S is **M**-invariant, we can define a right process $\mathbf{M}|_S := (\Omega', \mathcal{F}', (X_t|_{\Omega'})_{t \geq 0}, (P_z|_{\mathcal{F}'})_{z \in S_\Delta})$ with state space S where $\Omega' := \Omega \setminus \Omega_{E \setminus S}$ and $\mathcal{F}' = \mathcal{F} \cap \Omega'$. $\mathbf{M}|_S$ is called the *restriction* of **M** to S.
(ii) Let **M** be an m-special standard process associated with a Dirichlet form on $L^2(E; m)$. It follows by 5.28 (with h, g as in 5.4) that if S is **M**-invariant with $m(E \setminus S) = 0$, then $E \setminus S$ is $\mathcal{E}$-exceptional. A converse to this is contained in 6.5 below.

Definition 6.3. Two right processes **M** and $\mathbf{M}'$ with state space E and transition semigroups $(p_t)_{t > 0}$, $(p'_t)_{t > 0}$, are called m-*equivalent*, denoted $\mathbf{M} \sim \mathbf{M}'$, if there exists $S \in \mathcal{B}(E)$ such that

(i) $m(E\backslash S) = 0$.

(ii) S is both **M**-invariant and **M'**-invariant.

(iii) $p_t f(z) = p'_t f(z)$ for all $f \in \mathcal{B}_b(E)$, $t > 0$, and $z \in S$.

Two pairs of right processes $(\mathbf{M}, \hat{\mathbf{M}})$ and $(\mathbf{M'}, \hat{\mathbf{M}'})$ are called *m-equivalent* if $\mathbf{M} \sim \mathbf{M'}$ and $\hat{\mathbf{M}} \sim \hat{\mathbf{M}'}$.

The following is the main result of this section.

Theorem 6.4. *Let* $(\mathbf{M}, \hat{\mathbf{M}})$ *and* $(\mathbf{M'}, \hat{\mathbf{M}'})$ *be two pairs of right processes on* E. *Suppose that* $(\mathbf{M}, \hat{\mathbf{M}})$ *and* $(\mathbf{M'}, \hat{\mathbf{M}'})$ *are properly associated with the same quasi-regular Dirichlet form* $(\mathcal{E}, D(\mathcal{E}))$ *on* $L^2(E; m)$ *then* $(\mathbf{M}, \hat{\mathbf{M}})$ *and* $(\mathbf{M'}, \hat{\mathbf{M}'})$ *are m-equivalent.*

Corollary 6.5. *Let* **M** *be a right process associated with a quasi-regular Dirichlet form* $(\mathcal{E}, D(\mathcal{E}))$ *on* $L^2(E; m)$ *and let* $N \subset E$ *be an* $\mathcal{E}$-*exceptional set. Then there exists an* $\mathcal{E}$-*exceptional set* $N_0 \supset N$ *such that* $E\backslash N_0$ *is* **M**-*invariant.*

We postpone the proof of 6.4, 6.5 to Subsection b) below and first discuss its consequences. We say that a right process **M** on E with resolvent $(R_\alpha)_{\alpha>0}$ is *m-sectorial* if II.(5.3) is fulfilled. Let **M** be an m-sectorial right process with semigroup $(p_t)_{t>0}$ and resolvent $(R_\alpha)_{\alpha>0}$ such that m is p_t-supermedian for all $t > 0$. Then m is also αR_α-supermedian for all $\alpha > 0$. Moreover, II.(5.2) is fulfilled by virtue of 2.3. Hence it follows from the discussion of Chap.II, Subsection 5a) and 2.7 that there exists a Dirichlet form $(\mathcal{E}, D(\mathcal{E}))$ such that **M** is associated with it in the sense of 2.5. In particular, if $(\mathbf{M}, \hat{\mathbf{M}})$ is a pair of right processes on E *in duality w.r.t.* m (i.e., their transition semigroups are in duality w.r.t. m) such that **M** or equivalently $\hat{\mathbf{M}}$ is m-sectorial, then $(\mathbf{M}, \hat{\mathbf{M}})$ is associated with a Dirichlet form $(\mathcal{E}, D(\mathcal{E}))$ on $L^2(E; m)$. Obviously, a pair $(\mathbf{M'}, \hat{\mathbf{M}'})$ m-equivalent with $(\mathbf{M}, \hat{\mathbf{M}})$ is also associated with $(\mathcal{E}, D(\mathcal{E}))$. Thus we have constructed a map $\mathcal{A}$ from the set $\mathfrak{M}_m$ of all m-equivalence classes of pairs $(\mathbf{M}, \hat{\mathbf{M}})$ of m-sectorial right processes on E which are in duality w.r.t. m into the set $\mathcal{D}$ of all Dirichlet forms on $L^2(E; m)$. We know by 6.4 that $\mathcal{A}$ restricted to the set of those elements in $\mathfrak{M}_m$, which have a representative which is properly associated with its corresponding Dirichlet form, is one-to-one. By 5.1 $\mathcal{A}$ is hence, in particular, one-to-one on the set $\mathfrak{S}_m$ of all elements in $\mathfrak{M}_m$ which have a representative $(\mathbf{M}, \hat{\mathbf{M}})$ for which either **M** or $\hat{\mathbf{M}}$ is m-tight and m-special standard. If $\mathcal{D}_{q.r.}$ $(\subset \mathcal{D})$ denotes the set of all quasi-regular Dirichlet forms we know by 3.5 that $\mathcal{A}(\mathfrak{S}_m) = \mathcal{D}_{q.r.}$, i.e., we have the following

Theorem 6.6. *The map* $\mathcal{A}$ *specifies a one-to-one correspondence between* $\mathfrak{S}_m$ *and* $\mathcal{D}_{q.r.}$ *and each element in* $\mathfrak{S}_m$ *contains a representative which is (m-tight and) special standard.*

The family $\mathfrak{S}_m$ above is a little complicated, due to the generality of the state space E. If E is a metrizable Lusin space then we have a much more elegant correspondance (cf. 6.8 below), which will be a consequence of the following result.

Theorem 6.7. *Let E be a metrizable Lusin space. Then a Dirichlet form $(\mathcal{E}, \mathcal{D}(\mathcal{E}))$ on $L^2(E; m)$ is quasi-regular if and only if there exists a pair $(\mathbf{M}, \hat{\mathbf{M}})$ of right processes associated with $(\mathcal{E}, \mathcal{D}(\mathcal{E}))$. In this case $(\mathbf{M}, \hat{\mathbf{M}})$ is always properly associated with $(\mathcal{E}, \mathcal{D}(\mathcal{E}))$.*

For the proof of 6.7 see Subsection b) below. 6.6, 6.7 and 1.15 immediately imply that $\mathfrak{M}_m = \mathfrak{S}_m$ and we obtain

Theorem 6.8. *Let E be a metrizable Lusin space. Then the map $\mathcal{A}$ specifies a one-to-one correspondence between $\mathfrak{M}_m$ and $\mathcal{D}_{q.r.}$.*

b)   Proofs of 6.4, 6.5 and 6.7

Proof of 6.4. Let J_0 be as in Sect. 3, Step 2, and let $(F_k)_{k \in \mathbb{N}}$ be as in 3.10. We set

$$(6.2) \qquad N_1 := \bigcap_{k \geq 1} E \backslash F_k \ .$$

Let (p_t), $(\hat{p}_t)$, (p_t'), $(\hat{p}_t')$ be the transition semigroups of $\mathbf{M}$, $\hat{\mathbf{M}}$, $\mathbf{M}'$, $\hat{\mathbf{M}}'$ respectively. By 3.3(iii) and 5.30 we can find a Borel $\mathcal{E}$-exceptional set $N_2 \supset N_1$ such that for all $z \in E \backslash N_2$,

$$(6.3) \qquad \begin{aligned} p_t f(z) &= p_t' f(z) && \text{for all } f \in J_0, \ t \in \mathbb{Q}_+^* \\ \hat{p}_t f(z) &= \hat{p}_t' f(z) && \text{for all } f \in J_0, \ t \in \mathbb{Q}_+^* \end{aligned}$$

and

$$(6.4) \qquad \begin{aligned} P_z[\lim_{k \to \infty} \sigma_{E \backslash F_k} < \zeta] = 0, && P_z'[\lim_{k \to \infty} \sigma_{E \backslash F_k}' < \zeta'] = 0, \\ \hat{P}_z[\lim_{k \to \infty} \hat{\sigma}_{E \backslash F_k} < \hat{\zeta}] = 0, && \hat{P}_z'[\lim_{k \to \infty} \hat{\sigma}_{E \backslash F_k}' < \hat{\zeta}'] = 0. \end{aligned}$$

The meaning of the above notations are obvious, e.g., $\hat{\sigma}_{E \backslash F_k}'$ stands for the first hitting times of $E \backslash F_k$ relative to $\hat{\mathbf{M}}'$, etc. Since $J_0 \subset C(\{F_k\})$ by the right continuity of sample paths we get from (6.3) and (6.4)

$$(6.5) \qquad \begin{aligned} p_t f(z) &= p_t' f(z), && \text{for all } z \in E \backslash N_2, \ f \in J_0, \ t \in [0, \infty[\, , \\ \hat{p}_t f(z) &= \hat{p}_t' f(z), && \text{for all } z \in E \backslash N_2, \ f \in J_0, \ t \in [0, \infty[\, . \end{aligned}$$

We define

$$\begin{aligned} \Omega_{N_1} &:= \{ \lim_{k \to \infty} \sigma_{E \backslash F_k} < \zeta \}, & \Omega_{N_1}' &:= \{ \lim_{k \to \infty} \sigma_{E \backslash F_k}' < \zeta' \} \, , \\ \hat{\Omega}_{N_1} &:= \{ \lim_{k \to \infty} \hat{\sigma}_{E \backslash F_k} < \hat{\zeta} \}, & \hat{\Omega}_{N_1}' &:= \{ \lim_{k \to \infty} \hat{\sigma}_{E \backslash F_k}' < \hat{\zeta}' \} \, . \end{aligned}$$

Since each $E \backslash F_k$ is open and $N_1 = \bigcap_{k \geq 1}(E \backslash F_k)$, we can check that

$$(6.6) \qquad \Omega_{N_1} \supset \{\overline{X_0^t} \cap N_1 \neq \emptyset \text{ for some } 0 \leq t < \zeta\} \; .$$

The same assertions hold also for $\Omega'_{N_1}, \hat{\Omega}_{N_1}$ and $\hat{\Omega}'_{N_1}$. Thus we can find an increasing sequence of Borel $\mathcal{E}$-exceptional sets $(N_k)_{k \in \mathbb{N}}$ and corresponding increasing sequences $(\Omega_{N_k})_{k \in \mathbb{N}}, (\Omega'_{N_k})_{k \in \mathbb{N}}, (\hat{\Omega}_{N_k})_{k \in \mathbb{N}}, (\hat{\Omega}'_{N_k})_{k \in \mathbb{N}}$ such that for all $k \in \mathbb{N}$

$$(6.7) \qquad \Omega_{N_k} \supset \{\overline{X_0^t} \cap N_k \neq \emptyset \text{ for some } 0 \leq t < \zeta\}$$

$$(6.8) \qquad P_z[\Omega_{N_k}] = 0 \qquad \text{for all } z \in E \backslash N_{k+1} \; ,$$

and the same assertions hold for $(\Omega'_{N_k})_{k \in \mathbb{N}}, (\hat{\Omega}_{N_k})_{k \in \mathbb{N}}, (\hat{\Omega}'_{N_k})_{k \in \mathbb{N}}$. We now set $B := \bigcup_{k \geq 1} N_k$ and $S := E \backslash B$. Then B is again an $\mathcal{E}$-exceptional set, in particular $m(B) = 0$. If we set $\Omega_B := \bigcup_{k \geq 1} \Omega_{N_k}$, then from (6.7) and (6.8) we see that $S := E \backslash B$ is $\mathbf{M}$-invariant. The same argument holds for $\mathbf{M}', \hat{\mathbf{M}}$, and $\hat{\mathbf{M}}'$. Hence S is simultaneously invariant with respect to $\mathbf{M}, \mathbf{M}', \hat{\mathbf{M}}$, and $\hat{\mathbf{M}}'$. From (6.5) and the properties of J_0, by monotone class arguments we can prove that $p_t f(z) = p'_t f(z)$ and $\hat{p}_t f(z) = \hat{p}'_t f(z)$ for all $f \in \mathcal{B}_b(E), z \in S$ and $t \geq 0$. Thus $(\mathbf{M}, \hat{\mathbf{M}})$ and $(\mathbf{M}', \hat{\mathbf{M}}')$ are m-equivalent. $\qquad \square$

Proof of 6.5. We can always construct $(F_k)_{k \in \mathbb{N}}$ and J_0 in Sect.3, Step2, in such a way that $N \subset \bigcap_{k \geq 1} E \backslash F_k$. Then N_1 defined in (6.2) and hence the $\mathcal{E}$-exceptional set $B = \bigcup_{k \geq 1} N_k$ in the preceding proof contain N and is thus a desired set N_0. $\qquad \square$

For proving 6.7 we need a result of P.J. Fitzsimmons which we now restate.

Proposition 6.9. *Let E be a metrizable Lusin space and $(\mathbf{M}, \hat{\mathbf{M}})$ be a pair of right processes on E satisfying the following conditions:*

(i) $(\mathbf{M}, \hat{\mathbf{M}})$ are in duality w.r.t. m.

(ii) Both $\mathbf{M}$ and $\hat{\mathbf{M}}$ are transient (i.e., there exist strictly positive $f, \hat{f} \in \mathcal{B}_b(E)$ such that for all $z \in E$,

$$E_z \left[\int_0^\infty f(X_t) \, dt \right] < \infty \; , \qquad \hat{E}_z \left[\int_0^\infty \hat{f}(\hat{X}_t) \, dt \right] < \infty).$$

(iii) If L denotes the L^2-generator of $\mathbf{M}$, then $-L$ satisfies the sector condition I.(2.5) (or equivalently, the dual assertion holds for $\hat{\mathbf{M}}$).

Then both $\mathbf{M}$ and $\hat{\mathbf{M}}$ are m-special standard processes.

For the proof of 6.9 we refer to [Fi 89].

Proof of 6.7. The "only if" part follows directly from 3.5. For the "if" part and the last assertion, we assume that there exists a pair $(\mathbf{M}, \hat{\mathbf{M}})$ of right processes associated with $(\mathcal{E}, \mathcal{D}(\mathcal{E}))$, having transition semigroups $(p_t)_{t>0}$ and $(\hat{p}_t)_{t>0}$ respectively. We define $A_t = e^{-t} 1_{\{t<\zeta\}}$ (where ζ is the life time of $\mathbf{M}$). Then $(A_t)_{t \geq 0}$ is a decreasing right-continuous multiplicative functional of $\mathbf{M}$

with $E_A = E$ where $E_A := \{z \in E \mid P_z[A_0 = 1] = 1\}$. Thus by [Sh 88, (61.5)] there is a right process, say $\mathbf{M}'$, with state space E and transition semigroup $(p'_t)_{t>0}$ such that

$$(6.9) \qquad p'_t f(z) = E_z[f(X_t)A_t] = e^{-t}p_t f(z), \qquad z \in E, f \in \mathcal{B}_b(E) \,.$$

Similarly, there exists a right process $\hat{\mathbf{M}}'$ having transition semigroup $(\hat{p}'_t)_{t>0}$ such that

$$(6.10) \qquad \hat{p}'_t f(z) = e^{-t}\hat{p}_t f(z), \qquad z \in E, f \in \mathcal{B}_b(E) \,.$$

From (6.9) and (6.10) we see that $(\mathbf{M}', \hat{\mathbf{M}}')$ is associated with the Dirichlet form $(\mathcal{E}_1, \mathcal{D}(\mathcal{E}))$ (cf. I.(2.2)) We can check that $(\mathbf{M}', \hat{\mathbf{M}}')$ fulfills 6.9(i)-(iii), and hence both $\mathbf{M}'$ and $\hat{\mathbf{M}}'$ are m-special standard processes. Moreover, by 1.15 both $\mathbf{M}'$ and $\hat{\mathbf{M}}'$ are m-tight. Hence it follows from 5.1 that $(\mathcal{E}_1, \mathcal{D}(\mathcal{E}))$ is quasi-regular and $(\mathbf{M}', \hat{\mathbf{M}}')$ is properly associated with $(\mathcal{E}_1, \mathcal{D}(\mathcal{E}))$, which in turn implies that $(\mathcal{E}, \mathcal{D}(\mathcal{E}))$ is quasi-regular and $(\mathbf{M}, \hat{\mathbf{M}})$ is properly associated with $(\mathcal{E}, \mathcal{D}(\mathcal{E}))$.

$\square$

7   Notes/References

Section 1. In the literature there are many versions of the definition of a Markov process. Definition 1.5 is taken from [BG 68]. For the notions of right processes, special standard processes and Hunt processes we refer to [G 75] and [Sh 88], where the state spaces are assumed to be metrizable Radon spaces. The notion of m-special standard processes with state spaces being metrizable Lusinian were introduced in [GS 84]. Theorem 1.15 is taken from [AMR 91] where the result is proved also for E being metrizable co-Souslinian. The main idea is taken from [LR 90], but generalizes the corresponding result in [LR 90] considerably.

Section 2. Proposition 2.3 is new. Note that the goal of 2.3 is to keep this book up to Chap. IV, Sect. 4 reasonably self-contained (cf. Remark 2.2). Otherwise by employing Claim 1 of 5.14 we can show that (1.1) of [GG 83] is true and consequently

$$(7.1) \qquad \{R_1 f \mid f \in L^2(E; m)\} \text{ is dense in } L^2(E; m) \,,$$

which may be used as an alternative to (2.3). A root for the argument of $(ii) \Rightarrow (i)$ of Proposition 2.8 can be found in the proof of [F 80, Theorem 4.3.3]. Here again for the same reason as mentioned in Remark 2.2 we used (2.1) instead of (7.1).

Section 3. A different version of the conditions (i)-(iii) in Definition 3.1 was first formulated for symmetric Dirichlet forms in [AM 91 c,f]. But this book is the first discussing systematically the properties of (non-symmetric) quasi-regular Dirichlet forms and their connection to Markov processes. Theorem 3.5 is new. It together with Theorem 5.1 are fundamental results in this book. They generalize and improve the corresponding results obtained in [AM 91 c,f]. Here we used "Ray resolvent/compactification techniques" to construct the associated processes, which is a variant of the method used by Carillo-Menendez [Ca-Me 78]. (For a recent representation of this method we refer to [O 88]). It may be traced back to the original ideas of M. Fukushima [F 71]. The technique for proving Proposition 3.7 is taken from [AM 91e]. The short proof of Lemma 3.19 we learnt from M. Schweitzer.

Section 4. For the details of the theory of regular Dirichlet forms we recommend the fundamental work of [F 80] and [Si 74]. See also [O 88] for the non-symmetric case. The material of Subsections b), c) is taken from [RS 91 a], respectively from [AM 88], [AM 91 a,d] (see also [M 91]). The existence of the diffusion process for the Dirichlet forms discussed in Subsection b) was first proved in full generality by B. Schmuland in [S 90] extending earlier results in [Ku 82] and [AR 89b] (see also [ARZ 91]).

Section 5. Apart from the fact that the results of this section extend the corresponding results obtained in [AM 91 f] to the non-symmetric case, the arguments presented here are more complicated than those in [AM 91 f] because of the generality of the state spaces. For example, Proposition 5.14 is well-known when E is a metrizable Radon space (see e.g. [G 75] (7.2), (7.3) and (13.8)). But in the literature there seems to be no explicit statement of this result for general state spaces. In particular, some extra difficulties arise in proving 5.25-5.29 because we do not assume that E is metrizable. Lemma 5.6 can be found in [FO 88, (2.7)]. The semi-norm (5.15) is a modification of one introduced by Y. LeJan [Le 83] who defined a semi-norm $\| \ \|$ by $\|f\| := E \left[\sup_{t \geq 0} e^{-t}|f(X_t)| \vee e^{-t}|f(X_{t-})| \right]$ relative to a given process with cadlag paths on $[0, \infty)$. Theorem 5.29 (i) generalizes [F 80, Theorem 4.3.1] and [Le 77, Proposition I.5.4] (see also [Si 78]), and is relatively close to [F 76, Theorem 6(i)] and [Fi 89, (4.17)]. In the latter two papers the authors used "fine capacity" instead of the usual capacity, while the proof of [F 80, Theorem 4.3.1] uses that the underlying Dirichlet form is regular. Theorem 5.29(ii) is essentially due to Y. LeJan [Le 83, Theorem 1].

Section 6. The notion of m-equivalence was introduced in [F 76]. Theorem 6.4 extends [F 76, Theorem 5]. Theorems 6.6 and 6.8 are new. (A result corresponding to 6.8 for the symmetric case was stated in [AM 91 f] without proof.)

Chapter V

Characterization of Particular Processes

This chapter deals with further sample path properties of Markov processes associated with Dirichlet forms. In Section 1 we prove that the continuity of sample paths (up to lifetime) is equivalent with the "local property" of the Dirichlet form extending M. Fukushima's result for the locally compact case. On the way we prove a result on the existence of "localizing" functions in the domain of a Dirichlet form. We give applications to the examples of Chap.IV, Sect. 4 at the end. In Section 2 we prove a characterization of those Dirichlet forms associated with Hunt processes (also in non-locally compact cases). In particular, we show that every quasi-regular Dirichlet form which contains the function $f \equiv 1$ in its domain is associated with a Hunt process provided the cemetery is an isolated point. In all of this chapter we assume E to be a Hausdorff topological space such that $\mathcal{B}(E) = \sigma(C(E))$. We fix a σ-finite positive measure m on $(E, \mathcal{B}(E))$ and adjoin a point Δ to E as an isolated point of $E_\Delta := E \cup \{\Delta\}$. If E is locally compact, E_Δ is also taken to be the one point compactification of E (cf. Chap. IV, Sect. 1). We extend m to $(E_\Delta, \mathcal{B}(E_\Delta))$ by $m(\{\Delta\}) := 0$. Again every function f on E is considered as a function on E_Δ with $f(\Delta) := 0$. For $A \subset E$ we set $A^c := E \backslash A$. Let $(\mathcal{E}, D(\mathcal{E}))$ be a fixed Dirichlet form on $L^2(E; m)$ with continuity constant $K \geq 1$ and let $(T_t)_{t>0}$, $(\hat{T}_t)_{t>0}$ and $(G_\alpha)_{\alpha>0}$, $(\hat{G}_\alpha)_{\alpha>0}$ be the associated strongly continuous contraction (co)semigroups and (co)resolvents. Again $(\,,\,)$ denotes the usual inner product in $L^2(E; m)$.

1　Local property and diffusions

In this section we assume $(\mathcal{E}, D(\mathcal{E}))$ to be quasi-regular. Hence we may assume that E is a (topological) Lusin space (cf. IV.3.2(iii)). Note that since E is strongly Lindelöf, the support of a positive measure on $(E, \mathcal{B}(E))$ is defined. For a $\mathcal{B}(E)$-measurable function u on E we set

$$(1.1) \qquad \mathrm{supp}[u] := \mathrm{supp}[|u| \cdot m]$$

and call $\mathrm{supp}[u]$ the *support* of u. Note that $u = u'$ m-a.e. implies $\mathrm{supp}[u] := \mathrm{supp}[u']$. Hence $\mathrm{supp}[u]$ is well-defined by (1.1) for all $u \in L^2(E; m)$.

Definition 1.1. $(\mathcal{E}, \mathcal{D}(\mathcal{E}))$ is said to have the *local property* (or to be *local*) if

$$(1.2) \qquad \mathcal{E}(u, v) = 0 \qquad \text{for all } u, v \in D(\mathcal{E}) \text{ with } \mathrm{supp}[u] \cap \mathrm{supp}[v] = \emptyset$$
$$\text{and } \mathrm{supp}[u], \mathrm{supp}[v] \text{ compact.}$$

The following proposition shows that "$\mathrm{supp}[u], \mathrm{supp}[v]$ compact" in 1.1 can be dropped and gives another property which is equivalent with the local property.

Proposition 1.2. *The following are equivalent:*

(i) $(\mathcal{E}, D(\mathcal{E}))$ *has the local property.*

(ii) $\mathcal{E}(u, v) = 0$ *for all* $u, v \in D(\mathcal{E})$ *with* $\mathrm{supp}[u] \cap \mathrm{supp}[v] = \emptyset$.

(iii) $u'_U = u$ *for all* $u \in D(\mathcal{E})$ *and all* $U \subset E$, *U open, with* $\mathrm{supp}[u] \subset U$ *(where* u'_U *is as specified in III.1.4).*

Proof. $(ii) \Leftrightarrow (i)$: Suppose (ii) holds. Let $u, v \in D(\mathcal{E})$ with $\mathrm{supp}[u] \cap \mathrm{supp}[v] = \emptyset$. Since $u = u^+ - u^-$, $v = v^+ - v^-$ with $u^+, u^-, v^+, v^- \in D(\mathcal{E})$, we may assume that $u, v \geq 0$. Let $w := u_E$. Then w is 1-excessive, $w \geq u$ and if $(F_n)_{n \in \mathbb{N}}$ is an $\mathcal{E}$-nest of compacts, then by III.2.12 $w_{F_n^c} \to 0$ in $D(\mathcal{E})$ as $n \to \infty$. Let

$$(1.3) \qquad u'_n := (w - w_{F_n^c}) \wedge u, \ n \in \mathbb{N} .$$

Then $u'_n \to u$ in $L^2(E; m)$ as $n \to \infty$, $\sup_{n \in \mathbb{N}} \mathcal{E}_1(u'_n, u'_n) < \infty$, and $\mathrm{supp}[u'_n] \subset F_n \cap \mathrm{supp}[u]$, $n \in \mathbb{N}$. Consequently, by I.2.12 the Cesaro means u_n, $n \in \mathbb{N}$, of a subsequence of $(u'_n)_{n \in \mathbb{N}}$ converge to u in $D(\mathcal{E})$ and clearly, $\mathrm{supp}[u_n] \subset F_n \cap \mathrm{supp}[u]$, $n \in \mathbb{N}$. Similarly, we can find $v_n \in D(\mathcal{E})$, $n \in \mathbb{N}$, with compact support contained in $\mathrm{supp}[v]$, such that $v_n \to v$ in $D(\mathcal{E})$ as $n \to \infty$. Thus by (ii)

$$\mathcal{E}(u, v) = \lim_{n \to \infty} \mathcal{E}(u_n, v_n) = 0 .$$

The converse implication is trivial.

$(ii) \Leftrightarrow (iii)$:By IV.3.5 there exists a pair $(\mathbf{M}, \hat{\mathbf{M}})$ of m-tight special standard processes associated with $(\mathcal{E}, D(\mathcal{E}))$. Let $(R_\alpha)_{\alpha > 0}$, $(\hat{R}_\alpha)_{\alpha > 0}$ denote the corresponding resolvents. Let $u \in D(\mathcal{E})$. Then it follows by (the co-version of) IV.5.6 and III.1.4 that

$$(1.4) \qquad (u - u'_U, f) = \mathcal{E}_1(u - u'_U, \hat{R}_1^{U^c} f) = \mathcal{E}_1(u, \hat{R}_1^{U^c} f)$$
$$\text{for all } f \in L^2(E; m), U \subset E, U \text{ open} ,$$

(since $\hat{R}_1^{U^c} f \in D(\mathcal{E})_{U^c}$). By (1.4) we now see that

(1.5) $\mathcal{E}(u,v) = 0$ for all $v \in D(\mathcal{E})$ with $\mathrm{supp}[u] \cap \mathrm{supp}[v] = \emptyset$

clearly implies

(1.6) $u'_U = u$ for all open $U \subset E$ with $\mathrm{supp}[u] \subset U$.

But, indeed (1.5) $\Leftrightarrow$ (1.6), since if u, v are as in (1.5), then by (1.6) with $U := E \setminus \mathrm{supp}[v]$ we have $u = u'_U$, hence by III.1.4

$$\mathcal{E}(u,v) = \mathcal{E}(u'_U, v) = 0$$

because $v \in D(\mathcal{E})_{U^c}$. Since $u \in D(\mathcal{E})$ was arbitrary, the equivalence of (1.5) and (1.6) now implies that $(ii) \Leftrightarrow (iii)$. $\square$

The following lemma is a bit technical, but is useful in applications (cf. 1.4 below).

Lemma 1.3. *Let $\mathcal{D} \subset L^\infty(E; m)$ be a dense subset of $\mathcal{D}(\mathcal{E})$ and suppose that for each compact subset K of E there exists a family $\mathcal{U}_K$ of open subsets of E containing K such that*

(i) for every closed set $F \subset E$ with $K \cap F = \emptyset$ there is $U \in \mathcal{U}_K$ such that $F \subset (\overline{U})^c$,

(ii) for every $U \in \mathcal{U}_K$ there is $\chi_{K,U} \in D(\mathcal{E})$ with $1_K \leq \chi_{K,U} \leq 1_U$ on E.

Let $\mathcal{D}_0$ denote the set of all functions $\chi_{K,U}$, $K \subset E$, K compact, $U \in \mathcal{U}_K$, specified in (ii). If for all $u, v \in \mathcal{D}$ and all $\chi_{K,U}, \chi_{K',U'} \in \mathcal{D}_0$

$$\mathcal{E}(\chi_{K,U} \cdot u, \chi_{K',U'} \cdot v) = 0 \ \ \text{whenever} \ \ \mathrm{supp}[\chi_{K,U} \cdot u] \cap \mathrm{supp}[\chi_{K',U'} \cdot v] = \emptyset \ ,$$

then $(\mathcal{E}, D(\mathcal{E}))$ has the local property.

Proof. Let $u, v \in D(\mathcal{E}) \cap L^\infty(E; m)$ with $\mathrm{supp}[u] \cap \mathrm{supp}[v] = \emptyset$ and $\mathrm{supp}[u]$, $\mathrm{supp}[v]$ compact. Let $u_n, v_n \in \mathcal{D}$, $n \in \mathbb{N}$, such that $u_n \xrightarrow[n \to \infty]{} u$ and $v_n \xrightarrow[n \to \infty]{} v$ in $D(\mathcal{E})$ (w.r.t. $\tilde{\mathcal{E}}^{1/2}$). By I.4.17 we may assume that $(\|u_n\|_\infty \vee \|v_n\|_\infty)_{n \in \mathbb{N}}$ is bounded. Let $K := \mathrm{supp}[u]$, $K' := \mathrm{supp}[v]$ and $U \in \mathcal{U}_K$, $U' \in \mathcal{U}_{K'}$ such that $K' \subset (\overline{U})^c$, $\overline{U} \subset (\overline{U'})^c$. If we can show that

(1.7) $$\frac{1}{N} \sum_{k=1}^{N} u_{n_k} \cdot \chi_{K,U} \xrightarrow[N \to \infty]{} u \ \text{and} \ \frac{1}{N} \sum_{k=1}^{N} v_{n_k} \cdot \chi_{K',U'} \xrightarrow[N \to \infty]{} v \ \text{in} \ D(\mathcal{E})$$

for some subsequence $(n_k)_{k \in \mathbb{N}}$,

then

$$\mathcal{E}(u,v) = \lim_{N \to \infty} \frac{1}{N^2} \sum_{i,k=1}^{N} \mathcal{E}(u_{n_i} \cdot \chi_{K,U}, v_{n_k} \cdot \chi_{K',U'}) = 0$$

by assumption, since $\mathrm{supp}[u_i \cdot \chi_{K,U}] \cap \mathrm{supp}[v_k \cdot \chi_{K',U'}] = \emptyset$ for all $i, k \in \mathbb{N}$. But clearly $u_n \cdot \chi_{K,U} \xrightarrow[n \to \infty]{} u \cdot \chi_{K,U} = u$ in $L^2(E; m)$ and by I.4.15

$$\sup_{n\in\mathbb{N}} \mathcal{E}(u_n \cdot \chi_{K,U}, u_n \cdot \chi_{K,U})^{1/2}$$

$$\leq \sup_{n\in\mathbb{N}} \|u_n\|_\infty \, \mathcal{E}(\chi_{K,U}, \chi_{K,U})^{1/2} + \|\chi_{K,U}\|_\infty \sup_{n\in\mathbb{N}} \mathcal{E}(u_n, u_n)^{1/2} < \infty \; .$$

Hence by I.2.12, (1.7) follows for $(u_n)_{n\in\mathbb{N}}$ and in the same way for $(v_n)_{n\in\mathbb{N}}$. $\quad\square$

Exercise 1.4. Suppose E is a locally compact separable metric space and m a positive Radon measure on $(E, \mathcal{B}(E))$. Prove that if $(\mathcal{E}, D(\mathcal{E}))$ is regular and

$$\mathcal{E}(u,v) = 0 \text{ whenever } u,v \in C_0(E) \cap D(\mathcal{E}) \text{ with } \operatorname{supp}[u] \cap \operatorname{supp}[v] = \emptyset \; ,$$

then $(\mathcal{E}, D(\mathcal{E}))$ has the local property. (Hint: Show first that if $u \in C_0(E)$, then there exist $u_n \in C_0(E) \cap \mathcal{D}(\mathcal{E})$ with $\operatorname{supp}[u_n] \subset \{u \neq 0\}$, $n \in \mathbb{N}$, such that $u_n \underset{n\to\infty}{\longrightarrow} u$ w.r.t. uniform norm. Use also that for every compact set $K \subset E$ and every open set $U \supset K$ there exists $\chi \in C_0(E)$ such that $1_K \leq \chi \leq 1_U$.)

The aim of this section is to prove the following result:

Theorem 1.5. *Let* $\mathbf{M} = (\Omega, \mathcal{F}, (X_t)_{t\geq 0}, (P_z)_{z\in E_\Delta})$ *be an m-tight special standard process with life time ζ associated with $(\mathcal{E}, D(\mathcal{E}))$. Then $(\mathcal{E}, D(\mathcal{E}))$ has the local property if and only if*

$$(1.8) \qquad P_z\,[t \mapsto X_t \text{ is continuous on } [0, \zeta[\,] = 1 \text{ for } \mathcal{E}\text{-q.e. } z \in E \; .$$

In order to prove 1.5 we need some preparations which are also of their own interest. Note that the existence of $\mathbf{M}$ is equivalent with the quasi-regularity of $(\mathcal{E}, D(\mathcal{E}))$ by IV.3.5 and IV.5.1. Below we denote by $\tilde{D}(\mathcal{E})$ the set of all $\mathcal{B}(E)$-measurable funcions which are $\mathcal{E}$-quasi-continuous m-versions of elements in $D(\mathcal{E})$. We start with the previously announced generalization of IV.5.7.

Proposition 1.6. *Let $U \subset E$, U open, and $u \in \tilde{D}(\mathcal{E})$. Then $\tilde{u}_U$ defined by*

$$(1.9) \qquad\qquad \tilde{u}_U(z) := E_z[e^{-\sigma_U} u(X_{\sigma_U})], \; z \in E,$$

is an $\mathcal{E}$-quasi-continuous m-version of u'_U (cf. III.1.4).

Proof. Clearly, if suffices to consider the case $u \geq 0$. If $u = R_1 f$ for some $f \in \mathcal{B}_b^+(E) \cap L^2(E; m)$, then by IV.5.7 and III.1.6(i) we know that $\tilde{u}_U$ is an m-version of u'_U. Its $\mathcal{E}$-quasi-continuity follows by the same arguments as in the first part of the proof of IV.5.25(ii). The resolvent equation now implies the assertion if $u = R_\alpha f$ for some $\alpha > 0$, $f \in \mathcal{B}_b^+(E) \cap L^2(E; m)$. Now let $u \in \tilde{D}(\mathcal{E})$, u bounded. By I.2.13(ii) and III.3.5 we can find an $\mathcal{E}$-nest $(F_k)_{k\in\mathbb{N}}$ and a sequence $(\alpha_j)_{j\in\mathbb{N}}$ such that $\alpha_j \uparrow \infty$ as $j \to \infty$ and

$$\alpha_j R_{\alpha_j} u \underset{j\to\infty}{\longrightarrow} u \text{ both in } \tilde{\mathcal{E}}_1^{1/2}\text{-norm and uniformly on}$$

$$\text{each } F_k, k \in \mathbb{N} \; .$$

Setting $u_j := \alpha_j R_{\alpha_j} u$ we obtain from IV.5.30(i) that

$$u_j(X_{\sigma_U}) \xrightarrow[j\to\infty]{} u(X_{\sigma_U}) \quad P_z\text{-a.s. for } \mathcal{E}\text{-q.e. } z \in E,$$

hence, since $(u_j)_{j\in\mathbb{N}}$ are uniformly bounded by $\|u\|_\infty$,

$$(\tilde{u}_j)_U \xrightarrow[j\to\infty]{} \tilde{u}_U \quad \mathcal{E}\text{-q.e. .}$$

But clearly by III.1.4,

$$\mathcal{E}_1((u_j)'_U - u'_U, (u_j)'_U - u'_U) \leq K^2 \mathcal{E}_1(u_j - u, u_j - u) \xrightarrow[j\to\infty]{} 0$$

and thus III.3.5 implies that $\tilde{u}_U$ is an $\mathcal{E}$-quasi-continuous m-version of u'_U. Applying I.4.17, III.3.5, and the monotone convergence theorem we obtain the assertion for arbitrary $u \in \tilde{D}(\mathcal{E})$, $u \geq 0$. $\qquad\square$

The following existence result on "localizing" functions in $D(\mathcal{E})$ is crucial.

Proposition 1.7. *Let $U \subset E$, U open. Then there exist $u_n \in \tilde{D}(\mathcal{E})$, $n \in \mathbb{N}$, such that $0 \leq u_n \leq 1_U$, $\mathrm{supp}[u_n]$ is a compact subset of U, and $c_n \in]0,\infty[$, $n \in \mathbb{N}$, such that*

$$u := \sum_{n=1}^{\infty} c_n u_n \in \tilde{D}(\mathcal{E})$$

with $0 \leq u \leq 1_U$ and $u > 0$ $\mathcal{E}$-q.e. on U.

Proof. Let $(E_k)_{k\in\mathbb{N}}$ be as in IV.3.1(i). By IV.3.2(iii) we may assume that each E_k is metrizable. For each k we fix a metric ρ_k on E_k compatible with the trace topology inherited from E. Set for $k, j \in \mathbb{N}$

$$B_{k,j} := \left\{ z \in E_k \,\Big|\, \rho_k(z, E_k\backslash U) > \frac{1}{j} \right\} \cup E_k^c .$$

Then each $B_{k,j}$ is an open subset of E. Let $\varphi \in \mathcal{B}_b(E) \cap L^1(E;m)$ be such that $0 < \varphi \leq 1$. We define (cf. IV.(5.17))

$$h_k := R_1^{E_k}\varphi = E.\left[\int_0^{\sigma_{E_k^c}} e^{-s}\varphi(X_s)\,ds\right], \quad k \in \mathbb{N},$$

$$h_{k,j} := R_1^{B_{k,j}}\varphi = E.\left[\int_0^{\tau_{B_{k,j}^c}} e^{-s}\varphi(X_s)\,ds\right], \quad k, j \in \mathbb{N} .$$

By IV.5.25(ii) (and its proof) we know that $h_k, h_{k,j} \in \tilde{D}(\mathcal{E})$ and that

$$(1.10) \qquad\qquad \{h_{k,j} = 0\} = B_{k,j}^c \text{ and } \{h_k = 0\} \supset E_k^c .$$

Setting

$$g_{k,j} := h_k \wedge h_{k,j}$$

we have by (1.10) that

$$(1.11) \qquad\qquad U^c \subset B_{k,j}^c \cup E_k^c \subset \{g_{k,j} = 0\}$$

and thus

$$(1.12) \qquad \mathrm{supp}[g_{k,j}] \subset \{z \in E_k \mid \rho_k(z, E_k \backslash U) \geq \frac{1}{j}\} \subset U \ .$$

By IV.5.30 there exists an $\mathcal{E}$-exceptional set $N \subset E$ such that for all $z \in E \backslash N$

$$h_k(z) \uparrow E_z \left[\int_0^\infty e^{-s} \varphi(X_s)\, ds\right] \text{ as } k \to \infty \ .$$

Hence for every $z \in U \backslash N$ there exists $k \in \mathbb{N}$ such that $h_k(z) > 0$. Moreover, we can find $j \in \mathbb{N}$ such that $z \in B_{k,j}$ and consequently $h_{k,j}(z) > 0$. Let $\{u_i \mid i \in \mathbb{N}\} = \{h_{k,j} \mid k, j \in \mathbb{N}\}$ and define

$$u(z) := \sum_{n=1}^\infty c_n u_n(z), \ z \in E,$$

where $c_n := 2^{-n}(1 + \mathcal{E}_1^{1/2}(u_n, u_n)), n \in \mathbb{N}$. Then, $u > 0$ $\mathcal{E}$-q.e. on U and (by III.3.3) u is $\mathcal{E}$-quasi-continuous. Because of (1.11), (1.12) it remains to show that $u \in D(\mathcal{E})$. But this is an immediate consequence of I.2.12 since $u \in L^2(E; m)$ and for all $n \in \mathbb{N}$

$$\mathcal{E}(\sum_{i=1}^n c_i u_i, \sum_{i=1}^n c_i u_i) \leq \sum_{i,j=1}^n c_i c_j |\mathcal{E}(u_i, u_j)| \leq K \ .$$

$\square$

Lemma 1.8. *For $U \subset E$, U open, the following assertions are equivalent:*

(i) $P_z[X_{\sigma_U} \in U] = 0$ *for $\mathcal{E}$-q.e. $z \in U^c$.*

(ii) *For all $u \in \tilde{D}(\mathcal{E})$ with $\mathrm{supp}[u] \subset U$*

$$\tilde{u}_U(z) \left(:= E_z[e^{-\sigma_U} u(X_{\sigma_U})]\right) = 0 \text{ for } \mathcal{E}\text{-q.e. } z \in U^c \ .$$

Proof. $(i) \Rightarrow (ii)$ is obvious.
$(ii) \Rightarrow (i)$: Let $u = \sum_{n=1}^\infty c_n u_n \in \tilde{D}(\mathcal{E})$ be as specified in 1.7. Then, by the monotone convergence theorem for every $z \in E$

$$\tilde{u}_U(z) = E_z[e^{-\sigma_U} u(X_{\sigma_U})] = \sum_{n=1}^\infty c_n \, E_z[e^{-\sigma_U} u_n(X_{\sigma_U})] \ .$$

Hence $\tilde{u}_U = 0$ for $\mathcal{E}$-q.e. $z \in U^c$. By 1.6 and IV.3.3(iii) $\tilde{u}_U$ only changes on an $\mathcal{E}$-exceptional set if u is changed on an $\mathcal{E}$-exceptional. Hence we can assume that $u > 0$ everywhere on U and (i) follows. $\square$

Corollary 1.9. *The following assertions are equivalent.*

(i) $(\mathcal{E}, D(\mathcal{E}))$ *has the local property.*

(ii) $P_z[X_{\sigma_U} \in U] = 0$ *for $\mathcal{E}$-q.e. $z \in U^c$ for all $U \subset E$, U open.*

Proof. The assertion follows by 1.2 and 1.8 since for $u \in \tilde{D}(\mathcal{E})$ and $U \subset E$, U open, with $\mathrm{supp}[u] \subset U$, $u = u'_U$ is equivalent with $\tilde{u}_U = 0$ $\mathcal{E}$-q.e. on U^c by 1.6, III.1.4 and IV.3.3(iii). $\qquad\square$

Proof of 1.5. By 1.9 it suffices to prove that $(1.8) \Leftrightarrow 1.9(ii)$. But $(1.8) \Rightarrow 1.9(ii)$ is obvious.

$1.9(ii) \Rightarrow (1.8)$: Assume 1.9(ii) and let $(E_k)_{k \in \mathbb{N}}$ be an $\mathcal{E}$-nest consisting of metrizable compact sets. Let $\mathcal{U}$ be a countable family of open sets such that $\{U \cap E_k \mid U \in \mathcal{U}\}$ is a base for the topology on E_k for every $k \in \mathbb{N}$. By 1.9(ii) we can find an $\mathcal{E}$-exceptional set N such that for every $U \in \mathcal{U}$

$$P_z[X_{\sigma_U} \in U] = 0 \text{ for all } z \in U^c \backslash N .$$

By the Markov property (cf. also IV.1.9(v)) this yields that for all $s > 0$

$$(1.13) \quad \begin{aligned} &P_z[X_s \in U^c \backslash N , X_{\sigma_U} \circ \theta_s \in U] \\ &= E_z[1_{\{X_s \in U^c \backslash N\}} E_{X_s}[1_{\{X_{\sigma_U} \in U\}}]] = 0 \text{ for all } z \in E, \ U \in \mathcal{U} . \end{aligned}$$

Let $(F_k)_{k \in \mathbb{N}}$ be an $\mathcal{E}$-nest such that $F_k \subset E_k$ and $N \subset \bigcap F_k^c$. Let

$$\Omega_0 := \{ \lim_{k \to \infty} \sigma_{F_k^c} \geq \zeta \} .$$

By IV.5.30, $P_z[\Omega_0] = 1$ for $\mathcal{E}$-q.e. $z \in E$. It is easy to see that

$$\begin{aligned} \Omega_d &:= \{\omega \in \Omega_0 \mid X_{t-}(\omega) \neq X_t(\omega) \text{ for some } t \in]0, \zeta(\omega)[\} \\ &\subset \bigcup_{U \in \mathcal{U}} \bigcup_{s \in \mathbb{Q}_+^*} \{\omega \in \Omega_0 \mid X_s(\omega) \in U^c \backslash N, \ X_{\sigma_U}(\theta_s \omega) \in U\} . \end{aligned}$$

By (1.13) it follows that $\Omega_d \in \mathcal{F}$ and $P_z[\Omega_d] = 0$ for all $z \in E$. Consequently, $P_z[\Omega_0 \backslash \Omega_d] = 1$ for $\mathcal{E}$-q.e. $z \in E$ and (1.8) is proved, since $\mathbf{M}$ is right continuous. $\qquad\square$

Definition 1.10. Let $\mathbf{M} = (\Omega, \mathcal{F}, (X_t)_{t \geq 0}, (P_z)_{z \in E_\Delta})$ be an m-tight special standard process with state space E and life time ζ. We say that $\mathbf{M}$ is a *diffusion* if

$$P_z[t \mapsto X_t \text{ is continuous on } [0, \zeta[] = 1 \text{ for all } z \in E .$$

Theorem 1.11. *A quasi-regular Dirichlet form possesses the local property if and only if it is associated with a pair of diffusions* $(\mathbf{M}, \hat{\mathbf{M}})$.

Proof. The "if-part" follows from 1.5 and the "only if-part" follows from IV.3.5, 1.5, and IV.6.5 (by appropriate restriction and trivial extension). $\qquad\square$

Examples 1.12. (i) Obviously, in the classical case with maximal domain (cf. Chap. II, Subsection 1c)) we have a Dirichlet form with the local property, hence also in the case with "minimal" domain (cf. Chap.II, Subsection 1b)) and the non-regular case (cf. Chap.II, Subsection 2e)). The remaining examples in Chap.II, Sections 1,2 are all obtained as closures of bilinear forms with domain

$C_0^\infty(U)$, $U \subset \mathbb{R}^d$, U open, $n \in \mathbb{N}$. Since for all compact $K \subset U$ and open $V \subset U$ with $K \subset V$ there exists $\chi \in C_0^\infty(U)$ with $1_K \leq \chi \leq 1_V$, it follows by 1.3 that all these Dirichlet forms have the local property except the ones in Chap.II, Subsections 1d) if $\alpha \neq 1$, and 2c) if $J \not\equiv 0$.

(ii) Assume that E is a locally convex topological real vector space which is a (topological) Souslin space and that $\mu := m$ is a finite positive measure on $\mathcal{B}(E)$ with $\operatorname{supp}\mu = E$. Let $k \in E\backslash\{0\}$ be μ-admissible and let $(\mathcal{E}_k, D(\mathcal{E}_k))$ be the closure of

$$\mathcal{E}_k(u, v) = \int \frac{\partial u}{\partial k} \frac{\partial v}{\partial k}\, d\mu; \quad u, v \in \mathcal{F}C_b^\infty,$$

(cf. Chap.II, Subsection 3a)). To prove the local property of $(\mathcal{E}_k, D(\mathcal{E}_k))$ it is sufficient to show that

$$(1.14) \qquad \frac{\partial u}{\partial k} = 0 \ \mu\text{-a.e. on } E\backslash\operatorname{supp}[u] \text{ for all } u \in D(\mathcal{E}_k) \ .$$

To this end we first prove that if $v, w \in D(\mathcal{E}_k) \cap L^\infty(E; \mu)$ then

$$(1.15) \qquad \frac{\partial(v \cdot w)}{\partial k} = v\frac{\partial w}{\partial k} + w\frac{\partial v}{\partial k} \ .$$

Indeed, let $v_n, w_n \in \mathcal{F}C_b^\infty$, $n \in \mathbb{N}$, with $v_n \to v$, $w_n \to w$ in $D(\mathcal{E}_k)$ as $n \to \infty$. By I.4.17 we can assume that $(\|v_n\|_\infty \vee \|w_n\|_\infty)_{n \in \mathbb{N}}$ is bounded, hence $v_n \cdot w_n \to v \cdot w$ in $L^2(E; \mu)$ as $n \to \infty$. Since for all $n \in \mathbb{N}$

$$\frac{\partial(v_n w_n)}{\partial k} = v_n\frac{\partial w_n}{\partial k} + w_n\frac{\partial v_n}{\partial k} \ .$$

we see by the closability that $(v \cdot w \in D(\mathcal{E}_k)$ and $)$ (1.15) holds. Now let $u \in D(\mathcal{E}_k)$. By I.4.17 we may assume that u is bounded. By 1.7 there exists $v \in D(\mathcal{E}_k) \cap L^\infty(E; \mu)$ such that $0 \leq v \leq 1_{E\backslash\operatorname{supp}[u]}$ and $v > 0$ μ-a.e. on $E\backslash\operatorname{supp}[u]$. Hence, by (1.15)

$$0 = u\frac{\partial v}{\partial k} + v\frac{\partial u}{\partial k}$$

and thus

$$(u\frac{\partial v}{\partial k} =) \ v\frac{\partial u}{\partial k} = 0 \ .$$

Consequently, $\frac{\partial u}{\partial k} = 0$ μ-a.e. on $E\backslash\operatorname{supp}[u]$ and (1.14) is proven. By (1.14) it follows that all Dirichlet forms appearing in Chap.II, Sect.3 have the local property.

2   A new capacity and Hunt processes

a) $\operatorname{Cap}_{1,g}$ and $\operatorname{Cap}_{h,1}$

Below we denote by $\mathcal{S}$ (resp. $\hat{\mathcal{S}}$) the family of all 1-excessive (resp. 1-coexcessive) functions in $D(\mathcal{E})$ (cf. III.1.1). Note that for $u, v \in \mathcal{S}$ we have $(u \vee v)_E \in \mathcal{S}$,

$u \vee v \leq (u \vee v)_E$ and $(u \vee v)_E \leq 1$ if $u, v \leq 1$ (cf. III.1.5 (i)-(iii)) with the corresponding dual assertions for $\hat{S}$. In particular,

$$(2.1) \qquad \{u \in \mathcal{S} \,|\, u \leq 1\}, \ \{u \in \hat{\mathcal{S}} \,|\, u \leq 1\} \text{ are } \textit{upper directed}\,.$$

(A family of functions $\mathcal{S}$ is called upper directed if for any $u, v \in \mathcal{S}$ there exists $w \in \mathcal{S}$ such that $u \vee v \leq w$.)

Definition 2.1. Let $h \in \mathcal{S}$ and $g \in \hat{\mathcal{S}}$. We define for an open set $U \subset E$

$$(2.2) \qquad \begin{aligned} \mathrm{Cap}_{1,g}(U) &:= \sup\{\mathrm{Cap}_{u,g}(U) \,|\, u \in \mathcal{S}, \ u \leq 1\} \\ \mathrm{Cap}_{h,1}(U) &:= \sup\{\mathrm{Cap}_{h,u}(U) \,|\, u \in \hat{\mathcal{S}}, \ u \leq 1\} \end{aligned}$$

and for arbitrary $A \subset E$

$$(2.3) \qquad \begin{aligned} \mathrm{Cap}_{1,g}(A) &:= \inf\{\mathrm{Cap}_{1,g}(U) \,|\, A \subset U \subset E, \ U \text{ open}\} \\ \mathrm{Cap}_{h,1}(A) &:= \inf\{\mathrm{Cap}_{h,1}(U) \,|\, A \subset U \subset E, \ U \text{ open}\} \end{aligned}$$

Note that $\mathrm{Cap}_{1,g}$ and $\mathrm{Cap}_{h,1}$ are defined in accordance with III.2.4 if $1 \in D(\mathcal{E})$ or if $g = 1$ resp. $h = 1$. In what follows we fix $h \in \mathcal{S}$ and $g \in \hat{\mathcal{S}}$.

Lemma 2.2. *(i) Let U, W be open subsets of E. Then*

$$U \subset W \Rightarrow \mathrm{Cap}_{1,g}(U) \subset \mathrm{Cap}_{1,g}(W),$$

and

$$\mathrm{Cap}_{1,g}(U \cup W) + \mathrm{Cap}_{1,g}(U \cap W) \leq \mathrm{Cap}_{1,g}(U) + \mathrm{Cap}_{1,g}(W)\,.$$

(ii) Let $(U_n)_{n \in \mathbb{N}}$ be an increasing sequence of open subsets of E, then

$$\mathrm{Cap}_{1,g}\Big(\bigcup_{n \geq 1} U_n\Big) = \lim_{n \to \infty} \mathrm{Cap}_{1,g}(U_n)\,.$$

The corresponding dual assertions hold for $\mathrm{Cap}_{h,1}$.

Proof. (i): The first assertion is obvious by III.2.6(i). We now prove the second. By (2.1) and III.2.5(iv) we can find an increasing sequence $(u_n)_{n \in \mathbb{N}}$ in $\mathcal{S}$ such that

$$\mathrm{Cap}_{1,g}(U \cup W) = \lim_{n \to \infty} \mathrm{Cap}_{u_n, g}(U \cup W)$$

and

$$\mathrm{Cap}_{1,g}(U \cap W) = \lim_{n \to \infty} \mathrm{Cap}_{u_n, g}(U \cap W)\,.$$

Now the assertion is an immediate consequence of the strong subadditivity of $\mathrm{Cap}_{u_n, g}$ (cf. III.2.7(i)).

(ii):

$$\mathrm{Cap}_{1,g}(\bigcup_{n\geq 1} U_n) \;=\; \sup_{\substack{u\in\mathcal{S} \\ u\leq 1}} \mathrm{Cap}_{u,g}(\bigcup_{n\geq 1} U_n)$$

$$=\; \sup_{n\in\mathbb{N}} \sup_{\substack{u\in\mathcal{S} \\ u\leq 1}} \mathrm{Cap}_{u,g}(U_n)$$

$$=\; \sup_{n\in\mathbb{N}} \mathrm{Cap}_{1,g}(U_n)$$

where we used III.2.7(ii) in the third step. Now (ii) follows by the first part of (i). The dual assertions of (i), (ii) for $\mathrm{Cap}_{h,1}$ can be proved in the same way. $\square$

As a consequence of the above lemma we obtain for all $h \in S$, $g \in \hat{S}$

Proposition 2.3. $\mathrm{Cap}_{1,g}$ and $\mathrm{Cap}_{h,1}$ are Choquet capacities which are countably subadditive, i.e., they satisfy the properties stated in III.2.8.

Proof. See the proof of III.2.8. $\hspace{2cm}\square$

Lemma 2.4. Let $U \subset E$, U open. Then there exist e_U, $\hat{e}_U$ in $L^\infty(E;m)$ such that if $h = G_1\varphi$, $g = \hat{G}_1\varphi$ for some $\varphi \in L^1(E;m) \cap L^2(E;m)$, $\varphi > 0$, then

$$\mathrm{Cap}_{1,g}(U) \;=\; \int e_U\,\varphi\,dm \;,$$

$$\mathrm{Cap}_{h,1}(U) \;=\; \int \hat{e}_U\,\varphi\,dm \;.$$

Proof. We only prove the existence of e_U, the existence proof for $\hat{e}_U$ is similar. Let

$$\mathcal{S}_U := \{u_U \,|\, u \in \mathcal{S},\ u \leq 1\} \;.$$

Then $\mathcal{S}_U$ is upper directed. Define

$$e_U \;:=\; \mathrm{ess\,sup}\,\mathcal{S}_U$$

(i.e., e_U is the m-a.e. unique minimal $g \in \mathcal{B}(E)^+$ such that $g \geq f$ m-a.e. for all $f \in \mathcal{S}_U$). We claim that e_U (exists and) is as desired. To show this we fix a strictly positive function $\Phi \in L^1(E;m)$ and let

$$C := \sup\{\int f\,\Phi\,dm \,|\, f \in \mathcal{S}_U\} \;.$$

Then we can take $f_n \in \mathcal{S}_U$, $n \in \mathbb{N}$, $f_n \uparrow$, such that

$$\lim_{n\to\infty} \int f_n\Phi\,dm = C \;.$$

Set $f_\infty := \lim_{n\to\infty} f_n$. Then $f_\infty \geq f$ m-a.e. for all $f \in \mathcal{S}_U$, because otherwise, we could find $f \in \mathcal{S}_U$ such that $\int (f \vee f_\infty)\Phi\,dm > \int f_\infty\Phi\,dm = C$, and hence we could find $f'_n \in \mathcal{S}_U$, $f'_n \geq f \vee f_n$, $n \in \mathbb{N}$, $f'_n \uparrow$, such that

$$\lim_{n\to\infty} \int f'_n\Phi\,dm \geq \lim_{n\to\infty} \int (f \vee f_n)\Phi\,dm = \int (f \vee f_\infty)\Phi\,dm > C \;.$$

This would contradict the definition of C. Thus clearly, $f_\infty = e_U$ m-a.e. . Now for any $u \in \mathcal{S}$, $u \leq 1$, we have $u_U \leq e_U$ m-a.e. , and therefore,

$$
\begin{aligned}
\mathrm{Cap}_{u,g}(U) &= \mathcal{E}_1(u_U, g) = \int u_U\, \varphi\, dm \leq \int e_U\, \varphi\, dm \\
&= \lim_{n \to \infty} \int f_n \varphi\, dm = \lim_{n \to \infty} \mathcal{E}_1(f_n, g) \leq \mathrm{Cap}_{1,g}(U)
\end{aligned}
$$

Consequently, $\mathrm{Cap}_{1,g}(U) = \int e_U\, \varphi\, dm$. $\qquad\square$

We set

$$
(2.4) \quad
\begin{aligned}
\mathcal{S}_0 &:= \{h \in \mathcal{S} \mid h = G_1\varphi \text{ with } \varphi \in L^1(E; m) \cap L^2(E; m),\ \varphi > 0\} \\
\hat{\mathcal{S}}_0 &:= \{g \in \hat{\mathcal{S}} \mid g = \hat{G}_1\varphi \text{ with } \varphi \in L^1(E; m) \cap L^2(E; m),\ \varphi > 0\}
\end{aligned}
$$

Proposition 2.5. *(i) $\mathrm{Cap}_{1,g}(E) < \infty$ for all $g \in \hat{\mathcal{S}}_0$.*

(ii) If $g_1, g_2 \in \hat{\mathcal{S}}_0$, then Cap_{1,g_1} and Cap_{1,g_2} are equivalent, i.e., $\mathrm{Cap}_{1,g_1}(U_n) \downarrow 0$ if and only if $\mathrm{Cap}_{1,g_2}(U_n) \downarrow 0$ as $n \to \infty$ for any decreasing sequence of open sets $U_n \subset E$.
The dual assertions hold for $\mathrm{Cap}_{h,1}$.

Proof. Since $e_U \leq 1$, (i) is evident by 2.4. (ii) follows from the fact that for $g_1, g_2 \in \hat{\mathcal{S}}_0$, and open $U_n \subset E$, $n \in \mathbb{N}$, we have by 2.4 for $i = 1, 2$

$$
\mathrm{Cap}_{1,g_i}(U_n) \downarrow 0 \text{ if and only if } e_{U_n} \downarrow 0 \text{ m-a.e. as } n \to \infty .
$$

$\qquad\square$

Proposition 2.6. *Suppose that $(\mathcal{E}, D(\mathcal{E}))$ is quasi-regular and $\mathbf{M} = (\Omega, \mathcal{F}, (X_t)_{t \geq 0}, (P_z)_{z \in E_\Delta})$ is a right process properly associated with $(\mathcal{E}, D(\mathcal{E}))$. Then for $g := \hat{G}_1\varphi \in \hat{\mathcal{S}}_0$ and $U \subset E$, U open, we have*

$$
(2.5) \qquad\qquad \mathrm{Cap}_{1,g}(U) = E_{\varphi \cdot m}[e^{-\sigma_U}] .
$$

The dual assertion holds for $\mathrm{Cap}_{h,1}$ with $h \in \mathcal{S}_0$.

Proof. Let $u \in \mathcal{S}$, $u \leq 1$, and let $\tilde{u}$ be an $\mathcal{E}$-quasi-continuous m-version of u with $0 \leq \tilde{u} \leq 1$ everywhere on E. By 1.6 (which also holds in this situation)

$$
\mathrm{Cap}_{u,g}(U) = \mathcal{E}_1(u_U, g) = (u_U, \varphi) = E_{\varphi \cdot m}[e^{-\sigma_U} \tilde{u}(X_{\sigma_U})] \leq E_{\varphi \cdot m}[e^{-\sigma_U}] .
$$

Hence $\mathrm{Cap}_{1,g}(U) \leq E_{\varphi \cdot m}[e^{-\sigma_U}]$. Furthermore, if $h = R_1\varphi := E.[\int_0^\infty e^{-s}\varphi(X_s)\, ds]$, $u_n = (nh) \wedge 1$, $n \in \mathbb{N}$, then $u_n \in \mathcal{S}$ and $u_n \uparrow 1_E$ pointwise as $n \to \infty$. Note that $\{\sigma_U < \zeta\} = \{\sigma_U < \infty\}$ (where ζ is the life time of $\mathbf{M}$), hence

$$
E_{\varphi \cdot m}[e^{-\sigma_U}] = \lim_{n \to \infty} E_{\varphi \cdot m}[e^{-\sigma_U} u_n(X_{\sigma_U})] = \lim_{n \to \infty} \mathrm{Cap}_{u_n,g}(U) \leq \mathrm{Cap}_{1,g}(U) .
$$

$\qquad\square$

b)   Strict quasi-regularity and Hunt processes

Given an increasing sequence $(F_k)_{k\in\mathbb{N}}$ of closed sets of E, we define

$$(2.6)\qquad C_\infty(\{F_k\}) \; := \; \{f : A \to R \mid \bigcup_{k\geq 1} F_k \subset A \subset E, f_{|F_k\cup\{\Delta\}}$$
$$\text{is continuous for every } k \in \mathbb{N}\} \; .$$

Recall that $f(\Delta) = 0$ and note that $C_\infty(\{F_k\})$ only differs from $C(\{F_k\})$ (cf. III.(3.2)) if E is locally compact and E_Δ is the one point compactification, since otherwise Δ is an isolated point of E by our convention.

Definition 2.7. An increasing sequence $(F_k)_{k\in\mathbb{N}}$ of closed sets of E is called an *s.$\mathcal{E}$-nest (strict $\mathcal{E}$-nest)* if $\mathrm{Cap}_{1,g}(F_k^c) \downarrow 0$ as $n \to \infty$ for some (hence all) $g \in \hat{S}_0$. A subset $N \subset E$ is called *s.$\mathcal{E}$-exceptional (strictly $\mathcal{E}$-exceptional)* if $\mathrm{Cap}_{1,g}(N) = 0$ for some (hence all) $g \in \hat{S}_0$. We say that a property of points in E holds *s.$\mathcal{E}$-q.e. (strictly $\mathcal{E}$-quasi-everywhere)* if the property holds outside some s.$\mathcal{E}$-exceptional set. An s.$\mathcal{E}$-q.e. defined function f on E is called *s.$\mathcal{E}$-quasi-continuous (strictly $\mathcal{E}$-quasi-continuous)* if there exists an s.$\mathcal{E}$-nest $(F_k)_{k\in\mathbb{N}}$ such that $f \in C_\infty(\{F_k\})$.

Remark 2.8. We have that $\mathrm{Cap}_{u,g} \leq \mathrm{Cap}_{1,g}$ for $g \in \hat{S}_0$, $u \in \mathcal{S}$, $u \leq 1$. Hence it follows from III.2.11 that the above notions are indeed stricter than the corresponding notions defined in Chapter III. That is, every s.$\mathcal{E}$-nest is an $\mathcal{E}$-nest, every s.$\mathcal{E}$-exceptional set is $\mathcal{E}$-exceptional (and hence m-negligible), every s.$\mathcal{E}$-quasi-continuous function is $\mathcal{E}$-quasi-continuous.

Lemma 2.9. *Let* $u \in D(\mathcal{E})$ *such that it has an s.$\mathcal{E}$-quasi-continuous m-version* $\tilde{u}$. *Then for* $g \in \hat{S}_0$ *and* $\lambda > 0$

$$\mathrm{Cap}_{1,g}\{|\tilde{u}| > \lambda\} \leq \frac{K^2}{\lambda}\mathcal{E}_1(u,u)^{1/2}\mathcal{E}_1(g,g)^{1/2} \; .$$

Proof. Let $(F_k)_{k\in\mathbb{N}}$ be an s.$\mathcal{E}$-nest, such that $\tilde{u} \in C_\infty(\{F_k\})$. We set $U_k := \{|\tilde{u}| > \lambda\} \cup F_k^c$, $k \in \mathbb{N}$. Then each U_k is open. If $h \in \mathcal{S}$, $h \leq 1$, one shows as in the proof of III.3.4 that

$$\mathrm{Cap}_{h,g}(U_k) \leq K^2\lambda^{-1}\mathcal{E}_1(u,u)^{1/2}\mathcal{E}_1(g,g)^{1/2} + \mathrm{Cap}_{h,g}(F_k^c)$$

and consequently,

$$\mathrm{Cap}_{1,g}(U_k) \leq K^2\lambda^{-1}\mathcal{E}_1(u,u)^{1/2}\mathcal{E}_1(g,g)^{1/2} + \mathrm{Cap}_{1,g}(F_k^c) \; .$$

Letting $k \to \infty$ we get the required inequality. $\qquad\qquad\square$

We say that $(f_n)_{n\in\mathbb{N}}$ converges to f *s.$\mathcal{E}$-quasi-uniformly* if there exists an s.$\mathcal{E}$-nest $(F_k)_{k\in\mathbb{N}}$ such that $f_n \xrightarrow[n\to\infty]{} f$ uniformly on each $F_k \cup \{\Delta\}$.

Exercise 2.10. Show that III.3.3, III.3.5, III.3.8, III.3.9 remain true if we replace $C(\{F_k\})$ by $C_\infty(\{F_k\})$ and add the prefix "s." to $\mathcal{E}$-nest, $\mathcal{E}$-quasi-continuous, and $\mathcal{E}$-quasi-uniformly in the statements. In what follows we shall refer to these statements as "strict version" of III.3.3, III.3.5 etc..

Definition 2.11. A Dirichlet form $(\mathcal{E}, D(\mathcal{E}))$ on $L^2(E; m)$ is called *s.quasi-regular (strictly quasi-regular)* if:

(i) There exists an s.$\mathcal{E}$-nest $(E_k)_{k\in\mathbb{N}}$ such that $E_k \cup \{\Delta\}$ is compact in E_Δ for each k.

(ii) There exists an $\tilde{\mathcal{E}}_1^{1/2}$-dense subset of $D(\mathcal{E})$ whose elements have s.$\mathcal{E}$-quasi-continuous m-versions.

(iii) There exist $u_n \in D(\mathcal{E})$, $n \in \mathbb{N}$, having s.$\mathcal{E}$-quasi-continuous m-versions $\tilde{u}_n$, $n \in \mathbb{N}$, and an s.$\mathcal{E}$-exceptional set $N \subset E$ such that $\{\tilde{u}_n \,|\, n \in \mathbb{N}\}$ separates the points of $E_\Delta \backslash N$.

Note that the notion of s.quasi-regularity crucially depends on "how" we adjoin Δ to E.

Proposition 2.12.   *(i) If $(\mathcal{E}, D(\mathcal{E}))$ is s.quasi-regular, then it is quasi-regular.*

(ii) Suppose E is a locally compact separable metric space, E_Δ is the one point compactification, and m a positive Radon measure on $(E, \mathcal{B}(E))$. If $(\mathcal{E}, D(\mathcal{E}))$ is regular, then it is s.quasi-regular.

Proof. (i): It follows from 2.8 that 2.11(ii), (iii) imply IV.3.1(ii), (iii). Moreover, if E_Δ is not the one point compactification, then $E_k \cup \{\Delta\}$ is compact, if and only if E_k is compact. In this case 2.11(i) implies also IV.3.1(i) and hence $(\mathcal{E}, D(\mathcal{E}))$ is quasi-regular. We now prove that if E is locally compact and E_Δ is the one point compactification, then the conditions 2.11(i)-(iii) also imply condition IV.3.1(i). Applying the strict version of III.3.5 (cf.2.10), we conclude from 2.11(ii) that every element in $D(\mathcal{E})$ has an s.$\mathcal{E}$-quasi-continuous m-version. In particular, $g \in \hat{\mathcal{S}}_0$ has an s.$\mathcal{E}$-quasi-continuous m-version $\tilde{g}$, i.e., $\tilde{g} \in C_\infty(\{F_k\})$ for some s.$\mathcal{E}$-nest $(F_k)_{k\in\mathbb{N}}$. Let $k \in \mathbb{N}$ and $E_k' := F_k \cap \{\tilde{g} \geq \frac{1}{k}\} = (F_k \cup \{\Delta\}) \cap \{\tilde{g} \geq \frac{1}{k}\}$ (Recall $\tilde{g}(\Delta) = 0$ by convention). Then E_k' is a compact subset of E. Let $U_k := E \backslash E_k'$. Then $g \wedge \frac{1}{k} + \hat{g}_{F_k^c} \geq g$ m-a.e. on U_k. Hence it follows from III.1.5 and III.1.2 that for $h \in \mathcal{S}_0$,

$$\mathrm{Cap}_{h,g}(U_k) \;=\; \mathcal{E}_1(h_{U_k}, g) \leq \mathcal{E}_1(h_{U_k}, g \wedge \frac{1}{k}) + \mathcal{E}_1(h_{U_k}, \hat{g}_{F_k^c})$$

$$\leq \;\mathcal{E}_1(h, g \wedge \frac{1}{k}) + \mathrm{Cap}_{1,g}(F_k^c) \xrightarrow[k\to\infty]{} 0$$

Hence $(E_k')_{k\in\mathbb{N}}$ is an $\mathcal{E}$-nest satisfying IV.3.1(i) and the proof of (i) is complete. (ii): 2.11(i) is fulfilled by taking e.g. $E_k = E$ for each $k \in \mathbb{N}$. 2.11(ii) follows from the fact that $C_0(E) \cap D(\mathcal{E})$ is dense in $D(\mathcal{E})$ w.r.t. $\tilde{\mathcal{E}}_1^{1/2}$ and 2.11(iii) holds since $C_0(E) \cap D(\mathcal{E})$ is dense in $C_0(E)$ w.r.t. uniform norm and since E is separable. Hence $(\mathcal{E}, D(\mathcal{E}))$ is s.quasi-regular. $\qquad\square$

Remark. The converse of 2.12(i) is false. See 2.14(iii) below.

We need one more "strict" notion. Let $\mathbf{M} = (\Omega, \mathcal{F}, (X_t)_{t\geq 0}, (P_z)_{z\in E_\Delta})$ be a right process with state space E and life time ζ and ν a positive measure on $(E_\Delta, \mathcal{B}(E_\Delta))$. We say that $\mathbf{M}$ is *s.ν-tight (strictly ν-tight)* if there exists an increasing sequence $(K_k)_{k\in\mathbb{N}}$ of compact metrizable sets in E_Δ such that

$$(2.7) \qquad P_\nu[\lim_{k\to\infty} \sigma_{E_\Delta\setminus K_k} < \infty] = 0 \ .$$

Note that if E is a metrizable Lusin space and $\mathbf{M}$ is a Hunt process with state space E, then $\mathbf{M}$ is always s.ν-tight for any σ-finite positive measure ν. This is because in this case we can always consider $\mathbf{M}$ as a *conservative* process (i.e. with infinite life time) with state space E_Δ and a.s. cadlag paths on $[0, \infty)$, and then apply IV.1.15 to obtain s.ν-tightness.
We now state the main theorem of this section.

Theorem 2.13. *A necessary and sufficient condition for $(\mathcal{E}, D(\mathcal{E}))$ to be associated with an s.m-tight Hunt process is that $(\mathcal{E}, D(\mathcal{E}))$ is s.quasi-regular. In this case $(\mathcal{E}, D(\mathcal{E}))$ is s.properly associated (strictly properly associated) with $\mathbf{M}$ in the sense that*

$$(2.8) \qquad \begin{array}{l} p_t f \text{ is an s.}\mathcal{E}\text{-quasi-continuous m-version of } T_t f \\ \text{for all } t > 0, \ f \in L^2(E; m) \ , \end{array}$$

where $(p_t)_{t>0}$ is the transition semigroup of $\mathbf{M}$.

Remark 2.14. (i) For the reason mentioned before 2.13, the phrase "s.m-tight" can be dropped if E is a metrizable Lusin space.
(ii) Note that in contrast with quasi-regularity, s.quasi-regularity is not a "symmetric" notion, since $\mathrm{Cap}_{1,g}$ might not be equivalent with $\mathrm{Cap}_{h,1}$. Therefore, if $(\mathcal{E}, D(\mathcal{E}))$ is associated with a Hunt process it is not clear whether it is also coassociated with a Hunt process, though it is then always coassociated with a special standard process. At this point we may introduce the notion of *co-s.quasi-regularity* in the same manner as s.quasi-regularity but relative to $\mathrm{Cap}_{h,1}$, and then give a dual result characterizing those Dirichlet forms coassociated with Hunt processes.
(iii) Note that if Δ is isolated from E the Hunt process $\mathbf{M}$ in 2.13 if it exists has automatically left limits on $]0, \zeta]$ in E P_z-a.s. on $\{\zeta < \infty\}$ for all $z \in E$. Using this one can easily construct an example of a quasi-regular Dirichlet form which is not s.quasi-regular. Let

$$B := \{x \in \mathbb{R}^2 \,|\, |x| < 1\}$$

and

$$E := B\setminus\{(x_1, 0) \in \mathbb{R}^2 \,|\, x_1 \in\,] - 1, -\tfrac{1}{2}[\cup]\tfrac{1}{2}, 1[\} \ .$$

Then E is not locally compact and we adjoin Δ as an isolated point. Let $\mathbf{M}' = (\Omega, \mathcal{F}, (Y_t)_{t\geq 0}, (P_z)_{z\in\mathbb{R}^d})$ be classical Brownian motion on $\mathbb{R}^d$, but with $(-\tfrac{1}{2}, 0)$, $(\tfrac{1}{2}, 0)$ as traps. We set $\zeta := \inf\{t \geq 0 \,|\, Y_t \notin E\}$ and define

$$X_t := \begin{cases} Y_t & \text{if } t < \zeta \\ \Delta & \text{if } t \geq \zeta \end{cases}.$$

Then (cf. [F 80,§4.4]) $\mathbf{M} = (\Omega, \mathcal{F}, (X_t)_{t\geq 0}, (P_z)_{z \in E_\Delta})$ is an m-symmetric standard process with state space E and life time ζ where $m :=$ Lebesgue measure restricted to E. Hence by IV.6.8 $\mathbf{M}$ is properly associated with a quasi-regular Dirichlet form $(\mathcal{E}, D(\mathcal{E}))$ on $L^2(E; m)$. Since

$$P_z[\zeta < \infty] = 1 \quad \text{for all } z \in E$$

(cf. [Chu 82, p.149]), it is clear that for each $z \in E \setminus \{(-\frac{1}{2}, 0), (\frac{1}{2}, 0)\}$ $\mathbf{M}$ does not have all left limits on $]0, \zeta]$ in E P_z-a.s. . So, $\mathbf{M}$ is not a Hunt process. As we shall see in 2.25 below this means that $(\mathcal{E}, D(\mathcal{E}))$ is not s.quasi-regular. However, the fact that $(\mathcal{E}, D(\mathcal{E}))$ as above is not associated with a Hunt process is only true as long as we do not change the original topology of E (cf. VI.1.6 below).

(iv) Similarly to IV.2.8 (cf. also IV.2.9) one can show that (2.8) is equivalent with

$$(2.8)' \qquad \begin{array}{l} R_\alpha f \text{ is an s.}\mathcal{E}\text{-quasi-continuous } m\text{-version of } G_\alpha f \\ \text{for all } \alpha > 0, \ f \in \mathcal{B}_b(E) \cap L^2(E; m) , \end{array}$$

where $(R_\alpha)_{\alpha > 0}$ is the resolvent of $\mathbf{M}$.

The proof of 2.13 will be given in the next subsection. Now we discuss some consequences and examples related to 2.13.

Proposition 2.15. *Suppose that $(\mathcal{E}, D(\mathcal{E}))$ is a quasi-regular Dirichlet form on $L^2(E; m)$ such that $1 \in D(\mathcal{E})$ and Δ is adjoined to E as an isolated point of E_Δ. Then $(\mathcal{E}, D(\mathcal{E}))$ is always s.- and co-s.quasi-regular and hence associated with a pair of Hunt processes.*

Proof. Comparing IV.3.1(i)-(iii) and the above 2.11(i)-(iii) and taking the fact into account that Δ is an isolated point of E_Δ, we only need to show that $\text{Cap}_{1,g}$ (resp. $\text{Cap}_{h,1}$) is equivalent with $\text{Cap}_{h,g}$ for $h \in \mathcal{S}_0$, $g \in \hat{\mathcal{S}}_0$. We may take $h \in \mathcal{S}_0$, $g \in \hat{\mathcal{S}}_0$ such that $h \leq 1$, $g \leq 1$ and hence evidently $\text{Cap}_{h,g} \leq \text{Cap}_{1,g} \leq \text{Cap}_{1,1}$; $\text{Cap}_{h,g} \leq \text{Cap}_{h,1} \leq \text{Cap}_{1,1}$. Therefore, every increasing sequence $(F_k)_{k\in\mathbb{N}}$ of closed subsets of E with

$$\lim_{k\to\infty} \text{Cap}_{1,1}(F_k^c) = 0$$

is an $\mathcal{E}$-nest by III.2.11. But, if $(F_k)_{k\in\mathbb{N}}$ is an $\mathcal{E}$-nest, then by III.2.12 (and III.1.7)

$$\text{Cap}_{1,1}(F_k^c) = \mathcal{E}_1(1_{F_k^c}, 1_{F_k^c}) \xrightarrow[k\to\infty]{} 0$$

(where in this case $1_{F_k^c}$ means "reduced function" and not "indicator function"). Consequently, $\text{Cap}_{h,g}$ and $\text{Cap}_{1,1}$ are equivalent. $\qquad\square$

For example, suppose $(\mathcal{E}, D(\mathcal{E}))$ is a Dirichlet form on $L^2(E; m)$ with $1 \in D(\mathcal{E})$ such that: (i) $\mathrm{Cap}_{1,1}$ is tight; (ii) $C_b(E) \cap D(\mathcal{E})$ is $\tilde{\mathcal{E}}_1^{1/2}$-dense in $D(\mathcal{E})$; (iii) there is a countable family of $C_b(E) \cap D(\mathcal{E})$ separating the points of E. Then by 2.15 $(\mathcal{E}, D(\mathcal{E}))$ is always associated with a pair of Hunt processes provided Δ is assumed to be an isolated point of E_Δ. In the symmetric case the associated Hunt process can be constructed directly along the lines of the construction used in [F 80] and [AM 91 f]. This fact was first discovered by I.Shigekawa and T.Taniguchi ([ShiTa 91]) in the symmetric case.

We now turn to the case where E is locally compact and E_Δ is the one point compactification. As we already mentioned in the proof of 2.12, in this case condition 2.11(i) is always fulfilled. Hence 2.11(ii) and (iii) are necessary and sufficient for $(\mathcal{E}, D(\mathcal{E}))$ to be associated with a Hunt process. In particular, we have

Corollary 2.16. *Suppose that E is locally compact. Then a sufficient condition for $(\mathcal{E}, D(\mathcal{E}))$ to be associated with a Hunt process is given by*

$$
\begin{aligned}
&C_0(E) \cap D(\mathcal{E}) \text{ is } \tilde{\mathcal{E}}_1^{1/2}\text{-dense in} D(\mathcal{E}) \text{ and there is a countable set} \\
(2.9) \quad &\{u_n \mid n \in \mathbb{N}\} \subset C_0(E) \cap D(\mathcal{E}) \text{ separating the points of } E_\Delta \backslash N \\
&\text{for some s.}\mathcal{E}\text{-exceptional set } N \subset E\ .
\end{aligned}
$$

It is evident that every regular Dirichlet form on a locally compact separable metric space E fulfills condition (2.9). However, condition (2.9) is really weaker than assuming that $C_0(E) \cap D(\mathcal{E})$ is dense both in $D(\mathcal{E})$ w.r.t. $\tilde{\mathcal{E}}_1^{1/2}$ and in $C_0(E)$ w.r.t. uniform norm. Here is a simple example illustrating this point.

Example 2.17. Let $E := \mathbb{R}^d \cup \{\Delta\}$ be the one point compactification of $\mathbb{R}^d$ for some $d \geq 1$. Let m be the usual Lebesgue measure on $\mathbb{R}^d$ which is extended to E by setting $m(\{\Delta\}) = 0$. Let $(\mathcal{E}, D(\mathcal{E}))$ be the classical Dirichlet form on $L^2(\mathbb{R}^d; m)$ (cf. Chap.II, Subsection 1b)). We may regard $(\mathcal{E}, D(\mathcal{E}))$ as a Dirichlet form on $L^2(E; m)$. $(\mathcal{E}, D(\mathcal{E}))$ is not regular on $L^2(E; m)$ because $C(E) \cap D(\mathcal{E})$ is not dense in $C(E)(= C_0(E))$ w.r.t. the uniform norm (since 1_E cannot be approximated by functions in $L^2(E; m)$ w.r.t. the uniform norm.) But $(\mathcal{E}, D(\mathcal{E}))$ satisfies condition (2.9). Hence there exists a Hunt process with state space E associated with $(\mathcal{E}, D(\mathcal{E}))$. Indeed, if $\mathbf{M} := (\Omega, \mathcal{F}, (X_t)_{t \geq 0}, (P_z)_{z \in E_\Delta})$ is the Hunt process associated with $(\mathcal{E}, D(\mathcal{E}))$ such that Δ is a trap of $\mathbf{M}$ (i.e., $P_\Delta[X_t = \Delta,$ for all $t \geq 0] = 1$) and that the restriction of $\mathbf{M}$ to $\mathbb{R}^d$ is the classical Brownian motion, then $\mathbf{M}$ is associated with $(\mathcal{E}, D(\mathcal{E}))$. Note, however, that m is not a Radon measure on $(E, \mathcal{B}(E))$ in this case. More complicated examples of s.quasi-regular Dirichlet forms (where m is a Radon measure) which are not necessarily regular arise from the perturbation theory of Dirichlet forms. We shall turn to this topic in Subsection d) below.

c)   Proof of 2.13

Let ν be a σ-finite positive measure on $(E_\Delta, \mathcal{B}(E_\Delta))$. We call a right process $\mathbf{M} = (\Omega, \mathcal{F}, (X_t)_{t\geq 0}, (P_z)_{z\in E_\Delta})$ a ν-*Hunt process* if for one (and hence all) measures $\mu \in \mathcal{P}(E_\Delta)$ equivalent to ν, $\mathbf{M}$ satisfies IV.(M.8) and (M.9) with ζ replaced by ∞ and E by E_Δ. The proof of 2.13 follows by 2.18 (2.25, 2.26) and 2.27 below which contain a slightly stronger result than 2.13.

Proposition 2.18. *If* $(\mathcal{E}, D(\mathcal{E}))$ *is associated with an s.m-tight m-Hunt process* $\mathbf{M} = (\Omega, \mathcal{F}, (X_t)_{t\geq 0}, (P_z)_{z\in E_\Delta})$, *then* $(\mathcal{E}, D(\mathcal{E}))$ *is s.quasi-regular and* $\mathbf{M}$ *is s.properly associated with* $(\mathcal{E}, D(\mathcal{E}))$.

The proof of 2.18 is split into three claims below. Up to 2.22 below let $(\mathcal{E}, D(\mathcal{E}))$ and $\mathbf{M}$ be as in 2.18. By IV.5.1 we know that $(\mathcal{E}, D(\mathcal{E}))$ is quasi-regular. We fix $g \in \hat{\mathcal{S}}_0$ such that $g = \hat{G}_1\varphi$ with $\varphi \in L^1(E; m)$, $0 < \varphi \leq 1$ everywhere on E. We set $\mu := \varphi \cdot m$ and denote the resolvent of $\mathbf{M}$ by $(R_\alpha)_{\alpha>0}$.

Claim 1. $(\mathcal{E}, D(\mathcal{E}))$ satisfies 2.11(i).
Proof.   Let $(K_k)_{k\in\mathbb{N}}$ be an increasing sequence of compact metrizable subsets of E_Δ satisfying (2.7) (with $\nu = m$ or equivalently $\nu = \mu$). We set $E_k := K_k \backslash \{\Delta\}$, $k \in \mathbb{N}$. Then $(E_k)_{k\in\mathbb{N}}$ is an increasing sequence of closed subsets of E such that $E_k \cup \{\Delta\}$ is compact in E_Δ and $\sigma_{E_k^c} \geq \sigma_{E_\Delta \backslash K_k}$. Therefore, by 2.6 and (2.7)

$$(2.10) \qquad \mathrm{Cap}_{1,g}(E_k^c) = E_\mu[e^{-\sigma_{E_k^c}}] \leq E_\mu[e^{-\sigma_{E_\Delta \backslash K_k}}] \underset{k\to\infty}{\longrightarrow} 0 \ .$$

Thus $(E_k)_{k\in\mathbb{N}}$ is an s.$\mathcal{E}$-nest and Claim 1 is proved. $\square$

For proving 2.11(ii), we first strengthen 2.6 as follows.

Lemma 2.19. *Let* $B \in \mathcal{B}(E)$, *then*

$$\mathrm{Cap}_{1,g}(B) = E_\mu[e^{-D_B}] = E_\mu[e^{-S_B}],$$

where D_B *is defined by IV.(1.11) and*

$$S_B := \inf\{t \geq 0 \,|\, \overline{X_0^t} \cap B \neq \emptyset\} \ .$$

In this case $\overline{X_0^t}$ *denotes the closure of* $\{X_s \,|\, s \in [0,t]\}$ *in* E_Δ *(cf. IV.(5.14)). In particular,* $D_B = S_B \ P_\mu$-*a.e.* .

Proof.   First note that by [DM 78, IV.50] S_B is an $(\mathcal{F}_t)^{P_\mu}$-stopping time. If B is open, then $S_B = D_B = \sigma_B$. Hence in this case the assertion follows directly from 2.6. Since for $B \in \mathcal{B}(E)$ and $B \subset U \subset E$, U open, $S_U \leq S_B$ it follows that

$$(2.11) \qquad \mathrm{Cap}_{1,g}(B) \geq E_\mu[e^{-S_B}] \geq E_\mu[e^{-D_B}] \text{ for all } B \in \mathcal{B}(E) \ .$$

Now assume that B is closed. We set for $n, k \in \mathbb{N}$,

$$B_{n,k} := \{z \in E_k \mid \rho_k(z, B) < \frac{1}{n}\} \cup E_k^c ,$$

where E_k is as in the proof of Claim 1 and ρ_k is a metric on E_k compatible with the trace topology inherited from E. Let $D_{n,k} := D_{B_{n,k}}$ and $D_k := \lim_{n \to \infty} D_{n,k}$. Obviously, we have $D_{n,k} \leq D_B \wedge \sigma_{E_k^c}$ for all $n \in \mathbb{N}$ and hence $D_k \leq D_B \wedge \sigma_{E_k^c}$. But by IV.(M.9) with ζ replaced by ∞, we have that $X_{D_k} = \lim_{n \to \infty} X_{D_{n,k}}$ P_μ-a.s. and consequently, $X_{D_k} \in B \cap E_k$ P_μ-a.s. on $\{D_k < \sigma_{E_k^c}\}$. Therefore, $D_k = D_B \wedge \sigma_{E_k^c}$ P_μ-a.s.. Moreover, by (2.7) $\lim_{k \to \infty} \sigma_{E_k^c} \geq \lim_{k \to \infty} \sigma_{E_\Delta \setminus F_k} = \infty$ P_μ-a.s.. Thus we have

$$\lim_{k \to \infty} \lim_{n \to \infty} D_{n,k} = D_B \quad P_\mu\text{-a.s..}$$

Since $B \subset B_{n,k}$ and $B_{n,k}$ is open, we have by 2.6 (since $D_{n,k} = \sigma_{B_{n,k}}$)

$$\mathrm{Cap}_{1,g}(B) \leq \mathrm{Cap}_{1,g}(B_{n,k}) = E_\mu[e^{-D_{n,k}}] \quad \text{for all } n, k \in \mathbb{N} .$$

First letting $n \to \infty$ and then letting $k \to \infty$ we obtain $\mathrm{Cap}_{1,g}(B) \leq E_\mu[e^{-D_B}]$ and by (2.11) the assertion is proved for closed B. Let now $B \in \mathcal{B}(E)$ be arbitrary. For $k \in \mathbb{N}$, $\mathrm{Cap}_{1,g}$ restricted to E_k is a finite Choquet capacity (cf. 2.3, 2.5(i)), hence by III.2.9(i) there exist compact sets $K_n \subset B \cap E_k$, $n \in \mathbb{N}$, such that

$$\mathrm{Cap}_{1,g}(B \cap E_k) = \lim_{n \to \infty} \mathrm{Cap}_{1,g}(K_n) = \lim_{n \to \infty} E_\mu[e^{-D_{K_n}}] \leq E_\mu[e^{-D_B}] .$$

Therefore,

$$\mathrm{Cap}_{1,g}(B) \leq \mathrm{Cap}_{1,g}(B \cap E_k) + \mathrm{Cap}_{1,g}(E_k^c) \leq E_\mu[e^{-D_B}] + \mathrm{Cap}_{1,g}(E_k^c) ,$$

and by (2.10), (2.11) we get the desired assertion. $\qquad\square$

We also need the following variant of IV.5.14.

Lemma 2.20. *Let $\tilde{C}_s$ denote the set of all $f \in \mathcal{B}_b(E)$ such that $t \mapsto f(X_t)$ is right continuous on $[0, \infty[$ and $t \mapsto f(X_{t-})$ is left continuous on $]0, \infty[$ P_μ-a.s.. Then $R_\alpha f \in \tilde{C}_s$ for all $f \in \mathcal{B}_b(E)$, $\alpha > 0$.*

Proof. See the proof of IV.5.14 with ζ replaced by ∞. $\qquad\square$

Claim 2. $(\mathcal{E}, D(\mathcal{E}))$ satisfies 2.11(ii) and $\mathbf{M}$ is s.properly associated with $(\mathcal{E}, D(\mathcal{E}))$.

Proof. We define for $f \in \mathcal{B}_b(E)$.

$$\|f\|_s := E_\mu[\sup_{t \geq 0}(e^{-t}|f(X_t)| \vee e^{-t}|f(X_{t-})|)] ,$$

Then $\| \ \|_s$ is a seminorm on $\mathcal{B}_b(E)$. Using 2.19 it is obvious that $\mathrm{Cap}_{1,g}(B) = \|1_B\|_s$ for any $B \in \mathcal{B}(E)$. Let $C_\infty(E)$ be the set of all $f \in \mathcal{B}_b(E)$ such that f is continuous on E_Δ. (Recall that by convention $f(\Delta) = 0$). Following the argument of IV.5.23 we can show that $C_\infty(E)$ is dense in $\tilde{C}_s$ w.r.t. $\| \ \|_s$. As in the proof of IV.5.24 by 2.19 we can then conclude that each element of $\tilde{C}_s$ is s.$\mathcal{E}$-quasi-continuous. In particular, by 2.20 every $R_\alpha f$ is s.$\mathcal{E}$-quasicontinuous for $f \in \mathcal{B}_b(E)$ and $\alpha > 0$. Since $\{G_\alpha f \mid f \in L^2(E; m), \alpha > 0\}$ is dense in $D(\mathcal{E})$ by I.2.13(ii) and because of 2.14(iv) the claim is proven. $\qquad\square$

Lemma 2.21. *Let $(F_k)_{k\in\mathbb{N}}$ be an s.$\mathcal{E}$-nest. Then*

$$(2.12) \qquad P_z[\lim_{k\to\infty} \sigma_{F_k^c} = \infty] = 1 \text{ for s.}\mathcal{E}\text{-q.e. } z \in E$$

Proof. Without loss of generality we assume that μ is a probability measure. Let $(\mathcal{F}_t)_{t\geq 0}$ be the natural filtration of **M**. Replacing $(\mathcal{F}_t)$ by $(\mathcal{F}_t^{P_\mu})$ and $\mathcal{F}$ by $\mathcal{F}^{P_\mu}$ we may assume that $(\mathcal{F}_t)_{t\geq 0}$ satisfies the usual conditions on $(\Omega, \mathcal{F}, P_\mu)$. Let $f_k(z) := E_z[e^{-\sigma_{F_k^c}}]$ and $f(z) := \lim_{k\to\infty} f_k(z)$, $z \in E$. For an arbitrary bounded $(\mathcal{F}_t)$-stopping time τ we have for $k \in \mathbb{N}$,

$$\tau + \sigma_{F_k^c} \circ \theta_\tau = \inf\{t > \tau \mid X_t \in F_k^c\} \geq \sigma_{F_k^c} .$$

Hence by the strong Markov property (more precisely, by IV.1.9(v)) and 2.19

$$\begin{aligned} E_\mu[e^{-\tau} f_k(X_\tau)] &= E_\mu[e^{-\tau} E_{X_\tau}[e^{-\sigma_{F_k^c}}]] = E_\mu[e^{-(\tau + \sigma_{F_k^c}\circ\theta_\tau)}] \\ &\leq E_\mu[e^{-\sigma_{F_k^c}}] = \mathrm{Cap}_{1,g}(F_k^c) . \end{aligned}$$

Therefore,

$$0 \leq E_\mu[e^{-\tau} f(X_\tau)] = \lim_{k\to\infty} E_\mu[e^{-\tau} f_k(X_\tau)] \leq \lim_{k\to\infty} \mathrm{Cap}_{1,g}(F_k^c) = 0 .$$

It follows by [DM 82, IV.48] that $(e^{-t} f(X_t))_{t\geq 0}$ is right continuous P_μ-a.s., since by monotone class arguments it is optional. This in turn implies that

$$(2.13) \qquad P_\mu[\omega \in \Omega \mid e^{-t} f(X_t)(\omega) \neq 0 \text{ for some } t \geq 0] = 0 ,$$

since $e^{-t} f(X_t) = 0$ P_μ-a.s. for all rational t. We now set $B := \{z \in E \mid f(z) \neq 0\}$. Then (2.13) implies that $D_B = \infty$ P_μ-a.s. . Consequently, by 2.19 $\mathrm{Cap}_{1,g}(B) = 0$, i.e., $f = 0$ s.$\mathcal{E}$-q.e., which is equivalent with (2.12). $\qquad\square$

Claim 3. $(\mathcal{E}, D(\mathcal{E}))$ satisfies 2.11(iii).

Proof. Let $h := R_1\varphi$. By 2.20 and the proof of Claim 2 we know that h is s.$\mathcal{E}$-quasi-continuous. Define $\tilde{h}_F(z) := E_z[\int_{S_F}^\infty e^{-s}\varphi(X_s)\,ds]$ for $z \in E$ and $F \subset E$, F compact. Applying 2.10, 2.19-2.21 one shows following the arguments in the proof of IV.5.25(ii) that $\tilde{h}_F$ is an s.$\mathcal{E}$-quasi-continuous m-version of an element in $D(\mathcal{E})$ and that $F = \{\tilde{h}_F = h\}$. Let F_n, $n \in \mathbb{N}$, be a countable family of compact subsets of E such that $\bigcup_{n\geq 1} F_n^c = E$ and separating the points of $Y := \bigcup_{k\geq 1} E_k$ where E_k is as specified in the proof of Claim 1 (cf. Chap.IV, Subsection 5c)). Let $u_n := h - \tilde{h}_{F_n}$, $n \in \mathbb{N}$. Then each u_n is an s.$\mathcal{E}$-quasi-continuous m-version of an element in $D(\mathcal{E})$ and $\{u_n \mid n \in \mathbb{N}\}$ separates the points of Y. Moreover, since $h > 0$ on E, for each $z \in Y$ we can find u_n such that $u_n(z) > 0$. Therefore, $\{u_n \mid n \in \mathbb{N}\}$ separates also the points of $Y \cup \{\Delta\}$. $\qquad\square$

Thus the proof of 2.18 is complete. We now turn to the proof of the sufficiency part of 2.13. We first list several properties of s.quasi-regular Dirichlet forms whose proofs are very similar to the corresponding arguments in Chapter IV. We leave the details to the reader as exercises.

Exercise 2.22. Suppose $(\mathcal{E}, D(\mathcal{E}))$ is s.quasi-regular.

(i) We can assume that all E_k in 2.11(i) are metrizable (cf. IV. 3.2(iii))

(ii) Under the assumption of (i), we have $\sigma\{\tilde{u}_n \mid n \in \mathbb{N}\} \supset \mathcal{B}(E_\Delta \backslash N)$ where $N, \tilde{u}_n, n \in \mathbb{N}$, are as in 2.11(iii) (cf. IV.3.2(iv)).

(iii) Each element $u \in D(\mathcal{E})$ has an s.$\mathcal{E}$-quasi-continuous m-version denoted by $\tilde{u}$ (cf. IV. 3.3 (ii)).

(iv) If f is s.$\mathcal{E}$-quasi-continuous and $f \geq 0$ m-a.e. on an open subset U of E, then $f \geq 0$ s.$\mathcal{E}$-q.e. on U (cf. IV.3.3(iii)).

(v) If D_1 contains one s.$\mathcal{E}$-quasi-continuous m-versions of each element in an $(\tilde{\mathcal{E}}_1^{1/2}$-) dense subset of $D(\mathcal{E})$, then there exists an s.$\mathcal{E}$-exceptional set $N \subset E$ such that D_1 separates the points of $E_\Delta \backslash N$ (cf. IV.3.4(i)).

In what follows we assume that our fixed Dirichlet form $(\mathcal{E}, D(\mathcal{E}))$ on $L^2(E; m)$ is s.-quasi-regular. Let $Y := \bigcup_{k \geq 1} E_k$ with E_k, $k \in \mathbb{N}$, as in 2.11(i). By 2.22(i) we may assume that each E_k is metrizable. Then Y is a (topological) Lusin space and we may identify $L^2(E; m)$ with $L^2(Y; m)$ canonically (cf. IV.3.2(iii)). Following the argument of IV.3.7 we have the following result whose proof we leave as an exercise.

Lemma 2.23. *Let $\alpha > 0$. There exists a kernel $\tilde{R}_\alpha$ from $(E, \mathcal{B}(E))$ to $(Y, \mathcal{B}(Y))$ such that $\tilde{R}_\alpha(z, Y) \leq 1$ for all $z \in E$, and $\tilde{R}_\alpha f$ is an s.$\mathcal{E}$-quasi-continuous m-version of $G_\alpha f$ for each $f \in L^2(Y; m)$.*

Lemma 2.24. *Let $\tilde{R}_\alpha$, $\alpha > 0$, be as specified in 2.23. Then there exists a countable family D of bounded m-versions of a dense set in $D(\mathcal{E})$ and an s.$\mathcal{E}$-nest $(F_k)_{k \in \mathbb{N}}$ satisfying the following properties:*

(i) *D is a $\mathbb{Q}$-algebra and $u \in D$ implies $\tilde{R}_\alpha u \in D$ for any $\alpha \in \mathbb{Q}_+^*$.*

(ii) *$F_k \subset E_k$ for all $k \in \mathbb{N}$.*

(iii) *$u \in C_\infty(\{F_k\})$ for all $u \in D$ and there exist $\alpha_n \in \mathbb{Q}_+^*$, $n \in \mathbb{N}$, such that $\lim_{n \to \infty} \alpha_n \tilde{R}_{\alpha_n} u(z) = u(z)$ for all $u \in D$, $z \in Y_1 := \bigcup_{k \geq 1} F_k$.*

(iv) *$\{\tilde{R}_1 u \mid u \in D\}(\subset D)$ separates the points of $Y_\Delta := Y_1 \cup \{\Delta\}$.*

Proof. Let D_0 be a countable family of bounded functions in $L^2(E; m)$ such that D_0 is dense in $L^2(E; m)$. Let $D_1 := \{\tilde{R}_1 u \mid u \in D_0\}$. Then D_1 is dense in $D(\mathcal{E})$ (cf. the proof of IV.3.3(i)). Let D be the smallest family of bounded s.$\mathcal{E}$-quasi-continuous functions which is a $\mathbb{Q}$-algebra containing D_1 such that $\tilde{R}_\alpha u \in D$ for all $u \in D$, $\alpha \in \mathbb{Q}_+^*$. Then it follows from IV.3.9 that D is again a countable family. By definition D satisfies (i). By I.2.13(ii) and the strict version of III.3.5 (cf. 2.10), and applying the usual diagonal argument we can find an increasing sequence $(\alpha_n)_{n \in \mathbb{N}}$ in $\mathbb{Q}_+^*$ and an s.$\mathcal{E}$-exceptional set N_1 such that $\lim_{n \to \infty} \alpha_n \tilde{R}_{\alpha_n} u(z) = u(z)$ for all $z \in E \backslash N_1$ and all $u \in D$. Since $\{\tilde{R}_1 u \mid u \in D\}$

is dense in $D(\mathcal{E})$, by 2.22(v) there exists an s.$\mathcal{E}$-exceptional set N_2 such that $\{\tilde{R}_1 u \,|\, u \in D\}(\subset D)$ separates the points of $E_\Delta \backslash N_2$. Let $(F_{1k})_{k\in\mathbb{N}}$ be an s.$\mathcal{E}$-nest such that $(N_1 \cup N_2) \subset \cap F_{1k}^c$. By the strict version of III.3.3 we may take an s.$\mathcal{E}$-nest $(F_{2k})_{k\in\mathbb{N}}$ such that $u \in C_\infty(\{F_{2k}\})$ for all $u \in D$. Let $F_k := F_{1k} \cap F_{2k} \cap E_k$, then $(F_k)_{k\in\mathbb{N}}$ is again an s.$\mathcal{E}$-nest. D and $(F_k)_{k\in\mathbb{N}}$ are as desired. $\square$

Proposition 2.25. *Let* $\mathbf{M} = (\Omega, \mathcal{F}, (X_t)_{t\geq 0}, (P_z)_{z\in E_\Delta})$ *be a right process on* E *with life time* ζ *which is properly associated with the s.quasi-regular Dirichlet form* $(\mathcal{E}, D(\mathcal{E}))$. *Then* $\mathbf{M}$ *is an s.m-tight m-Hunt process and is s.properly associated with* $(\mathcal{E}, D(\mathcal{E}))$.

Proof. The last assertion follows by 2.18. To prove the first, let $(R_\alpha)_{\alpha>0}$ denote the resolvent of $\mathbf{M}$. Let φ, g and μ be as defined after 2.18. Taking $K_k := F_k \cup \{\Delta\}$, $k \in \mathbb{N}$, with $(F_k)_{k\in\mathbb{N}}$ as in 2.24, by 2.6 one can easily see that $\mathbf{M}$ is s.m-tight. We now verify that $\mathbf{M}$ is an m-Hunt process. Let D and $\tilde{R}_\alpha$ be as specified in 2.24 and let $u \in D$. Since every s.$\mathcal{E}$-quasi-continuous function is $\mathcal{E}$-quasi-continuous (cf. 2.8), $\tilde{R}_\alpha u$ is also an $\mathcal{E}$-quasi-continuous m-version of $G_\alpha u$ for all $u \in D$. Consequently, by IV.2.8 and IV.3.3(iii) we have for $\alpha \in \mathbb{Q}_+^*$ and $u \in D$,

$$\tilde{R}_\alpha u = R_\alpha u := E.[\int_0^\infty e^{-\alpha t} u(X_t)\, dt] \quad \mathcal{E}\text{-q.e.} \; .$$

By IV.6.5 there exists an $\mathcal{E}$-exceptional set B such that $E\backslash B$ is $\mathbf{M}$-invariant and $\tilde{R}_\alpha u(z) = R_\alpha u(z)$ for all $z \in E\backslash B$. Hence

$$(2.14) \qquad\qquad \tilde{R}_\alpha u(X_t) = R_\alpha u(X_t) \quad P_\mu\text{-a.s. on } \{t < \zeta\}$$

Since $\tilde{R}_\alpha u(X_t(\omega)) = R_\alpha u(X_t(\omega)) = 0$ for all $t \geq \zeta(\omega)$, from (2.14) we conclude that $\{\tilde{R}_\alpha u(X_t) \neq R_\alpha u(X_t)$ for some $t \geq 0\} \in \mathcal{F}^{P_\mu}$ and is of P_μ-measure zero. We now set

$$(2.15) \qquad \Omega_0 := \{\tilde{R}_\alpha u(X_t) = R_\alpha u(X_t) \text{ for all } t \geq 0, \alpha \in \mathbb{Q}_+^*, u \in D\}\; .$$

and

$$\Omega_1 := \{\omega \in \Omega_0 \,|\, t \mapsto R_1 u(X_t)(\omega) \text{ is right continuous on } [0, \infty[\text{ and}$$
$$\text{has left limits on }]0, \infty[\text{ for all } u \in D\}\; .$$

Then $\Omega_1 \in \mathcal{F}^{P_\mu}$ and $P_\mu[\Omega_1] = 1$, since $(e^{-t} R_1 u(X_t))_{t\geq 0}$ is cadlag for all $u \in D$ by Claim 1 in the proof of IV.5.14 (cf. also IV.5.13). We set $\sigma_\infty := \lim_{k\to\infty} \sigma_{F_k^c}$. By the s.$m$-tightness of $\mathbf{M}$ we have that $\sigma_\infty = \infty$ P_μ-a.s.. Let

$$(2.16) \qquad\qquad \Omega_2 := \{\omega \in \Omega_1 \,|\, \sigma_\infty(\omega) = \infty\}$$

Then $P_\mu[\Omega_2] = 1$. We shall show that

$$(2.17) \qquad \Omega_2 \subset \{\omega \in \Omega \,|\, X_{t-}(\omega) \text{ exists in } Y_\Delta \text{ for all } t \in]0, \infty[\}$$

where Y_Δ is as defined in 2.24(iv). Indeed, let $\omega \in \Omega_2$ and $t \in]0,\infty[$. Then $t < \sigma_{F_k^c}(\omega)$ for some $k \in \mathbb{N}$, and hence $X_s(\omega)$ is in the compact set $F_k \cup \{\Delta\}$ for all $0 < s \le t$. For an arbitrary increasing sequence $(t_n)_{n\in\mathbb{N}}$ in $]0,t[$, with limit t, there exists a subsequence $t'_n \uparrow t$ and a point $z \in F_k \cup \{\Delta\}$ such that $X_{t'_n}(\omega) \to z$ as $n \to \infty$. If $(s_n)_{n\in\mathbb{N}}$ is another sequence in $]0,t[$ satisfying $s_n \uparrow t$ and $X_{s_n}(\omega) \to y$ as $n \to \infty$ for some $y \in F_k \cup \{\Delta\}$, then, since $\tilde{R}_1 u$ is continuous on $F_k \cup \{\Delta\}$ for all $u \in D$, we have

$$\tilde{R}_1 u(z) = \lim_{n\to\infty} \tilde{R}_1 u(X_{t'_n}(\omega)) = R_1 u(X_t(\omega))_- = \lim_{n\to\infty} \tilde{R}_1 u(X_{s_n}(\omega)) = \tilde{R}_1 u(y) \ .$$

Thus $z = y$ by 2.24(iv), proving (2.17). Let $(\tau_n)_{n\in\mathbb{N}}$ be an increasing sequence of $(\mathcal{F}_t^{P_\mu})$-stopping times with limit τ. The proof will be complete if we can show

$$(2.18) \qquad P_\mu[\lim_{n\to\infty} X_{\tau_n} \ne X_\tau, \ \tau < \infty] = 0 \ .$$

To this end we first assume that τ is bounded and set $V := \lim_{n\to\infty} X_{\tau_n}$. Note that $V \in Y_1$ and $X_\tau \in Y_\Delta$ P_μ-a.s., since $\mathbf{M}$ is s.m-tight (where Y_1 is as in 2.24(iii)). Let $g \in C_b(Y_\Delta)$ and $f \in D$, then for $\alpha \in \mathbb{Q}_+^*$ we have by the strong Markov property that

$$
\begin{aligned}
E_\mu[g(V)\tilde{R}_\alpha f(X_\tau)] &= E_\mu[g(V)R_\alpha f(X_\tau)] = E_\mu[g(V)E_{X_\tau}[\int_0^\infty e^{-\alpha t} f(X_t)\,dt]] \\
&= E_\mu[g(V)e^{\alpha\tau}\int_\tau^\infty e^{-\alpha t} f(X_t)\,dt] = \lim_{n\to\infty} E_\mu[g(X_{\tau_n})e^{\alpha\tau_n}\int_{\tau_n}^\infty e^{-\alpha t} f(X_t)\,dt] \\
&= \lim_{n\to\infty} E_\mu[g(X_{\tau_n})R_\alpha f(X_{\tau_n})] = \lim_{n\to\infty} E_\mu[g(X_{\tau_n})\tilde{R}_\alpha f(X_{\tau_n})] \\
&= E_\mu[g(V)\tilde{R}_\alpha f(V)] \ ,
\end{aligned}
$$

where we used the s.m-tightness of $\mathbf{M}$ again in the last step. Let $\alpha_n, n \in \mathbb{N}$, be as specified in 2.24(iii). Then $\lim_{n\to\infty} \alpha_n \tilde{R}_{\alpha_n} f(z) = f(z)$ for all $z \in Y_1$. But evidently, $\lim_{n\to\infty} \alpha_n \tilde{R}_{\alpha_n} f(\Delta) = 0 = f(\Delta)$. Replacing α above by α_n and letting $n \to \infty$, we conclude that

$$(2.19) \quad E_\mu[g(V)f(X_\tau)] = E_\mu[g(V)f(V)] \ \text{ for all } g \in C_b(Y_\Delta), f \in D \ .$$

But if $f = 1_{Y_\Delta}$, then (2.19) holds trivially. Using 2.24(i) and (iv), we conclude by monotone class arguments that for any $\Phi \in \mathcal{B}_b(Y_\Delta \times Y_\Delta)$

$$E_\mu[\Phi(V, X_\tau)] = E_\mu[\Phi(V, V)] \ .$$

In particular, choosing $\Phi = $ indicator function of the diagonal in $Y_\Delta \times Y_\Delta$, we obtain (2.18) for bounded τ. But the general case follows easily from the equality

$$P_\mu[\lim_{n\to\infty} X_{\tau_n} \ne X_\tau, \ \tau < \infty] = P_\mu[\bigcup_{k\in\mathbb{N}} \{\lim_{n\to\infty} X_{\tau_n \wedge k} \ne X_{\tau \wedge k}\}]$$

$\square$

Proposition 2.26. *Let* $\mathbf{M} = (\Omega, \mathcal{F}, (X_t)_{t\geq 0}, (P_z)_{z\in E_\Delta})$ *be a right process properly associated with the s.quasi-regular Dirichlet form* $(\mathcal{E}, D(\mathcal{E}))$. *Then* $\mathbf{M}$ *is a Hunt process up to a modification on an s.$\mathcal{E}$-exceptional set. More precisely, there exists* $S \in \mathcal{B}(E)$ *such that* $E \backslash S$ *is s.$\mathcal{E}$-exceptional and if we set* $S_\Delta := S \cup \{\Delta\}$ *and*

$$(2.20) \qquad \begin{aligned} \tilde{\Omega} := \{&X_t \in S_\Delta \text{ for all } t \in [0, \infty[\\ &\text{and } X_{t-} \text{ exists in } S_\Delta \text{ for all } t \in]0, \infty[\} \,, \end{aligned}$$

then $P_z[\tilde{\Omega}] = 1$ *for all* $z \in S$ *and the trivial extension* $\overline{\mathbf{M}}$ *(cf. IV.3.23(i)) of*

$$(2.21) \qquad \mathbf{M}_{|S} := (\tilde{\Omega}, \mathcal{F} \cap \tilde{\Omega}, (\tilde{X}_t)_{t\geq 0}, (\tilde{P}_z)_{z\in S_\Delta})$$

to E *is an s.m-tight Hunt process s.properly associated with* $(\mathcal{E}, D(\mathcal{E}))$. *(Here* $\tilde{X}_t$, $t \geq 0$, $\tilde{P}_z$, $z \in S_\Delta$, *denote the respective restrictions to* $\tilde{\Omega}$ *and* $\mathcal{F} \cap \tilde{\Omega}$.)

Proof. By 2.25 $\mathbf{M}$ is an s.m-tight m-Hunt process s.properly associated with $(\mathcal{E}, D(\mathcal{E}))$. Let $\tilde{R}_\alpha$ and D be as specified in 2.24 and let $(R_\alpha)_{\alpha>0}$ be the resolvent of $\mathbf{M}$. We set

$$(2.22) \qquad N_1 := \{\tilde{R}_\alpha u \neq R_\alpha u \text{ for some } u \in D \text{ and some } \alpha \in \mathbb{Q}^*_+\} \,.$$

For $\omega \in \Omega_0$ (defined by (2.15)) we have $D_{N_1}(\omega) = \infty$. Consequently, $E_\mu[e^{-D_{N_1}}] = 0$ which implies $\mathrm{Cap}_{1,g}(N_1) = 0$ by 2.19. Let $(F_k)_{k\in\mathbb{N}}$ be as specified in 2.24. We take an s.$\mathcal{E}$-nest $(F_{1k})_{k\in\mathbb{N}}$ such that $F_{1k} \subset F_k$ for all $k \in \mathbb{N}$ and $N_1 \subset \bigcap F_{1k}^c$. Set $E_1 := \bigcup F_{1k}$. Define $\sigma_{1k} := \sigma_{F_{1k}^c}$. By 2.21 there exists $N_2 \in \mathcal{B}(E)$ with $\mathrm{Cap}_{1,g}(N_2) = 0$ such that $P_z[\lim_{k\to\infty} \sigma_{1k} = \infty] = 1$ for all $z \in E \backslash N_2$. Similarly as in the proof of 2.25 one shows that for all $z \in E \backslash N_2$, $\mathbf{M}$ is an ε_z-Hunt process (ε_z denotes the Dirac measure at z) and that

$$(2.23) \qquad \begin{aligned} P_z[&X_t \in E_1 \cup \{\Delta\} \text{ for all } t \in [0, \infty[\\ &\text{and } X_{t-} \in E_1 \cup \{\Delta\} \text{ for all } t \in]0, \infty[] = 1 \,. \end{aligned}$$

Let $(F_{2k})_{k\in\mathbb{N}}$ be an s.$\mathcal{E}$-nest such that $F_{2k} \subset F_{1k}$ for all $k \in \mathbb{N}$ and $N_2 \subset \bigcap F_{2k}^c$. Set $E_2 := \bigcup F_{2k}$. Repeating the above argument we may find $N_3 \in \mathcal{B}(E)$ with $\mathrm{Cap}_{1,g}(N_3) = 0$ such that for each $z \in N_3$, (2.23) holds with $E_2 \cup \{\Delta\}$ replacing $E_1 \cup \{\Delta\}$. In this way we can construct a decreasing sequence of Borel sets $E_n, n \in \mathbb{N}$, with $\mathrm{Cap}_{1,g}(E \backslash E_n) = 0$ for all $n \in \mathbb{N}$ such that for all $z \in E_{n+1}$

$$(2.24) \qquad \begin{aligned} P_z[&X_t \in E_n \cup \{\Delta\} \text{ for all } t \in [0, \infty[\\ &\text{and } X_{t-} \in E_n \cup \{\Delta\} \text{ for all } t \in]0, \infty[] = 1 \,. \end{aligned}$$

We now set $S := \bigcap_{n\in\mathbb{N}} E_n$. Then S satisfies all the properties stated in the assertion. $\qquad\square$

Corollary 2.27. *There exists an s.m-tight Hunt process s.properly associated with the s.quasi-regular Dirichlet form* $(\mathcal{E}, D(\mathcal{E}))$.

Proof. Since by 2.12 $(\mathcal{E}, D(\mathcal{E}))$ is quasi-regular, by IV.3.5 there is a right process properly associated with $(\mathcal{E}, D(\mathcal{E}))$. Applying 2.26 we obtain the assertion. $\qquad\square$

We conclude this subsection with a result corresponding to IV.5.30.

Proposition 2.28. *Let* $\mathbf{M} = (\Omega, \mathcal{F}, (X_t)_{t\geq 0}, (P_z)_{z\in E_\Delta})$ *be a right process on* E *properly associated with the s.quasi-regular Dirichlet form* $(\mathcal{E}, D(\mathcal{E}))$.

(i) *Let* $(F_k)_{k\in\mathbb{N}}$ *be an s.$\mathcal{E}$-nest, then*

$$P_z[\lim_{k\to\infty} \sigma_{E\setminus F_k} < \infty] = 0 \quad \text{for s.}\mathcal{E}\text{-q.e. } z \in E .$$

(ii) *Let* $u : E \to \mathbb{R}$ *be s.$\mathcal{E}$-quasi-continuous, then for s.$\mathcal{E}$-q.e.* $z \in E$, $u(X_{t-}) = u(X_t)_-$ *for all* $t \in]0,\infty[$ P_z-*a.s. and*

$$P_z[t \mapsto u(X_t) \text{ is right continuous on } [0,\infty[\text{ and}$$
$$t \mapsto u(X_{t-}) \text{ is left continuous on }]0,\infty[] = 1 .$$

Proof. By 2.25 we are in fact in the situation of 2.18, hence 2.21 applies to prove (i). (ii) is then an immediate consequence by virtue of 2.26. $\qquad\square$

Examples 2.29. (i) As a consequence of 2.12(ii) and 2.13 we obtain that all the Dirichlet forms in Chap.II, Sections 1,2, where E was an open subset of $\mathbb{R}^d$ and $C_0^\infty(U)$ was dense in $D(\mathcal{E})$ w.r.t. $\tilde{\mathcal{E}}_1^{1/2}$, are associated with Hunt processes as regular Dirichlet forms.

(ii) All Dirichlet forms in Chap.II, Section 3 are s.quasi-regular by 2.15, hence are associated with Hunt processes. This, however, already follows from the results of Chap.IV, Subsection 4b). Indeed, clearly 1 is in the kernel of the corresponding generators. Hence $T_t 1 = 1$ for all $t > 0$ for the respective semigroups $(T_t)_{t>0}$. Therefore, the life times of the associated m-special standard processes, which exist according to Chap.IV., Subsection 4b), are infinite, so they are trivially Hunt processes.

d)   Perturbation of Dirichlet forms

In this subsection we assume that E is a locally compact separable metric space and m is a positive Radon measure on E. Let $(\mathcal{E}, D(\mathcal{E}))$ be a symmetric regular Dirichlet form on $L^2(E; m)$ and $\mathbf{M} = (\Omega, \mathcal{F}, (X_t)_{t\geq 0}, (P_z)_{z\in E_\Delta})$ a Hunt process associated with $(\mathcal{E}, D(\mathcal{E}))$. Let ν be a smooth measure of $(\mathcal{E}, D(\mathcal{E}))$ corresponding to a PCAF (A_t^ν) (cf. IV.(4.7)). We consider the perturbation $(\mathcal{E}^\nu, D(\mathcal{E}^\nu))$ of $(\mathcal{E}, D(\mathcal{E}))$ by ν (cf. IV.(4.11), (4.12)). To avoid confusion, all notations relative to $(\mathcal{E}^\nu, D(\mathcal{E}^\nu))$ will be marked by a superscript ν, e.g. $\mathrm{Cap}_{1,g}^\nu$ stands for the capacity (cf.(2.3)) relative to $(\mathcal{E}^\nu, D(\mathcal{E}^\nu))$, etc.. A function u is called *strictly* $\mathrm{Cap}_{1,1}$-*quasi-continuous* if it is quasi-continuous in the restricted sense relative to $(\mathcal{E}, D(\mathcal{E}))$ in the sense of [F 80, p.64]. In what follows we fix a function $\varphi \in \mathcal{B}_b(E) \cap L^1(E; m)$ such that $0 < \varphi \leq 1$ everywhere on E. Let

$$(2.25) \qquad \begin{aligned} g &:= E_{\cdot}\left[\int_0^\infty e^{-t}\varphi(X_t)\,dt\right] , \\ g^\nu &:= E_{\cdot}\left[\int_0^\infty e^{-t-A_t^\nu}\varphi(X_t)\,dt\right] , \end{aligned}$$

provided the integrals make sense. Then $g \in \hat{\mathcal{S}}_0$ and $g^\nu \in \hat{\mathcal{S}}_0^\nu$. (The latter can be seen by IV.4.4.)

Lemma 2.30. *Let g and g^ν be as above. Then for any $A \subset E$*

$$(2.26) \qquad \mathrm{Cap}_{1,g^\nu}^\nu(A) \leq \mathrm{Cap}_{1,g}(A) \leq \mathrm{Cap}_{1,1}(A) .$$

Proof. Since $D(\mathcal{E}^\nu) \subset D(\mathcal{E})$ (cf. IV.(4.11)), every element of $D(\mathcal{E}^\nu)$ admits a strictly $\mathrm{Cap}_{1,1}$-quasi-continuous m-version. For a $\mathrm{Cap}_{1,1}$-quasi-continuous function $u \in D(\mathcal{E}^\nu)$ and an open set $U \subset E$, we introduce the notation

$$(2.27) \qquad \tilde{u}_U^\nu(z) = E_z[e^{-\sigma_U - A_{\sigma_U}^\nu} u(X_{\sigma_U})] ,$$

provided the integral makes sense. Now let $u = R_1^\nu f := E_{\cdot}[\int_0^\infty e^{-t-A_t^\nu} f(X_t)\,dt]$ for some positive $f \in L^2(E;m)$. Then by IV.4.4 we see that $u \in D(\mathcal{E}^\nu)$ and u is $\mathrm{Cap}_{1,1}$-quasi-continuous. Moreover, by the strong Markov property one easily checks that $\tilde{u}_U^\nu = R_1^\nu f - R_1^{\nu,U^c} f$ (for the notations see IV.(4.10)). Thus, employing IV.(4.14) we conclude that $\tilde{u}_U^\nu$ is an m-version of $(u^\nu)_U'$ where $(u^\nu)_U'$ is as specified in III.1.4 with $\mathcal{E}$ replaced by $\mathcal{E}^\nu$. This enables us to conclude by a similar argument as in the proof of 1.6, that $\tilde{u}_U^\nu$ is an m-version of $(u^\nu)_U'$ for any $\mathrm{Cap}_{1,1}$-quasi-continuous function $u \in D(\mathcal{E}^\nu)$. Consequently, a simple modification of the proof of 2.6 yields

$$(2.28) \qquad \mathrm{Cap}_{1,g^\nu}^\nu(U) = E_{\varphi \cdot m}[e^{-\sigma_U - A_{\sigma_U}^\nu}] \text{ for all } U \subset E, U \text{ open} .$$

Comparing (2.28) with (2.5) we get the first inequality in (2.26), while the second one in (2.26) is evident (cf. 2.8). $\qquad\square$

As an immediate consequence we obtain

Corollary 2.31. *Any strictly $\mathrm{Cap}_{1,1}$-quasi-continuous function is s.$\mathcal{E}^\nu$-quasi-continuous.*

Theorem 2.32. *$(\mathcal{E}^\nu, D(\mathcal{E}^\nu))$ is an s.quasi-regular Dirichlet form on $L^2(E;m)$ and hence there is a Hunt process associated with $(\mathcal{E}^\nu, D(\mathcal{E}^\nu))$.*

Proof. We have to check that $(\mathcal{E}^\nu, D(\mathcal{E}^\nu))$ satisfies conditions 2.11(i)-(iii). But 2.11(i) is trivially fulfilled by taking $E_k = E$ for all $k \in \mathbb{N}$. 2.11(ii) follows directly from 2.31 since $D(\mathcal{E}^\nu) \subset D(\mathcal{E})$ and any element of $D(\mathcal{E})$ admits a strictly $\mathrm{Cap}_{1,1}$-quasi-continuous m-version (cf. [F 80, Theorem 3.1.3]). Hence we only need to check 2.11(iii). To this end let $\{f_k \mid k \in \mathbb{N}\}$ be a countable family of functions in $C_0(E)$ separating the points of E_Δ. We denote by N the exceptional set of (A_t^ν) (cf. IV.(4.6)). Then $\mathrm{Cap}_{1,1}(N) = 0$ and consequently by 2.30 N is an s.$\mathcal{E}^\nu$-exceptional set. We now define for $n, k \in \mathbb{N}$

$$(2.29) \quad u_{n,k}(z) = \begin{cases} n\, E_z[\int_0^\infty e^{-nt - A_t^\nu} f_k(X_t)\, dt] =: n\, R_n^\nu f_k(z), & z \in E\backslash N\ . \\ 0\ , & z \in N\ . \end{cases}$$

Then $\lim\limits_{n\to\infty} u_{n,k}(z) = f_k(z)$ for all $z \in E\backslash N$. Hence $\{u_{n,k}\,|\,n, k \in \mathbb{N}\}$ separates the points of $E_\Delta\backslash N$. On the other hand by IV.4.4 $u_{n,k} \in D(\mathcal{E}^\nu)$ and $u_{n,k}$ is $\mathrm{Cap}_{1,1}$-quasi-continuous for all $n, k \in \mathbb{N}$. But it follows from [F 80, Theorem 3.1.3] and [F 80, Lemma 3.1.4] that any $\mathrm{Cap}_{1,1}$-quasi-continuous function in $D(\mathcal{E})$ is also strictly $\mathrm{Cap}_{1,1}$-quasi-continuous. This fact together with (2.26) implies that $u_{n,k}$ is s.$\mathcal{E}$-quasi-continuous for all $n, k \in \mathbb{N}$. Thus, 2.11(iii) is verified. $\qquad\square$

Remark. Recall that by Chap.IV, Subsection 4c) $(\mathcal{E}^\nu, D(\mathcal{E}^\nu))$ in 2.31 is in general not regular.

3   Notes/References

Section 1. The notion "local property" is taken from [F 80]. Theorems 1.5 and 1.11 generalize [F 80, Theorem 4.5.1]. The corresponding results for symmetric quasi-regular Dirichlet forms on metrizable spaces were announced in [AM 91a] without proof.

Section 2. The results in this section are basically new. A different version of the characterization of symmetric Dirichlet forms associated with Hunt processes on locally compact separable metric spaces was given in [AM 91b]. In this connection we mention that a characterization of semigroups associated with (non-symmetric) Hunt processes on locally compact separable metric spaces was first given by LeYan [Le 83]. In particular, the norm used in proving Claim 2 in Subsection 2c) was introduced in [Le 83] (cf. Notes/References of Chap.IV, Sect.5). Proposition 2.6 shows that $\mathrm{Cap}_{1,g}$ coincides with the capacity for Hunt processes defined in [Le 83], provided the Hunt process is associated with the corresponding Dirichlet form. A sufficient condition for a symmetric Dirichlet form on a metrizable Lusin space to be associated with a Hunt process was first discovered by Shigekawa and Taniguchi [ShiTa 91] (see the example after 2.15).

Chapter VI

Regularization

In this chapter we present a general "local compactification" method that enables us to associate to a quasi-regular Dirichlet form on an arbitrary topological space a regular Dirichlet form on a locally compact separable metric space. This is done in such a way that we can transfer results obtained in the latter "classical" framework to our more general situation. The "local compactification" is constructed in Section 1 and it is shown that "without loss of generality" one can also restrict to Hunt processes. In Section 2 the "transfer method" is described in general and subsequently illustrated by several examples. In all of this chapter E is supposed to be a Hausdorff topological space with $\mathcal{B}(E) = \sigma(C(E))$. We fix a σ-finite positive measure m on $(E, \mathcal{B}(E))$ and a quasi-regular Dirichlet form $(\mathcal{E}, D(\mathcal{E}))$ on $L^2(E; m)$.

1 Local compactification

Let $(E^\#, \mathcal{B}^\#)$ be a measurable space and let $i : E \to E^\#$ be a $\mathcal{B}(E)/\mathcal{B}^\#$-measurable map. Let $m^\# := m \circ i^{-1}$, i.e., the image measure of m under i. Define an isometry $i^* : L^2(E^\#; m^\#) \to L^2(E; m)$ by defining $i^*(u^\#)$ to be the m-class represented by $\tilde{u} \circ i$ for any $\mathcal{B}^\#$-measurable $m^\#$-version $\tilde{u}$ of $u^\# \in L^2(E^\#; m^\#)$. Note that the range of i^* is always closed, but, of course, in general strictly smaller than $L^2(E; m)$. Define

$$(1.1) \qquad D(\mathcal{E}^\#) := \{u^\# \in L^2(E^\#; m^\#) | i^*(u^\#) \in D(\mathcal{E})\}$$

$$(1.2) \qquad \mathcal{E}^\#(u^\#, v^\#) := \mathcal{E}(i^*(u^\#), i^*(v^\#)) ; \quad u^\#, v^\# \in D(\mathcal{E}^\#) .$$

Then clearly, $(\mathcal{E}^\#, D(\mathcal{E}^\#))$ is a positive definite bilinear form on $L^2(E^\#; m^\#)$ satisfying the weak sector condition, called the *image of* $(\mathcal{E}, D(\mathcal{E}))$ under i.

Exercise 1.1. (i) Show that the symmetric part of $(\mathcal{E}^\#, D(\mathcal{E}^\#))$ is closed. Show that if $D(\mathcal{E}^\#)$ is dense in $L^2(E^\#; m^\#)$ then $(\mathcal{E}^\#, D(\mathcal{E}^\#))$ is a Dirichlet form on $L^2(E^\#; m^\#)$, called the *image Dirichlet form of* $(\mathcal{E}, D(\mathcal{E}))$ under i.

(ii) Show that if i^* is onto then $D(\mathcal{E}^{\#})$ is dense in $L^2(E^{\#}; m^{\#})$.

Now we can prove the following

Theorem 1.2. *There exists an $\mathcal{E}$-nest $(E_k)_{k \in \mathbb{N}}$ consisting of compact metrizable sets in E and a locally compact separable metric space $Y^{\#}$ such that:*

(i) *$Y^{\#}$ is a local compactification of $Y := \cup E_k$ in the following sense: $Y^{\#}$ is a locally compact space containing Y as a dense subset and $\mathcal{B}(Y) = \{A \in \mathcal{B}(Y^{\#}) | A \subset Y\}$.*

(ii) *The trace topologies on E_k induced by E, $Y^{\#}$ respectively coincide for every $k \in \mathbb{N}$.*

(iii) *The image $(\mathcal{E}^{\#}, D(\mathcal{E}^{\#}))$ of $(\mathcal{E}, D(\mathcal{E}))$ under the inclusion map $i : Y \subset Y^{\#}$ is a regular Dirichlet form on $L^2(Y^{\#}; m^{\#})$ where $m^{\#} := m \circ i^{-1}$ is a positive Radon measure on $Y^{\#}$.*

Remark 1.3. Note that in 1.2 the terminology on image Dirichlet forms applies in the following sense. Since $E \setminus Y$ is $\mathcal{E}$-exceptional we may assume that $E = Y$ by just restricting every element in $L^2(E; m)$ to Y. Observe also that 1.2(i) implies that i^* is onto.

Proof of 1.2. By IV.3.4(ii) there exists a set $\mathcal{D}_0^+$ consisting of bounded positive $\mathcal{E}$-quasi-continuous $\mathcal{B}(E)$-measurable functions separating the points of $E \setminus N$ for some $\mathcal{E}$-exceptional set $N \in \mathcal{B}(E)$ such that the m-classes corresponding to $\mathcal{D}_0^+ - \mathcal{D}_0^+$ are dense in $\mathcal{D}(\mathcal{E})$ w.r.t. $\tilde{\mathcal{E}}_1^{1/2}$. Fix $\varphi_0 \in L^2(E; m)$, $0 < \varphi_0 \leq 1$ and let h_0 be an $\mathcal{E}$-quasi-continuous m-version of $G_1 \varphi_0$. Then $h_0 > 0$ $\mathcal{E}$-q.e. by III.3.6. For $u \in \mathcal{D}_0^+$, $n, m \in \mathbb{N}$, let $\tilde{G}_n(u \wedge m h_0)$ denote a fixed $\mathcal{E}$-quasi-continuous m-version of $G_n(u \wedge m h_0)$. Let $\mathcal{D}_1^+ := \{n \tilde{G}_n(u \wedge m h_0) | u \in \mathcal{D}_0^+, n, m \in \mathbb{N}\} \cup \{h_0\}$. Then $\mathcal{D}_1^+$ is countable and since $u \wedge m h_0 \uparrow u$ $\mathcal{E}$-q.e. on E as $m \to \infty$, by I.2.9(i) and I.2.13(ii) it follows that

$$(1.3) \qquad \mathcal{D}_1^+ - \mathcal{D}_1^+ \text{ is dense in } D(\mathcal{E}) \text{ w.r.t. } \tilde{\mathcal{E}}_1^{1/2} \ .$$

By I.4.2(i), IV.3.3(iii), and since h_0 is 1-excessive we also have that

$$(1.4) \qquad \text{for all } u \in \mathcal{D}_1^+ \text{ there exists } c \in]0, \infty[\text{ with } u \leq c\, h_0 \ \mathcal{E}\text{-q.e.} \ .$$

By III.3.3 and since $\mathcal{E}$ is quasi-regular we can find an $\mathcal{E}$-nest of compact metrizable sets such that $\mathcal{D}_1^+ \subset C(\{E_k\})$. By making E_k smaller we can achieve that for

$$(1.5) \qquad Y := \cup E_k$$

and for the $\mathbb{Q}$-algebra $\mathcal{D}$ generated by $\mathcal{D}_1^+$ we have

$$(1.6) \qquad h_0 \in \mathcal{D} \text{ and } h_0 > 0 \text{ on } Y \ ,$$

$$(1.7) \qquad \text{for all } u \in \mathcal{D} \text{ there exists } c, n \in \mathbb{N} \text{ with } |u| \leq c\, h_0^n \text{ on } Y \ ,$$

(1.8) $\mathcal{D}$ separates the points of Y ,

(1.9) $\mathcal{D}$ is dense in $D(\mathcal{E})$ w.r.t. $\tilde{\mathcal{E}}_1^{1/2}$.

By [Sch 73, Corollary 2, p.102] Y is a (topological) Lusin space. Let $\mathcal{D} = \{u_n \,|\, n \in \mathbb{N}\}$, set $g_n := \frac{2}{\pi} \arctan u_n$, $n \in \mathbb{N}$, and define a metric on Y by

$$\rho(x,y) := \sum_{n=1}^{\infty} \frac{1}{2^n} |g_n(x) - g_n(y)| \; ; \;\; x, y \in Y \; .$$

Since $\mathcal{D}$ separates the points of Y, (Y, ρ) is isometric to a subset of $[-1,1]^{\mathbb{N}}$, and hence the completion $(\overline{Y}, \rho)$ is a compact metric space. Furthermore, all g_n, u_n have unique continuous extensions $\overline{g}_n, \overline{u}_n$ to $\overline{Y}$ and clearly, $\{\overline{g}_n \,|\, n \in \mathbb{N}\}$ separates the points of $\overline{Y}$, hence so does $\overline{\mathcal{D}} := \{\overline{u}_n \,|\, n \in \mathbb{N}\}$. Let

(1.10) $Y^{\#} := \{\overline{h}_0 > 0\}$.

Then $(Y^{\#}, \rho)$ is a locally compact separable metric space with $Y \subset Y^{\#}$ by (1.6). Let $\mathcal{D}^{\#}$ denote the set of all restrictions to $Y^{\#}$ of functions in $\overline{\mathcal{D}}$. Then $\mathcal{D}^{\#}$ is a $\mathbb{Q}$-algebra of continuous functions which by (1.7) belongs to $C_\infty(Y^{\#})$, i.e., the set of all continuous functions f on $Y^{\#}$ *vanishing at infinity* (i.e., $\{f \geq \epsilon\}$ is a compact subset of $(Y^{\#}, \rho)$ for all $\epsilon > 0$). Since $\overline{h}_{0|Y^{\#}} \in \mathcal{D}^{\#}$ and $\mathcal{D}^{\#}$ separates the points of $Y^{\#}$ (by (1.6) resp. (1.8)),

(1.11) $\mathcal{D}^{\#}$ is dense in $C_\infty(Y^{\#})$ w.r.t. the uniform norm

by the Stone-Weierstraß theorem (cf. e.g. [Ch 69a, p.28]). In particular, $\mathcal{B}(Y^{\#}) = \sigma(\mathcal{D}^{\#})$ and the inclusion $i : Y \subset Y^{\#}$ is $\mathcal{B}(Y)/\mathcal{B}(Y^{\#})$-measurable. Hence by [P 67, Theorem 2.4, p.135]

(1.12) $\mathcal{B}(Y) = \{A \in \mathcal{B}(Y^{\#}) | A \subset Y\}$

and (i) is proved. (ii) follows since E_k is compact and $i_{|E_k}$ is continuous for all $k \in \mathbb{N}$ as $\mathcal{D} \subset C(\{E_k\})$.

Now let us prove (iii). Because of 1.3 we may assume that $E = Y$. For the inclusion map $i : Y \subset Y^{\#}$ let $m^{\#} := m \circ i^{-1}$ be the corresponding image measure on $(Y^{\#}, \mathcal{B}(Y^{\#}))$. By (1.12), $i^* : L^2(Y^{\#}; m^{\#}) \to L^2(Y; m)$ is onto. Hence by 1.1 the image $(\mathcal{E}^{\#}, D(\mathcal{E}^{\#}))$ of $(\mathcal{E}, D(\mathcal{E}))$ under i is a Dirichlet form on $L^2(Y^{\#}; m^{\#})$. Since $h_0 \in L^2(E; m)$ and hence $\overline{h}_0 \in L^2(Y^{\#}; m^{\#})$, $m^{\#}$ is a Radon measure on $Y^{\#}$. By (1.9) $\mathcal{D}^{\#}$ is a dense subset of $D(\mathcal{E}^{\#})$ w.r.t. $\tilde{\mathcal{E}}^{\#}{}_1^{1/2}$ and by (1.11) $\mathcal{D}^{\#}$ is also dense in $C_\infty(Y^{\#})$ w.r.t. the uniform norm. Realizing that for each $u \in \mathcal{D}^{\#}$, $u_\epsilon := u - (u \vee (-\epsilon)) \wedge \epsilon \in C_0(Y^{\#})$ for all $\epsilon > 0$ and $u_\epsilon \underset{\epsilon \searrow 0}{\longrightarrow} u$

uniformly and w.r.t. $\tilde{\mathcal{E}}^{\#}{}_1^{1/2}$ (cf. I.4.17(ii)) we see that $(\mathcal{E}^{\#}, D(\mathcal{E}^{\#}))$ is regular. $\square$

For the remainder of this section we fix $(E_k)_{k \in \mathbb{N}}$, Y, $Y^{\#}$ with metric ρ, $(\mathcal{E}^{\#}, D(\mathcal{E}^{\#}))$, $m^{\#}$ as in 1.2 and assume $E = Y$, hence write $E^{\#} := Y^{\#}$, which is no restriction by 1.3. We fix $\varphi \in L^2(E; m)$, $0 < \varphi \leq 1$ and set $h := G_1 \varphi$, $g := \hat{G}_1 \varphi$ and use the same letters for the corresponding elements in $L^2(E^{\#}; m^{\#})$. Let $\mathrm{Cap}_{h,g}$, $\mathrm{Cap}^{\#}_{h,g}$ denote the capacities corresponding to $(\mathcal{E}, D(\mathcal{E}))$, $(\mathcal{E}^{\#}, D(\mathcal{E}^{\#}))$ respectively (cf. III.2.4).

Corollary 1.4.   *(i) If $(F_k)_{k \in \mathbb{N}}$ is an $\mathcal{E}^{\#}$-nest, then $(F_k \cap E_k)_{k \in \mathbb{N}}$ is an $\mathcal{E}$-nest and vice versa.*

(ii) $N^{\#} \subset E^{\#}$ is $\mathcal{E}^{\#}$-exceptional if and only if $N^{\#} \cap E$ is $\mathcal{E}$-exceptional. In particular, $\mathrm{Cap}_{h,g}^{\#}(E^{\#} \setminus E) = 0$.

(iii) A function $u^{\#} : E^{\#} \to \mathbb{R}$ is $\mathcal{E}^{\#}$-quasi-continuous if and only if $u_{|E}^{\#}$ is $\mathcal{E}$-quasi-continuous.

(iv) $\mathrm{Cap}_{h,g}^{\#}(A^{\#}) = \mathrm{Cap}_{h,g}(A^{\#} \cap E)$ for all $A^{\#} \subset E^{\#}$.

Proof.   (i): Observe that by 1.2(ii) for $F \subset E_k$, $E \setminus F$ is open if and only if $E^{\#} \setminus F$ is open in $(E^{\#}, \rho)$. Hence in this case, since $m^{\#}(E^{\#} \setminus E) = 0$, the reduction of functions on $E^{\#} \setminus F$ w.r.t. $(\mathcal{E}^{\#}, D(\mathcal{E}^{\#}))$, respectively on $E \setminus F$ w.r.t. $(\mathcal{E}, D(\mathcal{E}))$ coincides. Consequently, (by III.2.6(i))

$$
(1.13) \qquad \mathrm{Cap}_{h,g}(E \setminus F) = \mathrm{Cap}_{h,g}^{\#}(E^{\#} \setminus F)
$$
$$
\text{for all closed } F \text{ with } F \subset E_k \text{ for some } k \in \mathbb{N}.
$$

If $(F_k)_{k \in \mathbb{N}}$ is an $\mathcal{E}^{\#}$-nest then by (1.13) for all $k \in \mathbb{N}$

$$
\begin{aligned}
\mathrm{Cap}_{h,g}(E \setminus (F_k \cap E_k)) &= \mathrm{Cap}_{h,g}^{\#}(E^{\#} \setminus (F_k \cap E_k)) \\
&\leq \mathrm{Cap}_{h,g}^{\#}(E^{\#} \setminus F_k) + \mathrm{Cap}_{h,g}(E \setminus E_k)
\end{aligned}
$$

which converges to zero. Hence since each $F_k \cap E_k$ is closed in E, $(F_k \cap E_k)_{k \in \mathbb{N}}$ is an $\mathcal{E}$-nest by III.2.11. The converse is analogous.

(i) immediately implies (ii) and also (iii) by 1.2(ii).

(iv): By (ii), (iii) it follows that for $A^{\#} \subset E^{\#}$ the reduction of functions on $A^{\#}$ w.r.t. $(\mathcal{E}^{\#}, D(\mathcal{E}^{\#}))$, respectively on $A^{\#} \cap E$ w.r.t. $(\mathcal{E}, D(\mathcal{E}))$ coincides. Hence the assertion follows by III.3.11. $\square$

Now we prove a comparison result between capacities on $E^{\#}$.

Proposition 1.5. *Let $A \subset E^{\#}$. Then:*

(i) $\mathrm{Cap}_{h,g}^{\#}(A) \leq \mathrm{Cap}_{1,1}^{\#}(A)$.

(ii) $\mathrm{Cap}_{h,g}^{\#}(A) = 0$ if and only if $\mathrm{Cap}_{1,1}^{\#}(A) = 0$.

Proof.   (i): Since $h, g \leq 1$ we have for $U \subset E^{\#}$, U open, with $\mathcal{L}_{1,U} \neq \emptyset$ by III.1.6(iii) that $h_U \leq 1_U$ and $\hat{g}_U \leq \hat{1}_U$ (where 1_U denotes the reduced function rather than the indicator function). Hence

$$
\mathrm{Cap}_{h,g}^{\#}(U) = \mathcal{E}_1^{\#}(h_U, \hat{g}_U) \leq \mathcal{E}_1^{\#}(1_U, \hat{1}_U) = \mathrm{Cap}_{1,1}^{\#}(U)
$$

which extends to arbitrary sets $A \subset E^{\#}$.

(ii): By (i) one implication is clear. To show the converse, suppose that $\mathrm{Cap}_{h,g}^{\#}(A) = 0$. We may assume that $A \in \mathcal{B}(E^{\#})$ and that A is relatively compact (since $E^{\#}$ is σ-compact) and by III.2.11 that $h = h_0$ (where h_0 is as in the proof of 1.2). Let $W \subset E^{\#}$ be a relatively compact open set such that $h \geq \alpha \in]0, \infty[$ on W. By assumption there exist open sets $U_n \subset E^\#$ with $A \subset U_n \subset W$, $n \in \mathbb{N}$, and

$$\lim_{n \to \infty} \mathrm{Cap}^\#_{h,g}(U_n) = 0 \ ,$$

hence by III.2.12,

$$\lim_{n \to \infty} \mathcal{E}^\#_1(h_{U_n}, h_{U_n}) = 0 \ .$$

But using III.1.5 we have (since $h_{U_n} = \hat{h}_{U_n} = h \geq \alpha$ $m^\#$-a.e. on U_n) for all $n \in \mathbb{N}$

$$\mathrm{Cap}^\#_{1,1}(A) \leq \mathrm{Cap}^\#_{1,1}(U_n) = \mathcal{E}^\#_1(1_{U_n}, \hat{1}_{U_n}) \leq \alpha^{-2} \mathcal{E}^\#_1(h_{U_n}, \hat{h}_{U_n}) \ .$$

By III.1.7 the latter is equal to $\alpha^{-2} \mathcal{E}^\#_1(h_{U_n}, h_{U_n})$ and hence $\mathrm{Cap}^\#_{1,1}(A) = 0$. $\quad\square$

Before we describe the general "transfer method" in the next section we first prove the following important consequence of 1.2, 1.4, 1.5.

Theorem 1.6. *Let* $\mathbf{M} = (\Omega, \mathcal{F}, (X_t)_{t \geq 0}, (P_z)_{z \in E_\Delta})$ *be a right process properly associated with the quasi-regular Dirichlet form* $(\mathcal{E}, D(\mathcal{E}))$ *on* $L^2(E; m)$. *Then there exists* $N \subset E$, N $\mathcal{E}$-*exceptional, such that* $S := E \backslash N$ *is* $\mathbf{M}$-*invariant and if* $\overline{\mathbf{M}}$ *is the trivial extension to* $E^\#$ *(cf. IV.3.23(i)) of* $\mathbf{M}$ *restricted to* S, *then* $\overline{\mathbf{M}}$ *is a Hunt process properly associated with the regular Dirichlet form* $(\mathcal{E}^\#, D(\mathcal{E}^\#))$ *on* $L^2(E^\#; m^\#)$, *where* $E^\#_\Delta$ *is taken as the one point compactification of* $E^\#$.

Proof. Recall that $E^\# = Y^\#$ where Y is as specified in 1.2. Since $E \backslash Y$ is $\mathcal{E}$-exceptional, by IV.6.5 there exists $S_1 \subset Y$, $E \backslash S_1$ being $\mathcal{E}$-exceptional such that S_1 is $\mathbf{M}$-invariant. Let $\mathbf{M}_{|S_1}$ be the restriction of $\mathbf{M}$ to S_1 and let $\mathbf{M}^\#$ be the trivial extension of $\mathbf{M}_{|S_1}$ to $E^\#$. Then $\mathbf{M}^\#$ is a right process properly associated with the regular Dirichlet form $(\mathcal{E}^\#, D(\mathcal{E}^\#))$ on $L^2(E^\#; m^\#)$. By V.2.12(ii) $(\mathcal{E}^\#, D(\mathcal{E}^\#))$ is s.quasi-regular. Note that if N_0 is any s.$\mathcal{E}^\#$-exceptional set we can take $(F_{1k})_{k \in \mathbb{N}}$ in the proof of V.2.26 such that $N_0 \cup N_1 \subset \cap F^c_{1k}$, consequently $S \subset E \backslash N_0$. Hence using that by 1.5 for a subset of $E^\#$ s.$\mathcal{E}^\#$- and $\mathcal{E}^\#$-exceptionality are equivalent and exploiting 1.4, IV.6.5 we can construct a sequence $(S_n)_{n \in \mathbb{N}}$ of $\mathbf{M}$-invariant sets with $E^\# \backslash S_n$ s.$\mathcal{E}^\#$-exceptional and a sequence $(\tilde{S}_n)_{n \in \mathbb{N}}$ of Borel subsets of $E^\#$, each $\tilde{S}_n$ having the property of S in V.2.26 (for $\mathbf{M}^\#$, $(\mathcal{E}^\#, D(\mathcal{E}^\#))$), such that for all $n \in \mathbb{N}$

$$\tilde{S}_{n+1} \subset S_{n+1} \subset \tilde{S}_n \subset S_n \ .$$

It is easy to check that if $S := \cap S_n \cap \tilde{S}_n$, then S is $\mathbf{M}$-invariant, $E \backslash S$ is $\mathcal{E}$-exceptional, and S satisfies the requirements in V.2.26 (for $\mathbf{M}^\#$, $(\mathcal{E}^\#, D(\mathcal{E}^\#))$). Hence if $\mathbf{M}^\#|_S$ is as in V.(2.21), then its trivial extension $\overline{\mathbf{M}}$ to $E^\#$ is a Hunt process properly associated with $(\mathcal{E}^\#, D(\mathcal{E}^\#))$ such that $\overline{\mathbf{M}}|_S = \mathbf{M}|_S$. This completes the proof. $\quad\square$

2   Consequences – the transfer method

The purpose of this section is to illustrate the general "transfer method" mentioned in the summary of this chapter by a few important examples. We fix $(E_k)_{k\in\mathbb{N}}$, Y, $Y^\#$, $(\mathcal{E}^\#, D(\mathcal{E}^\#))$, h, g, as in the last section and as before we assume without loss of generality that $Y = E$ and set $E^\# := Y^\#$. We start with an example from the analytic part of the theory of Dirichlet forms.

Proposition 2.1. *Let $u \in D(\mathcal{E})$. Then u is 1-excessive if and only if there exists a (unique) σ-finite positive measure μ on $(E, \mathcal{B}(E))$ such that μ does not charge $\mathcal{E}$-exceptional sets, $\tilde{D}(\mathcal{E}) \subset L^1(E; \mu)$ and*

$$(2.1) \qquad \mathcal{E}_1(u, v) = \int \tilde{v}\, d\mu \text{ for all } v \in D(\mathcal{E})$$

(where $\tilde{v}$ as before denotes an $\mathcal{E}$-quasi-continuous m-version of v).

Proof. Clearly, if μ as in the assertion exists, then by III.1.2 and IV.3.3(iii) u is 1-excessive. Now suppose u is 1-excessive. Then u is the restriction of a function $u^\# \in D(\mathcal{E}^\#)$ which is 1-excessive w.r.t. $(\mathcal{E}^\#, D(\mathcal{E}^\#))$. Hence by [O 88, 2.3.1 and 2.3.6] there exists a positive Radon measure $\mu^\#$ on $(E^\#, \mathcal{B}(E^\#))$ charging no $\mathrm{Cap}_{1,1}^\#$-zero sets such that each $\mathrm{Cap}_{1,1}^\#$-quasi-continuous (cf. [O 88]) $m^\#$-version $\tilde{v}^\#$ of an element $v^\# \in D(\mathcal{E}^\#)$ is $\mu^\#$-integrable and that

$$\mathcal{E}_1^\#(u^\#, v^\#) = \int \tilde{v}^\#\, d\mu^\# \ .$$

Noting that by 1.5, [O 88, Theorem 2.2.3] and IV.3.3(iii) an $m^\#$-version of an element in $D(\mathcal{E}^\#)$ is $\mathcal{E}^\#$-quasi-continuous if and only if it is $\mathrm{Cap}_{1,1}^\#$-quasi-continuous, we conclude by 1.2, 1.4, 1.5, that $\mu := \mu_{|\mathcal{B}(E)}^\#$ is a positive σ-finite measure on $(E, \mathcal{B})$ which obviously has the desired properties. By monotone class arguments the uniqueness of μ is then an immediate consequence of IV.3.2(iv) since $h_0 \geq c_k$ for some $c_k \in]0, \infty[$ on E_k for each $k \in \mathbb{N}$ (where h_0 is as in the proof of 1.2). $\qquad\square$

Of course, there is also a "co-version" of 2.1. In a similar way one can transfer all results obtained within the analytic theory of regular Dirichlet forms on locally compact separable metric spaces to quasi-regular Dirichlet forms on arbitrary topological spaces like for example *sweeping out of measures*, existence of *equilibrium potentials*, *spectral synthesis*, the *first Beurling-Deny formula* etc. (cf. [Si 74], [F 80], [O 88],[Bl 71]).

We now want to illustrate the "transfer method" for an example connecting the probabilistic and analytic part of the theory of Dirichlet forms. Below, we fix a right process $\mathbf{M} = (\Omega, \mathcal{F}, (X_t)_{t\geq 0}, (P_z)_{z\in E_\Delta})$ with life time ζ and state space E properly associated with $(\mathcal{E}, D(\mathcal{E}))$.

Definition 2.2. A family $(A_t)_{t\geq 0}$ of functions on Ω is said to be an *additive functional* (abbreviated AF) of $\mathbf{M}$ if:

(i) $A_t(\cdot)$ is $\mathcal{F}_t$-measurable for all $t \geq 0$.

(ii) There exists a *defining set* $\Lambda \in \mathcal{F}$ and an *exceptional set* $N \subset E$ which is
$\mathcal{E}$-exceptional such that $P_z[\Lambda] = 1$ for all $z \in E \backslash N$, $\theta_t(\Lambda) \subset \Lambda$ for all $t > 0$
and for each $\omega \in \Lambda$, $t \mapsto A_t(\omega)$ is right continuous on $[0, \infty[$ and has left
limits on $]0, \zeta(\omega)[$, $A_0(\omega) = 0$, $|A_t(\omega)| < \infty$ for $t < \zeta(\omega)$, $A_t(\omega) = A_\zeta(\omega)$
for $t \geq \zeta(\omega)$, and $A_{t+s}(\omega) = A_t(\omega) + A_s(\theta_t\omega)$ for $s, t \geq 0$.

Two AF's $A := (A_t)_{t \geq 0}$, $B := (B_t)_{t \geq 0}$ are called *equivalent* and we write $A = B$
if they have a common defining set Λ and a common exceptional set N such
that $A_t(\omega) = B_t(\omega)$ for all $\omega \in \Lambda$ and $t \geq 0$. An AF is called a *continuous
additive functional* (abbreviated CAF) if $t \mapsto A_t(\omega)$ is continuous on $[0, \infty[$ and
a *positive continuous additive functional* (abbreviated PCAF) if $A_t(\omega) \geq 0$ for
all $t \geq 0$, $\omega \in \Lambda$.

Note that the additivity property of a PCAF $(A_t)_{t \geq 0} =: (A(t, \cdot))_{t \geq 0}$, expressed
in terms of the random measure $A(ds, \cdot)$ associated with it, reads

$$(2.2) \qquad\qquad A(ds + t, \omega) = A(ds, \theta_t\omega) \,,$$

where the measure on the left hand side is defined to be the translation of
$A(ds, \omega)$ by t.

Definition 2.3. A positive measure μ on $(E, \mathcal{B}(E))$ is called *smooth* (w.r.t.
$(\mathcal{E}, D(\mathcal{E}))$) if $\mu(N) = 0$ whenever $N \in \mathcal{B}(E)$ is $\mathcal{E}$-exceptional and there exists
an $\mathcal{E}$-nest $(F_k)_{k \in \mathbb{N}}$ of compact subsets of E such that

$$\mu(F_k) < \infty \text{ for all } k \in \mathbb{N} \,.$$

Note that in 2.3 we may always assume that $F_k \subset E_k$ for all $k \in \mathbb{N}$, and that if
$(\mathcal{E}, D(\mathcal{E}))$ is regular and E is a locally compact separable metric space, then by
IV.4.5 our notion of smooth measures coincide with that in [F 80], [O 88].

Theorem 2.4. *There is a one to one correspondence between smooth measures
μ of $(\mathcal{E}, D(\mathcal{E}))$ and PCAF's $(A_t)_{t \geq 0}$ of* $\mathbf{M}$ *which is specified by*

$$(2.3) \qquad \lim_{t \downarrow 0} E_m \left[\frac{1}{t} \int_0^t f(X_s) \, dA_s \right] = \int f \, d\mu \text{ for all } f \in \mathcal{B}^+(E) \,.$$

Proof. Observe that by 1.6 we can trivially extend $\mathbf{M}$ to a Hunt process
$\mathbf{M}^\#$ on $E^\#$. Every PCAF $(A_t)_{t \geq 0}$ of $\mathbf{M}$ can be extended (e.g. by zero) to a
PCAF $(A_t^\#)_{t \geq 0}$ of $\mathbf{M}^\#$ and vice versa. The assertion is now immediate by [O
88, 4.1.10, 4.1.19] and 1.2, 1.4 and 1.5 (including its proof). The details are left
to the reader. $\square$

Similarly, one can transfer other results from the classical framework. We only
formulate the following respective extension of the famous *Fukushima decom-
position* (cf. [F 80], and [O 88] for the non-symmetric case). We need some

preparations.

For an additive functional $A := (A_t)_{t\geq 0}$ we define its *energy*

$$(2.4) \qquad\qquad e(A) := \lim_{t\to 0} \frac{1}{2t} E_m[A_t^2]$$

if this limit exists in $[0,\infty]$. Define

$$(2.5) \qquad \mathcal{M} := \{M \mid M \text{ is an AF of } \mathbf{M}, \ E_z[M_t^2] < \infty, \ E_z[M_t] = 0$$
$$\text{for } \mathcal{E}\text{-q.e. } z \in E \text{ and all } t \geq 0\} .$$

Let $M \in \mathcal{M}$. Then if $t, s \geq 0$ we have by IV.1.7(iii), the Markov property, and IV.6.5 that for $\mathcal{E}$-q.e. $z \in E$

$$E_z[M_{t+s} \mid \mathcal{F}_t] = E_z[M_t + M_s \circ \theta_t \mid \mathcal{F}_t] = M_t \ P_z\text{-a.s.}$$

(because $\zeta = 0 \ P_\Delta$-a.s.). Since M is right continuous on its defining set it follows that M is a square integrable martingale. In particular, M has a defining set Λ such that $t \mapsto M_t(\omega)$ has left limits on $]0,\infty[$ for all $\omega \in \Lambda$. Furthermore, $t \mapsto E_m[M_t^2]$ is subadditive in t and therefore, $e(M)$ exists. Define

$$(2.6) \qquad\qquad \overset{\circ}{\mathcal{M}} := \{M \in \mathcal{M} \mid e(M) < \infty\} .$$

$M \in \overset{\circ}{\mathcal{M}}$ is called a *martingale additive functional of finite energy* (abbreviated MAF). Furthermore, define

$$(2.7) \qquad \mathcal{N}^c := \{N \mid N \text{ is a CAF}, \ e(N) = 0, \ E_z[|N_t|] < \infty$$
$$\text{for } \mathcal{E}\text{-q.e. } z \in E \text{ and all } t \geq 0\} .$$

Finally, note that if $\tilde{u} : E \to \mathbb{R}$ is an $\mathcal{E}$-quasi-continuous m-version of $u \in D(\mathcal{E})$, then by 1.4, 1.6, and V.2.28(ii) $(\tilde{u}(X_t) - \tilde{u}(X_0))_{t\geq 0}$ is an AF of $\mathbf{M}$. Note that one obtains an equivalent AF if one chooses a different $\mathcal{E}$-quasi-continuous m-version of u. Therefore, we may set

$$(2.8) \qquad\qquad A^{[u]} := (\tilde{u}(X_t) - \tilde{u}(X_0))_{t\geq 0}$$

Theorem 2.5. *If $u \in D(\mathcal{E})$, then there exist unique $M^{[u]} \in \overset{\circ}{\mathcal{M}}$, $N^{[u]} \in \mathcal{N}^c$ such that*

$$A^{[u]} = M^{[u]} + N^{[u]} .$$

Proof. As in the proof of 2.4 we can "lift" the problem to $E^{\#}$ and then restrict it again to E. In order to apply the corresponding result in the classical case (cf. [O 88, 5.1.3]) one only has to realize (cf. the proof of 2.1) that by 1.5 an m-version of an element in $D(\mathcal{E}^{\#})$ is $\mathcal{E}^{\#}$-quasi-continuous if and only if it is $\mathrm{Cap}_{1,1}^{\#}$-quasi-continuous. But then by V.2.28(ii) $(\tilde{u}^{\#}(X_t) - \tilde{u}^{\#}(X_0))_{t\geq 0}$ is an AF of $\mathbf{M}^{\#}$ in the sense of [O 88, §4.1] and [O 88, 5.13] applies. The remaining details are again left to the reader. $\qquad\square$

Remark 2.6. (i) Note that by 2.5 (or by 1.4, 1.6, and V.2.28(ii)) for all $u \in D(\mathcal{E})$ there exists a defining set Λ for $A^{[u]}$ such that $t \mapsto A_t^{[u]}(\omega)$ has left limits on $]0, \infty[$ for all $\omega \in \Lambda$.

(ii) Along the same line of arguments as above one can also reprove the existence of an m-special standard process $\mathbf{M}$ properly associated with $(\mathcal{E}, D(\mathcal{E}))$ using the corresponding results in [F 80, §6], and [O 88, §3.3].

3   Notes/References

Section 1. Theorem 1.2, Corollary 1.4, and Proposition 1.5 are new. Though they can also be derived from the method in [AMR 90] by h-transformation. Also Theorem 1.6 is new. But all these results have been announced in [AMR 92b]. As mentioned in Remark 2.6, using Theorem 1.2 one can also give a direct proof of the existence of the process, but the techniques would be very similar to the ones presented in Chap.IV, Sect.3.

Section 2. All consequences are new (see however [AMR 92b]), but they are based on the classical results in [O 88] (see also [F 80], [Si 74]). In our exposition of Theorem 2.5 and its preparation we benefitted a lot from that in [O 88, §4.1]. Proposition 2.1 and Theorem 2.4 can also be proved directly without the transfer method (see [DoM 91]). The result mentioned in Remark 2.6(i) was obtained independently by Z. Qian (private communication).

Appendix A

Some Complements

1   Adjoint operators

Let $\mathcal{H}$ be a real Hilbert space with inner product $(\,,\,)$ and norm $\|\,\| := (\,,\,)^{1/2}$. Let $(T, D(T))$ be a densely defined linear operator on $\mathcal{H}$. Its *adjoint operator* $(\hat{T}, D(\hat{T}))$ is defined by

$$(1.1) \qquad D(\hat{T}) := \{ g \in \mathcal{H} \,|\, f \mapsto (Tf, g) \text{ is continuous on } D(T) \}$$

For $g \in D(\hat{T})$, $\hat{T}g$ is the unique element in $\mathcal{H}$ such that

$$(1.2) \qquad\qquad (Tf, g) = (f, \hat{T}g) \text{ for all } f \in D(T) \, .$$

$\hat{T}g$ exists by the Riesz-representation theorem or I.2.6. Clearly, $(\hat{T}, D(\hat{T}))$ is a linear operator on $\mathcal{H}$. If $(S, D(S))$ is a linear operator such that

$$(Tf, g) = (f, Sg) \text{ for all } f \in D(T), g \in D(S) \, ,$$

then obviously, $D(S) \subset D(\hat{T})$ and $S = \hat{T}$ on $D(S)$, i.e., $(\hat{T}, D(\hat{T}))$ is the "maximal" operator in this class.

Proposition 1.1. *Suppose T is a densely defined bounded linear operator on $\mathcal{H}$. Then $D(\hat{T}) = \mathcal{H}$ and $\hat{T}$ is bounded with $\|T\| = \|\hat{T}\|$.*

Proof. For all $f \in D(T)$, $g \in \mathcal{H}$,

$$|(Tf, g)| \leq \|T\| \|f\| \|g\| \, .$$

Hence $D(\hat{T}) = \mathcal{H}$ and

$$
\begin{aligned}
\|\hat{T}\| = \sup_{\|g\| \leq 1} \|\hat{T}g\| \;&=\; \sup_{\|g\| \leq 1} \sup_{\|f\| \leq 1} |(f, \hat{T}g)| \, . \\
&=\; \sup_{\|f\| \leq 1} \sup_{\|g\| \leq 1} |(Tf, g)| \\
&=\; \sup_{\|f\| \leq 1} \|Tf\| = \|T\| \, ,
\end{aligned}
$$

where without loss of generality we assumed that $D(T) = \mathcal{H}$.   □

A densely defined linear operator $(T, D(T))$ on $\mathcal{H}$ is called *symmetric* if $D(T) \subset D(\hat{T})$ and $T = \hat{T}$ on $D(T)$; it is called *self-adjoint* if $(T, D(T)) = (\hat{T}, D(\hat{T}))$; it is called *essentially self-adjoint* if its *closure* (that is, its smallest closed extension) $(\overline{T}, D(\overline{T}))$ is self-adjoint. Clearly, if T is bounded and $D(T) = \mathcal{H}$ then T is self-adjoint if and only if it is symmetric.

For more details on adjoint operators the reader may consult any book on functional analysis (like e.g. [ReS 72]).

Exercise 1.2. Let $(G_\alpha)_{\alpha>0}$ be a strongly continuous contraction resolvent on $\mathcal{H}$ with corresponding strongly continuous contraction semigroup $(T_t)_{t>0}$ on $\mathcal{H}$ (cf. Chap.I, Sect. 1). Let $\hat{G}_\alpha, \alpha > 0$, and $\hat{T}_t$, $t > 0$, denote the respective adjoint operators. Prove that $(\hat{G}_\alpha)_{\alpha>0}, (\hat{T}_t)_{t>0}$ is a strongly continuous contraction resolvent resp. semigroup on $\mathcal{H}$ which correspond to each other. (Hint: Use I.1.3(ii) in a similar way as in the proof of I.2.8 to prove the strong continuity of $(\hat{G}_\alpha)_{\alpha>0}$.)

2   The Banach/Alaoglu and Banach/Saks theorems

Theorem 2.1. (Banach/Alaoglu) *Let B be a Banach space with norm $\|\ \|$ and B' its dual (with operator norm). Then the unit ball B'_1 in B' is compact in the weak*-topology (i.e., the topology generated by the maps $l \mapsto l(u)$, $l \in B'$, and u running through B).*

Proof. (cf. [ReS 72, Theorem IV.21]) Let

$$X := \prod_{u \in B} [-\|u\|, \|u\|] = \{b : B \to \mathbb{R} \mid |b(u)| \leq \|u\|\}$$

equipped with the product topology. Then X is compact by Tychonov's theorem and B'_1 consists exactly of all linear maps in X. Clearly, on B'_1 the weak*-topology coincides with the product topology induced by X. It remains to show that B'_1 is closed in X. So, let $(l_i)_{i \in I}$ be a net in B'_1 such that $\lim_{i \in I} l_i = b$ in X, then for $\alpha, \beta \in \mathbb{R}$; $u, v \in B$,

$$\begin{aligned}
b(\alpha u + \beta v) &= \lim_{i \in I} l_i(\alpha u + \beta v) = \alpha \lim_{i \in I} l_i(u) + \beta \lim_{i \in I} l_i(v) \\
&= \alpha b(u) + \beta b(v).
\end{aligned}$$

Consequently, $b \in B'_1$.   □

Theorem 2.2. (Banach/Saks) *Let $\mathcal{H}$ be a real Hilbert space with inner product $(\ ,\ )$ and norm $\|\ \| := (\ ,\ )^{1/2}$. Let $u, u_n \in \mathcal{H}, n \in \mathbb{N}$, with $u_n \xrightarrow[n \to \infty]{} u$ weakly in $\mathcal{H}$, then there exists a subsequence $(n_k)_{k \in \mathbb{N}}$ such that the Cesaro mean*

$$v_N := \frac{1}{N} \sum_{k=1}^{N} u_{n_k}, \qquad N \in \mathbb{N},$$

converges (strongly) to u in $\mathcal{H}$.

Proof. (cf. [RiN 55, Sect. 38]) We may assume that $u = 0$. Define a subsequence $(n_k)_{k\in\mathbb{N}}$, by $n_1 := 1$ and if $n_1,\ldots,n_N$ are already chosen, let $n_{N+1} > n_N$ such that

$$|(u_{n_1}, u_{n_{N+1}})| \leq \frac{1}{N}, \ldots, |(u_{n_N}, u_{n_{N+1}})| \leq \frac{1}{N}$$

which is possible since $u_n \xrightarrow[n\to\infty]{} 0$ weakly in $\mathcal{H}$. By the uniform boundedness principle there exists $c \in {]}0, \infty{[}$ with $\|u_n\| \leq c$ for all $n \in \mathbb{N}$. Hence

$$\|\frac{1}{N}\sum_{k=1}^{N} u_{n_k}\|^2 = \frac{1}{N^2}\sum_{k=1}^{N}\|u_{n_k}\|^2 + \frac{2}{N^2}\sum_{\substack{i,k=1\\i<k}}^{N}(u_{n_i}, u_{n_k})$$

$$\leq \frac{c^2}{N} + \frac{2}{N^2}\sum_{k=1}^{N}\frac{k-1}{k} \leq \frac{c^2+2}{N} \xrightarrow[N\to\infty]{} 0.$$

$\square$

3 Supplement on Ray resolvents and right processes

a) Proof of IV.1.20

Recall that in the situation of IV.1.20 the state space E is assumed to be a compact metric space. The proof of IV.1.20 presented below is a slight modification of [G 75, Section 3] (see also [BlHa 86, II.8.1, II.8.2]). We need a few preparations.

Theorem 3.1. (Hausdorff-Bernstein-Widder) *For a function $f : {]}0, \infty{[} \to \mathbb{R}$ the following two conditions are equivalent:*

(i) f is completely monotone, that is f is infinitely differentiable and $(-1)^n D^n f(\alpha) \geq 0$ for all $\alpha \in {]}0, \infty{[}$ and $n \in \mathbb{N} \cup \{0\}$. Here $D := \frac{d}{d\alpha}$.

(ii) $f(\alpha)$ is the Laplace transform of some positive measure μ on $([0, \infty{[}, \mathcal{B}([0, \infty{[}))$, i.e., $f(\alpha) = \int_0^\infty e^{-\alpha t}\mu(dt)$ for all $\alpha > 0$.

Proof. We confine ourselves to give only a sketch of the proof and refer to [Fe 71, Chap.XIII, Sect. 4] for details. $(ii) \Rightarrow (i)$ follows by differentiation of the Laplace transform. For proving $(i) \Rightarrow (ii)$, let f be completely monotone and consider the function $s \mapsto f(a - as)$ on $[0, 1{[}$ for some fixed $a > 0$. Its derivatives are all nonnegative and hence it admits a Taylor expansion (cf. [Fe 71, Chap.VII, §2, Theorem 2])

$$f(a - as) = \sum_{n=0}^{\infty}\frac{(-a)^n D^n f(a)}{n!}s^n, \qquad s \in [0, 1{[}\,.$$

Then the measure $\lambda_a := \sum_{n=0}^{\infty}\frac{(-a)^n D^n f(a)}{n!}\varepsilon_{a^{-1}n}$ (where $\varepsilon_{a^{-1}n}$ is the Dirac measure at $a^{-1}n$) has Laplace transform

$$f_a(\alpha) = \sum_{n=0}^{\infty} \frac{(-a)^n D^n f(a)}{n!} e^{-\alpha a^{-1} n} = f(a(1 - e^{-\alpha/a})), \qquad \alpha > 0 \; .$$

But obviously, $f_a(\alpha) \xrightarrow[a \to \infty]{} f(\alpha)$ for all $\alpha > 0$. Thus by the extended continuity theorem (cf. [Fe 71, Chap. XIII.§1, Theorem 2a]) there exists a nonnegative measure λ on $([0, \infty[, \mathcal{B}([0, \infty[))$ such that $\lambda_a \longrightarrow_{a \to \infty} \lambda$ (i.e., $\lambda_a(I) \to \lambda(I)$ for each bounded interval of continuity of λ) and f is its Laplace transform. $\qquad \Box$

Lemma 3.2. *Let g be a function on $[0, \infty[\times E$ such that $t \mapsto g(t, z)$ is right continuous on $[0, \infty[$ and $\int_0^{\infty} |g(t, z)| \, dt < \infty$ for each fixed $z \in E$. Suppose that $z \mapsto \int_0^{\infty} e^{-\alpha t} g(t, z) \, dt$ is $\mathcal{B}(E)$ measurable for all $\alpha > 0$, then g is $\mathcal{B}([0, \infty[) \otimes \mathcal{B}(E)$-measurable.*

Proof. By a monotone class argument (cf. e.g. [Sh 88, A.06]) we obtain that $z \mapsto \int_0^{\infty} f(t) g(t, z) \, dt$ is $\mathcal{B}(E)$-measurable for all $f \in \mathcal{B}_b([0, \infty[)$. Since by the right continuity of $t \mapsto g(t, z)$

$$g(s, z) = \lim_{n \to \infty} \frac{1}{n} \int_s^{s+n} g(t, z) \, dt, \qquad s \geq 0 \; ,$$

$z \mapsto g(s, z)$ is $\mathcal{B}(E)$-measurable for all $s \geq 0$. Using the right-continuity again the assertion now follows by a routine argument. $\qquad \Box$

Lemma 3.3. *Let $(R_\alpha)_{\alpha > 0}$ be a sub-Markovian resolvent of kernels on $(E, \mathcal{B}(E))$ (cf. Chap.II, Sect.5 for the definition) and $f \in \mathcal{B}_b(E)$. Then $\alpha \mapsto R_\alpha f(z)$ is infinitely differentiable on $]0, \infty[$ for all $z \in E$ and for every $n \in \mathbb{N}$*

(i) $D^n(R_\alpha f) = (-1)^n n! R_\alpha^{n+1} f$

(ii) $D^n(\alpha R_\alpha f) = (-1)^{n+1} n! R_\alpha^n (I - \alpha R_\alpha) f$

where $D := \frac{d}{d\alpha}$, $R_\alpha^0 := I$ is the identity, and $R_\alpha^n := R_\alpha \circ R_\alpha^{n-1}$, $n \in \mathbb{N}$.

Proof. (ii) clearly follows from (i) by Leibnitz's rule.
(ii): For $n \geq 2$ we have

$$R_\alpha^n f - R_\beta^n f = (R_\alpha - R_\beta) \sum_{k=0}^{n-1} R_\alpha^{n-1-k} R_\beta^k f \; .$$

Consequently, by II.(5.1)

$$(3.1) \qquad \frac{R_\alpha^n f - R_\beta^n f}{\alpha - \beta} = -R_\alpha R_\beta \sum_{k=0}^{n-1} R_\alpha^{n-1-k} R_\beta^k f \; .$$

Since $(R_\alpha)_{\alpha > 0}$ is sub-Markovian, (3.1) implies that

$$|R_\alpha^n f - R_\beta^n f| \leq n(\alpha \wedge \beta)^{-n-1} |\alpha - \beta| \sup_{z \in E} |f(z)| \; ,$$

hence in particular, $\alpha \mapsto R_\alpha^n f(z)$ is continuous on $[\alpha_0, \infty[$ uniformly in z for all $n \in \mathbb{N}$ and $\alpha_0 > 0$. Therefore, letting $\beta \to \alpha$ in (3.1) we obtain

$$D(R_\alpha^n f) = -n R_\alpha^{n+1} f \; , \qquad n \in \mathbb{N} \; .$$

Now (i) follows by induction. $\qquad \Box$

Lemma 3.4. *Let $(R_\alpha)_{\alpha>0}$ be a Ray resolvent. Then for each fixed $\alpha > 0$, $C(E) \cap S^\alpha$ separates the points of E and if we set*

$$(3.2) \quad \overline{S^\alpha} := \{f \in C(E) \mid f = u - v \text{ for some } u, v \in C(E) \cap S^\alpha\}, \quad \alpha > 0 ,$$

then $\overline{S^\alpha}$ is uniformly dense in $C(E)$.

Proof. Let $\alpha > 0$ be fixed. If $x, y \in E$, $x \neq y$, then by (R.2) of IV.1.19 there exists $\beta > 0$ and $f \in C(E) \cap S^\beta$ such that $f(x) \neq f(y)$. If $\beta \leq \alpha$, then $f \in S^\beta \subset S^\alpha$. If $\beta > \alpha$, let $g := R_\alpha f$ and $h := f + (\beta - \alpha)g$. Then either $g(x) \neq g(y)$ or $h(x) \neq h(y)$. But $g \in C(E) \cap S^\alpha$, thus it suffices to show that $h \in C(E) \cap S^\alpha$. From (R.1) of IV.1.19 we know that $h \in C(E)$. From the resolvent equation we have for $\lambda > 0$,

$$\lambda R_{\lambda+\alpha} f = \lambda R_{\lambda+\beta} f + \lambda(\beta - \alpha)R_{\lambda+\alpha}R_{\lambda+\beta}f \leq f + (\beta - \alpha)R_{\lambda+\alpha}f .$$

which yields

$$\lambda R_{\lambda+\alpha} h \leq f + (\beta - \alpha)R_{\lambda+\alpha}f + (\beta - \alpha)(R_\alpha f - R_{\lambda+\alpha}f) = h .$$

Thus $h \in S^\alpha$, proving that $C(E) \cap S^\alpha$ separates the points of E. For the last assertion we note that $\overline{S^\alpha}$ is a linear space, separates the points of E and contains the constants. Hence by the lattice form of the Stone-Weierstrass theorem $\overline{S^\alpha}$ is uniformly dense in $C(E)$ provided $\overline{S^\alpha}$ is a lattice. But if $f, g, h, k \in C(E) \cap S^\alpha$, then

$$\begin{aligned}
(f - g) \wedge (h - k) &= [(f + k) - (g + k)] \wedge [(h + g) - (g + k)] \\
&= [(f + k) \wedge (h + g)] - (g + k)
\end{aligned}$$

and hence $(f - g) \wedge (h - k) \in \overline{S^\alpha}$. $\qquad\qquad\square$

Proof of IV.1.20. A semigroup $(p_t)_{t \geq 0}$ satisfying IV.1.20(i) and (ii) is obviously unique by the uniqueness of the Laplace-transform. For the existence of $(p_t)_{t \geq 0}$ let R_0 be the potential kernel of $(R_\alpha)_{\alpha>0}$ (i.e., $R_0 f := \sup_{\alpha>0} R_\alpha f$ for $f \in \mathcal{B}(E)^+$). Let S^0 be the set of all supermedian functions for $(R_\alpha)_{\alpha>0}$, that is,

$$S^0 := \{f \in \mathcal{B}(E)^+ \mid \beta R_\beta f \leq f \text{ for all } \beta > 0\} .$$

For the moment we only assume that each αR_α is sub-Markovian, but impose the additional assumption

$$(3.3) \quad R_0 \text{ is a bounded kernel and } C(E) \cap S^0 \text{ separates the points of } E .$$

Under the above assumption, $\overline{S^0} := \{f \in C(E) \mid f = u - v \text{ for some } u, v \in C(E) \cap S^0\}$ is uniformly dense in $C(E)$, as can be seen by the same argument as in the last part of the proof of 3.4. Fix $f \in C(E) \cap S^0$, then by 3.3(ii) $D(\alpha R_\alpha f) = R_\alpha(I - \alpha R_\alpha)f \geq 0$ and consequently $\alpha \mapsto \alpha R_\alpha f$ is increasing. We set $\hat{f} := \lim_{\alpha \to \infty} \alpha R_\alpha f$. Fix $z \in E$ and define $g(\alpha) := \hat{f}(z) - \alpha R_\alpha f(z)$, $\alpha > 0$. Then $g \geq 0$ and by 3.3(ii) for $\alpha > 0$

$$(-1)^n D^n g(\alpha) = n! R_\alpha^n (I - \alpha R_\alpha) f(z) \geq 0 \text{ for all } n \in \mathbb{N} .$$

Thus by 3.1 there exists a positive measure $\lambda_z(f, \cdot)$ on $[0, \infty[$ such that

$$\hat{f}(z) - \alpha R_\alpha f(z) = \int_{[0,\infty[} e^{-\alpha t} \lambda_z(f, dt) \text{ for all } \alpha > 0 .$$

Since $R_0 f(z) < \infty$, by the monotone convergence theorem we have that

$$(3.4) \qquad \lambda_z(f, [0, \infty[) = \lim_{\alpha \downarrow 0} (\hat{f}(z) - \alpha R_\alpha f(z)) = \hat{f}(z) .$$

and

$$(3.5) \qquad \lambda_z(f, \{0\}) = \lim_{\alpha \to \infty} [\hat{f}(z) - \alpha R_\alpha f(z)] = 0 .$$

We now define

$$(3.6) \qquad p_t f(z) := \lambda_z(f,]t, \infty[) , \qquad t \geq 0 .$$

Then $t \mapsto p_t f(z)$ is decreasing and right continuous, and $p_t f(z) \uparrow \hat{f}(z) = p_0 f(z)$ as $t \downarrow 0$ by (3.4), (3.5). Furthermore, for all $\alpha > 0$

$$(3.7) \qquad \begin{aligned} \int_0^\infty e^{-\alpha t} p_t f(z)\, dt &= \int_0^\infty e^{-\alpha t} \int 1_{]t,\infty[}(s)\, \lambda_z(f, ds)\, dt \\ &= \int \int_0^s e^{-\alpha t}\, dt\, \lambda_z(f, ds) \\ &= \frac{1}{\alpha} \int (1 - e^{-\alpha s}) \lambda_z(f, ds) \\ &= \frac{1}{\alpha} [\hat{f}(z) - (\hat{f}(z) - \alpha R_\alpha f(z))] \\ &= R_\alpha f(z) . \end{aligned}$$

We have

$$D^n\left(\frac{f}{\alpha} - R_\alpha f\right) = \frac{(-1)^n n!}{\alpha^{n+1}} f - n!(-1)^n R_\alpha^{n+1} f = \frac{(-1)^n n!}{\alpha^{n+1}} (f - (\alpha R_\alpha)^{n+1} f) .$$

Since $\alpha R_\alpha f \leq f$, we see that $D^n\left(\frac{f}{\alpha} - R_\alpha f\right) \geq 0$. But (3.7) shows that $\frac{f}{\alpha} - R_\alpha f$ is the Laplace transform of $f - p_t f$. Employing 3.1 and the uniqueness of the Laplace transform we see that $(f - p_t f)\, dt$ is a nonnegative measure, and consequently $p_t f \leq f$ for all $t \geq 0$ since $t \mapsto p_t f$ is right continuous. In particular letting $f = 1$ we obtain

$$(3.8) \qquad p_t 1 \leq 1 \text{ for all } t \geq 0 .$$

By the uniqueness of the Laplace transform and (3.7) $f \mapsto p_t f(z)$, $f \in S^0$, is additive and positive homogeneous, hence can be extended to a linear map $f \mapsto p_t f(z)$ on $\overline{S^0}$; in particular, we have

$$(3.9) \qquad \int_0^\infty e^{-\alpha t} p_t f(z)\, dt = R_\alpha f(z) \text{ for all } \alpha > 0, f \in \overline{S^0} .$$

Now let $f \in \overline{S^0}$, $f \geq 0$, then by 3.3(i),

$$(-1)^n D^n(R_\alpha f) = n! R_\alpha^{n+1} f \geq 0 \;,$$

hence by 3.1, $\alpha \mapsto R_\alpha f(z)$ is the Laplacian transform of a positive measure on $([0, \infty[, \mathcal{B}([0, \infty[))$. Consequently, by the uniqueness of the Laplace transform and right continuity, (3.9) implies that $p_t f(z) \geq 0$ for all $t \geq 0$. We have thus proved that for $t \geq 0$, $f \mapsto p_t f(z)$ is a positive linear functional on $\overline{S^0}$ which because of (3.8) is bounded by 1. Hence, as $\overline{S^0}$ is uniformly dense in $C(E)$, by the Riesz representation theorem there exists a subprobability measure $p_t(z, \cdot)$ on $(E, \mathcal{B}(E))$ such that

$$(3.10) \qquad p_t f(z) = \int_E f(y) p_t(z, dy) \text{ for all } f \in \overline{S^0}, t \geq 0$$

(cf. IV.3.15). We now define $p_t f(z) := \int_E f(y) p_t(z, dy)$ for all $f \in \mathcal{B}_b(E)$. It coincides with our previous definition when $f \in \overline{S^0}$. Applying (3.9) and 3.2 we see that $(t, z) \mapsto p_t f(z)$ is $\mathcal{B}([0, \infty[) \times \mathcal{B}(E)$-measurable for all $f \in \overline{S^0}$ and this fact extends easily to all $f \in \mathcal{B}_b(E)$ by monotone class arguments. In particular, p_t is a sub-Markovian kernel on $(E, \mathcal{B})$ for each $t \geq 0$. Hence employing monotone class arguments we see that (3.9) holds for all $f \in \mathcal{B}_b(E)$ and all $\alpha > 0$. Letting $\alpha \downarrow 0$ and using the boundedness of R_0, (3.9) holds even for $\alpha = 0$. Since $\overline{S^0}$ is uniformly dense in $C(E)$ and $\sup_{z \in E} |p_t f(z)| \leq \sup_{z \in E} |f(z)|$ for all $t \geq 0$, it follows from the right continuity of $t \mapsto p_t f(z)$ for $f \in \overline{S^0}$ that $t \mapsto p_t f(z)$ is right continuous on $[0, \infty[$ for all $f \in C(E)$. Now we prove that $(p_t)_{t \geq 0}$ is a semigroup. It suffices to show that $p_t p_s f = p_{t+s} f$ for all $f \in C(E)$, $f \geq 0$. To this end, fix $f \in C(E)$, $f \geq 0$. Then both $p_t p_s f$ and $p_{t+s} f$ are right continuous in s, and so it suffices to show that they have the same Laplace transform, that is, after using Fubini's theorem and (3.9),

$$(3.11) \qquad p_t(R_\alpha f)(z) = \int_0^\infty e^{-\alpha s} p_{t+s} f(z)\, ds = e^{\alpha t} \int_t^\infty e^{-\alpha s} p_s f(z)\, ds$$

Observing that both sides of (3.11) are right continuous in t and employing again the uniqueness of Laplace transforms and (3.9), the proof of the semigroup property reduces to verifying

$$(3.12) \qquad R_\beta R_\alpha f = \int_0^\infty e^{-\beta t} e^{\alpha t} \Big(\int_t^\infty e^{-\alpha s} p_s f \Big)\, ds\, dt \;.$$

But the right hand side of (3.12) is equal to

$$
\begin{aligned}
\int_0^\infty e^{-\alpha s} p_s f \Big(\int_0^s e^{-(\beta - \alpha)t} dt \Big) ds &= (\beta - \alpha)^{-1} \int_0^\infty (e^{-\alpha s} - e^{-\beta s}) p_s f\, ds \\
&= (\beta - \alpha)^{-1} (R_\alpha - R_\beta) f = R_\beta R_\alpha f
\end{aligned}
$$

provided $\beta \neq \alpha$. If $\beta = \alpha$ the same computation shows that the right side of (3.12) equals

$$\int_0^s s e^{-\alpha s} p_s f \, ds = -D\left(\int_0^\infty e^{-\alpha s} p_s f \, ds\right) = -D(R_\alpha f) = R_\alpha R_\alpha f$$

where in the last step we used 3.3(i). Up to now we proved the existence of a semigroup of sub-Markovian kernels satisfying IV.1.20(i) and (ii) under the additional assumption (3.3).

For the general case we let $\beta > 0$ and define $R_\alpha^\beta := R_{\alpha+\beta}$. Then $(R_\alpha^\beta)_{\alpha>0}$ is sub-Markovian Ray resolvent with $R_0^\beta = R_{\alpha+\beta}$. By virtue of 3.4, $(R_\alpha^\beta)_{\alpha>0}$ satisfies the additional assumption (3.7). Thus by what we have proved there exists for each $\beta > 0$ a semigroup (p_t^β) of sub-Markovian kernels on $(E, \mathcal{B}(E))$ such that $t \mapsto p_t^\beta f(z)$ is right continuous on $[0, \infty[$ for all $f \in C(E)$, $z \in E$, and

$$(3.13) \qquad R_{\alpha+\beta} f(z) = \int_0^\infty e^{-\alpha t} p_t^\beta f(z) \, dt \text{ for all } \alpha > 0 \; .$$

Fix $z \in E$ and let

$$g(\alpha) := \int_0^\infty e^{-\alpha t}(e^{-\beta t} - p_t^\beta 1(z)) \, dt = \frac{1}{\alpha + \beta} - R_{\alpha+\beta} 1(z), \quad \alpha > 0 \; .$$

As before we can check that $(-1)^n D^n g \geq 0$, and again by virtue of 3.1 and the right continuity of $t \mapsto p_t^\beta 1$ we see that $p_t^\beta 1 \leq e^{-\beta t}$ for all $t \geq 0$. We now define $p_t := e^{\beta t} p_t^\beta$, $t \geq 0$. It is clear that $(p_t)_{t\geq 0}$ is a sub-Markovian semigroup. By (3.13) we have for $f \in C(E)$

$$(3.14) \qquad \int_0^\infty e^{-\alpha t} p_t f \, dt = \int_0^\infty e^{-\alpha t} e^{\beta t} p_t^\beta f \, dt = R_\alpha f \text{ for all } \alpha > \beta$$

Thus by the uniqueness of Laplace transforms $p_t f$ does not depend on β, and since $\beta > 0$ is arbitrary in (3.14) we see that $(p_t)_{t\geq 0}$ satisfies IV.1.20 (i) and (ii). Since $(R_\alpha)_{\alpha>0}$ is Markovian, it follows from IV.1.20(i) that

$$\int_0^\infty e^{-\alpha t}(1 - p_t 1) \, dt = \frac{1}{\alpha} - R_\alpha 1 = 0 \text{ for all } \alpha > 0 \; .$$

Thus by the right continuity of $t \mapsto p_t 1$ and the uniqueness of Laplace transforms we see that $p_t 1(z) = 1$ for all $t > 0$, $z \in E$, i.e., $(p_t)_{t\geq 0}$ is a semigroup of Markovian kernels.

To complete the proof it remains to check IV.1.20(iii). To this end, let $\beta > 0$ be fixed. For $f \in C(E) \cap S^\beta$, we put $\hat{f} := \lim_{\alpha\to\infty} \alpha R_{\alpha+\beta} f = \sup_{\alpha>0} \alpha R_{\alpha+\beta} f$. Then

$$(3.15) \quad p_0 f = \lim_{\alpha\to\infty} \alpha \int_0^\infty e^{-\alpha t} p_t f \, dt = \lim_{\alpha\to\infty} \alpha R_\alpha f = \lim_{\alpha\to\infty} \alpha R_{\alpha+\beta} f = \hat{f} \; .$$

Since $C(E)$ equipped with the uniform norm is a separable Banach space, we may take a countable dense subset $\{g_n \mid n \in \mathbb{N}\}$ of $C(E) \cap S^\beta$. Fix $z \in E$ and suppose that $g_n(z) = \hat{g}_n(z)$ for all $n \in \mathbb{N}$. Then $g(z) = \hat{g}(z)$ for all $g \in C(E) \cap S^\beta$ and if we set $\mathcal{K} := \{f \in C(E) \mid \lim_{\alpha\to\infty} \alpha R_\alpha f(z) = f(z)\}$, then $\mathcal{K}$ contains $C(E) \cap S^\beta$. But clearly $\mathcal{K}$ is a closed linear subspace of $C(E)$, consequently $\mathcal{K} = C(E)$ because of 3.4. Thus we have shown that

$$(3.16) \qquad D = \bigcap_{n \geq 1} \{g_n = \hat{g}_n\}$$

where D is as specified in IV.(1.14). Consequently, $D \in \mathcal{B}(E)$. Let $z \in E$. If $p_0(z, \cdot) = \varepsilon_z$, then by (3.15) $g_n(z) = p_0 g_n(z) = \hat{g}_n(z)$ for all $n \in \mathbb{N}$; thus by (3.16) $z \in D$. Conversely, if $z \in D$, then by (3.15) $p_0 f(z) = f(z)$ for all $f \in C(E) \cap S^\beta$. Thus by linearity and 3.4, $p_0(z, \cdot) = \varepsilon_z$. Let $n \in \mathbb{N}$.It follows from (3.15) that $p_t g_n = p_t p_0 g_n = p_t \hat{g}_n$, and hence $p_t(g_n - \hat{g}_n) = 0$. Since $\hat{g}_n \leq g_n$, we get $p_t(\cdot, \{g_n \neq \hat{g}_n\}) = 0$. It follows from (3.16) that $p_t(z, E \backslash D) = 0$ for all $t \geq 0$ and $z \in E$, which completes the proof. $\qquad \square$

b)   Proof of IV.1.22 (cf. [Sh 88, (9.13)]).

Let $W := \mathbb{R}^{[0,\infty)}$ be the space of all maps $\omega : [0, \infty[\rightarrow E$, and let $Y_t(\omega) := \omega(t)$ denote the coordinate process on W. Let $\mathcal{F}^0 := \sigma\{Y_s \mid s \in [0, \infty[\}$ and $\mathcal{F}_t^0 := \sigma\{Y_s \mid s \leq t\}$ be the σ-algebra and the filtration generated by $(Y_t)_{t \geq 0}$. Let $\mathcal{P}$ be the set of all probability measures on $(E, \mathcal{B}(E))$ and fix $\mu \in \mathcal{P}$. We can apply the Kolmogorov extension theorem to obtain a probability measure P_μ on $(W, \mathcal{F}^0)$ such that $(W, \mathcal{F}^0, (\mathcal{F}_t^0)_{t \geq 0}, (Y_t)_{t \geq 0}, P_\mu)$ is a Markov process with transition semigroup $(p_t)_{t \geq 0}$ and initial measure $\mu p_0 := \int p_0(z, \cdot) \mu(dz)$ (cf. e.g. [B 78, 65.5]). That is, for $0 \leq t_1 \leq t_2 \leq \ldots \leq t_n$ and $x_0, x_1, \ldots, x_n \in E$.

$$(3.17) \qquad \begin{aligned} &P_\mu[Y_0 \in dx_0, \ldots, Y_{t_n} \in dx_n] \\ &= \mu p_0(dx_0) p_{t_1}(x_0, dx_1) \ldots p_{t_n - t_{n-1}}(x_{n-1}, dx_n) \,. \end{aligned}$$

If $f \in C(E) \cap S^1$, then $e^{-t} p_t f \leq f$ for all $t \geq 0$. Indeed, it is easy to check that $g(\alpha) := \alpha^{-1} f - R_{\alpha+1} f$ is the Laplace transform of $f - e^{-t} p_t f$. Thus, if $f \in C(E) \cap S^1$ we see as in the proof of IV.1.20 that g is completely monotone and because of the right continuity of $t \mapsto e^{-t} p_t f$ that hence $e^{-t} p_t f \leq f$ for all $t \geq 0$. Let $f \in C(E) \cap S^1$. We have by the Markov property that

$$E_\mu[e^{-(t+s)} f(Y_{t+s}) \mid \mathcal{F}_s^0] = e^{-s} e^{-t} p_t f(Y_s) \leq e^{-s} f(Y_s) \,.$$

Therefore, $(e^{-t} f(Y_t), \mathcal{F}_t^0, P_\mu)_{t \in [0,\infty[}$ is a nonnegative supermartingale. By [DM 82, VI.2] for P_μ-almost all $\omega \in \Omega$, the mapping $t \mapsto Y_t(\omega)$ restricted to $\mathbb{Q}_+$, has left and right limits at each point $t \in [0, \infty[$. Since $\overline{S^1}$ (cf. (3.2)) is uniformly dense in $C(E)$ by 3.4 and $C(E)$ is separable, it follows that for every $\mu \in \mathcal{P}$, the set W_1 defined by

$$W_1 := \{\omega \in W \mid t \mapsto f(Y_t)(\omega) \text{ restricted to } \mathbb{Q}_+ \text{ has left and right limits}$$
$$\text{at each } t \in [0, \infty[\text{ for all } f \in C(E)\} \,.$$

is an element of $(\mathcal{F}^0)^{P_\mu}$ and $P_\mu[W_1] = 1$. We have that

$$(3.18) \qquad \begin{aligned} W_1 = \{\omega \in W \mid t \mapsto Y_t(\omega) \text{ restricted to } \mathbb{Q}_+ \text{ has left and right limits} \\ \text{at each } t \in [0, \infty[\} \,. \end{aligned}$$

Indeed, if we let W' denote the set specified on the right side of (3.18), then obviously $W' \subset W_1$. Conversely, if $t \mapsto Y_t(w)$ has no right limit for some $t \in [0, \infty[$ along $\mathbb{Q}_+$, then we can find two sequences $(t_n)_{n \in \mathbb{N}}$ and $(s_n)_{n \in \mathbb{N}}$ of rational numbers and two different points $x, y \in E$ such that $t_n \downarrow t$, $s_n \downarrow t$, $\lim_{n \to \infty} Y_{t_n}(\omega) = x$ and $\lim_{n \to \infty} Y_{s_n}(\omega) = y$, because $\hat{E}$ is compact. Let $f \in C(E)$ with $f(x) \neq f(y)$. Then obviously $t \mapsto f(Y_t)(\omega)$ has no right limit at t, proving that $\omega \notin W_1$. The same argument works also when $t \mapsto Y_t(\omega)$ has no left limit at t along $\mathbb{Q}_+$. Thus $W' = W_1$. We now restrict Y_t to W_1, and replace P_μ, $\mathcal{F}^0$ and $\mathcal{F}_t^0$ by their respective traces on W_1. The formula (3.17) remains valid since $P_\mu(W_1) = 1$. For $\omega \in W_1$ we define

$$(3.19) \qquad X_t(\omega) := \lim_{s \in \mathbb{Q}_+,\, s \downarrow t} Y_s(\omega), \quad t \geq 0 .$$

We are going to show that

$$(3.20) \qquad P_\mu[X_t = Y_t] = 1 \quad \text{for all } t \geq 0 ,$$

and

$$(3.21) \qquad P_\mu[X_t \in D \text{ for all } t \geq 0] = 1 .$$

If $f \in C(E) \cap S^1$ and $g \in C(E)$, then for $t \geq 0$, by the Markov property

$$(3.22) \qquad \begin{aligned} E_\mu[g(Y_t)f(X_t)] &= \lim_{s \in \mathbb{Q}_+,\, s \downarrow t} E_\mu[g(Y_t)f(Y_s)] \\ &= \lim_{s \in \mathbb{Q}_+,\, s \downarrow t} E_\mu[g(Y_t)p_{s-t}f(Y_t)] \\ &= E_\mu[g(Y_t)\hat{f}(Y_t)] \end{aligned}$$

where in the last step we used (3.15) with $\hat{f} := \lim_{\alpha \to \infty} \alpha R_{\alpha+1} f$. But we have that

$$E_\mu[f(Y_t) - \hat{f}(Y_t)] = \int (p_t f - p_t \hat{f}) d\mu = \int (p_t f - p_t p_0 f) d\mu = 0 .$$

Therefore, $f(Y_t) = \hat{f}(Y_t)$ P_μ-a.s. since $0 \leq \hat{f} \leq f$. Thus (3.22) yields

$$(3.23) \qquad E_\mu[g(Y_t)f(X_t)] = E_\mu[g(Y_t)f(Y_t)] .$$

But by 3.4, $\overline{S^1}$ is uniformly dense in $C(E)$, therefore (3.23) holds in fact for all $f \in C(E)$ and $g \in C(E)$. It follows by a routine monotone class argument that for every bounded $\mathcal{B}(E) \times \mathcal{B}(E)$-measurable function h,

$$E_\mu[h(Y_t, X_t)] = E_\mu[h(Y_t, Y_t)] .$$

Taking $h(x, y) = 1_{\{x \neq y\}}$ we obtain (3.20). For proving (3.21), let $g \in C(E) \cap S^1$ and set $h_n := n(g - nR_{n+1}g)$ for $n \in \mathbb{N}$, so that $R_1 h_n = nR_{n+1}g \uparrow \hat{g}$ as $n \to \infty$. Since $R_1 h_n \in C(E) \cap S^1$, $(e^{-t}R_1 h_n(X_t), \mathcal{F}_t^0, P_\mu)_{t \in [0, \infty[}$ is a right continuous supermartingale, it is then easy to see that $P_\mu[t \mapsto e^{-t}\hat{g}(X_t)$ is right continuous for all $t \geq 0] = 1$ (cf. [DM 82 VI.18]). As we observed above, $g(X_t) = \hat{g}(X_t)$

$P_\mu - a.s.$ for each fixed $t \geq 0$ and $g \in C(E) \cap S^1$. Therefore, by virtue of the right continuity

$$(3.24) \qquad P_\mu[g(X_t) = \hat{g}(X_t) \text{ for all } t \geq 0] = 1 \text{ for all } g \in C(E) \cap S^1 .$$

Let $\{g_n \mid n \in \mathbb{N}\} \subset C(E) \cap S^1$ be as specified in (3.16). It follows from (3.24) that

$$(3.25) \qquad P_\mu[g_n(X_t) = \hat{g}_n(X_t) \text{ for all } t \geq 0 \text{ and } n \geq 1] = 1 .$$

Combining (3.25) and (3.16) we obtain (3.21). We now set

$$\Omega := \{\omega \in W_1 \mid X_t(\omega) \in D \text{ for all } t \geq 0\} .$$

It follows from (3.21) that $\Omega \in \bigcap_{\mu \in \mathcal{P}}(\mathcal{F}^0)^{P_\mu}$ and $P_\mu(\Omega) = 1$ for all $\mu \in \mathcal{P}$. We now restrict X_t to Ω and replace $\mathcal{F}^0, \mathcal{F}^0_t$ by their respective traces on Ω. For $z \in D$, let P_z denote the restriction of P_{ε_z} on $(\Omega, \mathcal{F}^0)$. Then $(\Omega, \mathcal{F}^0, (\mathcal{F}^0)_{t \geq 0}, (X_t)_{t \geq 0}, (P_z)_{z \in D})$ is a Markov process with state space D, transition semigroup $(p_t)_{t \geq 0}$, such that for all $\omega \in \Omega$, $t \mapsto X_t(\omega)$ is right continuous and has left limits in E for all $t \in]0, \infty[$. Moreover, by virtue of IV.1.20(iii) we have

$$P_z[X_0 = z] = p_0(z, \{z\}) = 1 \text{ for all } z \in D .$$

Let $\mathcal{P}(D)$ denote the set of all probability measures on $(D, \mathcal{B}(D))$. We set

$$\mathcal{F} := \bigcap_{\mu \in \mathcal{P}(D)} (\mathcal{F}^0)^{P_\mu}, \qquad \mathcal{F}'_t := \bigcap_{\mu \in \mathcal{P}(D)} (\mathcal{F}^0_t)^{P_\mu}, \qquad \mathcal{F}_t := \bigcap_{s > t} \mathcal{F}'_s, .$$

We claim that $(\Omega, \mathcal{F}, (\mathcal{F}_t)_{t \geq 0}, (X_t)_{t \geq 0}, (P_z)_{z \in D})$ is a desired right process satisfying the conditions specified in IV.1.22. Taking into account what we have proved so far, it remains to show that $(X_t)_{t \geq 0}$ has the strong Markov property relative to $(\mathcal{F}_t)_{t \geq 0}$. To this end we denote by $C_u(D)$ the set of all uniformly continuous functions f on D. Then each $f \in C_u(D)$ is the restriction to D of some $\overline{f} \in C(E)$. Since $\sigma(C_u(D)) = \mathcal{B}(D)$ and by monotone class arguments, the strong Markov property (see Chap.IV., Sect.1, (M.7)) is equivalent with the property that for every $(\mathcal{F}_t)$-stopping time σ and all $A \in \mathcal{F}_\sigma$, $z \in D$,

$$(3.26) \qquad E_z[f(X_{\sigma+t})1_A] = E_z[p_t f(X_\sigma)1_A] \text{ for all } t \geq 0, f \in C_u(D) .$$

Here again we use the convention that $X_\infty = \Delta$ and $f(\Delta) = 0$ for any real function f on D, where Δ is some isolated point. Since both sides of (3.26) are right continuous in t and by the uniqueness of the Laplace transform, Fubini's theorem and IV.1.20(i) imply that (3.26) is equivalent with

$$(3.27) \quad E_z[\int_0^\infty e^{-\alpha t} f(X_{\sigma+t})1_A \, dt] = E_z[1_A R_\alpha f(X_\sigma)] \text{ for all } \alpha > 0, f \in C_u(D) .$$

For a given $(\mathcal{F}_t)$-stopping time σ, we define

$$\sigma_n := \sum_{k=1}^\infty \frac{k+1}{2^n} 1_{\{\frac{k-1}{2^n} \leq \sigma < \frac{k}{2^n}\}}, \quad n \in \mathbb{N} .$$

For $A \in \mathcal{F}_\sigma$, we have

$$A \cap \{\frac{k-1}{2^n} \le \sigma < \frac{k}{2^n}\} \in \mathcal{F}_{\frac{k}{2^n}} \subset \mathcal{F}'_{\frac{k+1}{2^n}} \ .$$

Hence the Markov property of $(X_t)_{t\ge 0}$ (relative to $(\mathcal{F}'_t)_{t\ge 0}$) implies for $f \in C_u(D)$, $z \in D$,

$$
\begin{aligned}
E_z[\int_0^\infty e^{-\alpha t} f(X_{\sigma_n+t}) 1_A \, dt] &= \sum_{k=1}^\infty \int_0^\infty e^{-\alpha t} E_z[f(X_{\frac{k+1}{2^n}+t}) 1_{A \cap \{\frac{k-1}{2^n} \le \sigma < \frac{k}{2^n}\}}] \, dt \\
&= \sum_{k=1}^\infty \int_0^\infty e^{-\alpha t} E_z[1_{A \cap \{\frac{k-1}{2^n} \le \sigma < \frac{k}{2^n}\}} E_{X_{\frac{k+1}{2^n}}}[f(X_t)]] \, dt \\
&= \int_0^\infty e^{-\alpha t} E_z[1_A E_{X_{\sigma_n}}[f(X_t)]] \, dt \\
&= E_z[1_A \int_0^\infty e^{-\alpha t} p_t f(X_{\sigma_n}) \, dt] \\
&= E_z[1_A R_\alpha f(X_{\sigma_n})]
\end{aligned}
$$

Using the right continuity of $(X_t)_{t\ge 0}$ as well as the continuity of f and $R_\alpha f$, and letting $n \to \infty$ we obtain (3.27) by the dominated convergence theorem, which completes the proof. $\qquad\square$

Bibliography

[ABR 89]   Albeverio, S., Brasche, J., Röckner, M.: Dirichlet forms and
              generalized Schrödinger operators. In: Proc. Summer School
              Schrödinger operators, ed. H. Holden and A. Jensen. Lecture
              Notes in Physics **345**, 1-42. Berlin: Springer 1989.

[AFHMR 90]   Albeverio, S., Fukushima, M., Hansen, W., Ma, Z.M., Röckner,
              M.: An invariance result for capacities on Wiener space. Preprint
              (1990). J. Funct. Anal. **106**, 35-49 (1992).

[AFHMR 91]   Albeverio, S., Fukushima, M., Hansen, W., Ma, Z.M., Röckner,
              M.: Capacities on Wiener space: tightness and invariance. C.R.
              Acad. Sci. Paris, t. **312**, Serie I, 931-935 (1991).

[AHPRS 90a]   Albeverio, S., Hida, T., Potthoff, J., Streit, L., Röckner, M.:
              Dirichlet forms in terms of white noise analysis I: construction
              and QFT examples. Reviews in Math. Phys. **1**, 291-312 (1990).

[AHPRS 90b]   Albeverio, S., Hida, T. Potthoff, J., Streit, L., Röckner, M.:
              Dirichlet forms in terms of white noise analysis II: closability and
              diffusion processes. Reviews in Math. Phys. **1**, 313-323 (1990).

[AH-K 75]   Albeverio, S.,Høegh-Krohn, R.: Quasi-invariant measures, sym-
              metric diffusion processes and quantum fields. In: Les méthodes
              mathématiques de la théorie quantique des champs, Colloques
              Internationaux du C.R.N.S., no. **248**, Marseille, 23-27 juin 1975,
              C.N.R.S., 1976.

[AH-K 77a]   Albeverio, S., Høegh-Krohn, R.: Dirichlet forms and diffusion
              processes on rigged Hilbert spaces. Z. Wahrsch. verw. Geb. **40**,
              1-57 (1977).

[AH-K 77b]   Albeverio, S., Høegh-Krohn, R.: Hunt processes and analytic
              potential theory on rigged Hilbert spaces. Ann. Inst. Henri
              Poincaré Sect. **B 13**, 269-291 (1977).

[AKR 90]   Albeverio, S., Kusuoka, S., Röckner, M.: On partial integration in infinite dimensional space and applications to Dirichlet forms. J. London Math. Soc. **42**, 122-136 (1990).

[AM 88]   Albeverio, S., Ma, Z.M.: Additive functionals, nowhere Radon and Kato class smooth measures associated with Dirichlet forms. BiBoS **395** (1988), to appear in Osaka, J. Math.

[AM 89]   Albeverio, S., Ma, Z.M.: Nowhere Radon smooth measures, perturbation of Dirichlet forms and singular quadratic forms. In Lecture Notes in Control and Information Sciences **126**, Berlin: Springer 1989.

[AM 91a]   Albeverio, S., Ma, Z.M.: Diffusion processes associated with singular Dirichlet forms. In Proc. Stochastic Analysis and Applications, ed. by A.B. Cruzeiro, J.C. Zambrini. New York: Birkhäuser 1991.

[AM 91b]   Albeverio, S., Ma, Z.M.: Characterization of Dirichlet forms associated with Hunt processes, SFB-preprint **237**, Bochum (1991). To appear in Proc. Swansea Conference, ed. by A. Truman.

[AM 91c]   Albeverio, S., Ma, Z.M.: A general correspondence between Dirichlet forms and right processes. Bull. Am. Math. Soc. **26**, 245-252 (1991).

[AM 91d]   Albeverio, S., Ma, Z.M.: Perturbation of Dirichlet forms – lower semiboundedness, closability and form cores. J. Funct. Anal. **99**, 332-356, 1991.

[AM 91e]   Albeverio, S., Ma, Z.M.: A note on quasicontinuous kernels representing quasi-linear positive maps. Forum Math. **3**, 889-400, 1991.

[AM 91f]   Albeverio, S., Ma, Z.M.: Necessary and sufficient condition for the existence of m-perfect processes associated with Dirichlet forms. In: Seminaire de Probabilités XXV. Lecture Notes in Math. **1485**, 374-406, Berlin: Springer 1991.

[AMR 90]   Albeverio, S., Ma, Z.M., Röckner, M.: A Beurling-Deny type structure theorem for Dirichlet forms on general state space. Preprint (1990). In: Memorial Volume for R. Høegh-Krohn, Vol. I. Ideas and methods in mathematical analysis, stochastics and applications. Ed. S. Albeverio, J.E. Fenstad, H. Holden, T. Lindstrøm. Cambridge: Cambridge University Press (1992).

[AMR 91] Albeverio, S., Ma, Z.M., Röckner, M.: A remark on the support of cadlag processes. Preprint (1991). To appear in: Festschrift G. Kallianpur.

[AMR 92a] Albeverio, S., Ma, Z.M., Röckner, M.: Non-symmetric Dirichlet forms and Markov processes on general state space. C.R. Acad. Sci. Paris, t. **314**, Série I, 77-82 (1992).

[AMR 92b] Albeverio, S., Ma, Z.M., Röckner, M.:Regularization of Dirichlet spaces and applications. To appear in C.R. Acad. Sci. Paris, Série I, (1992).

[AR 89a] Albeverio, S., Röckner, M.: Dirichlet forms, quantum fields and stochastic quantization. In: Stochastic analysis, path integration, and dynamics. Research Notes in Math. **200**, 1-21. Editors: K.D. Elworthy, J.C. Zambrini, Harlow: Longman 1989.

[AR 89b] Albeverio, S., Röckner, M.: Classical Dirichlet forms on topological vector spaces – construction of an associated diffusion process. Probab. Th. Rel. Fields **83**, 405-434 (1989).

[AR 90a] Albeverio, S., Röckner, M.: Classical Dirichlet forms on topological vector spaces – closability and a Cameron-Martin formula. J. Funct. Anal. **88**, 395-436 (1990).

[AR 90b] Albeverio, S., Röckner, M.: New developments in the theory and application of Dirichlet forms. In: Stochastic processes, physics and geometry, 27-76, Ascona/Locarno, Switzerland, 4-9 July 1988, Editors: S. Albeverio et al., Singapore: World Scientific 1990.

[AR 91] Albeverio, S., Röckner, M.: Stochastic differential equations in infinite dimensions: solutions via Dirichlet forms. Probab. Th. Rel. Fields **89**, 347-386 (1991).

[ARZ 91] Albeverio, S., Röckner, M., Zhang, T.S.: Girsanov transform for symmetric diffusions with infinite dimensional state space. Preprint (1991). Ann. Prob., in press.

[An 72] Ancona, A.: Continuité des contractions dans les espaces de Dirichlet. C.R. Acad. Sci. Paris, t. **282**, Série I, 871-873 (1976).

[An 75] Ancona, A.: Sur les espaces de Dirichlet: principes, fonctions de Green. J. Maths pures et appl. **54**, 75-124 (1975).

[An 76] Ancona, A.: Continuité des contractions dans les espaces de Dirichlet. In: Sem. Th. Potentiel. Paris no.2. Lecture Notes in Math. **563**, 1-26. Berlin: Springer 1976.

[Ba 70]   Badrikian, A.: Séminaire sur les fonctions aléatoires linéaires et les mesures cylindriques. Lecture Notes in Math. **139**. Berlin: Springer 1970.

[B 78]   Bauer, H.: Wahrscheinlichkeitstheorie und Grundzüge der Maßtheorie. Berlin-New York: de Gruyter 1978.

[Be De 58]   Beurling, A., Deny, J.: Espaces de Dirichlet. Acta Math. **99** , 203-224 (1958).

[Be De 59]   Beurling, A., Deny, J.: Dirichlet spaces. Proc. Nat. Acad. Sci. U.S.A. **45**, 208-215 (1959).

[Bl 71]   Bliedtner, J.: Dirichlet forms on regular functional spaces. In: Seminar on Potential Theory II. Lecture Notes in Math. **226**, 1-61, Berlin: Springer 1971.

[Bl Ha 86]   Bliedtner, J., Hansen, W.: Potential Theory. Berlin: Springer 1986.

[BG 68]   Blumenthal, R., Getoor, R.: Markov processes and potential theory. New York - London: Academic Press 1968.

[BH 86a]   Bouleau, N., Hirsch, F.: Propriétés d'absolue continuité dans les espaces de Dirichlet et application aux équations differentielles stochastiques. In: Séminaire de Probabilités, no. XX. Lecture Notes in Math. **1204**, 131-160, Berlin: Springer 1986.

[BH 86b]   Bouleau, N., Hirsch, F.: Formes de Dirichlet générales et densité des variables alétoires réelles sur l'espaces de Wiener. J. Funct. Anal. **69**, 229-259 (1986).

[BH 91]   Bouleau, N., Hirsch, F.: Dirichlet forms and analysis on Wiener space. Berlin - New York: de Gruyter 1991.

[Bou 74]   Bourbaki, N.: Topologie générale. Chapitres 5 à 10. Paris: Hermann 1974.

[C 75]   Cannon, J.T.: Convergence criteria for a sequence of semi-groups. Appl. Anal. **5**, 23-31 (1975).

[Ca-Me 75]   Carillo - Menendez, S.: Processus de Markov associé à une forme de Dirichlet non symétrique. Z. Wahrsch. verw. Geb. **33**, 139-154 (1975).

[Ch 69a]   Choquet, G.: Lectures on Analysis I: Integration and topological vector spaces. London: Benjamin 1969.

[Ch 69b]   Choquet, G.: Lectures on Analysis II: Representation Theory. London: Benjamin 1969.

[Chu 82]   Chung, K.L.: Lectures from Markov processes to Brownian Motion. Berlin: Springer 1982.

[Da 89]   Davies, E.B.: Heat kernels and spectral theory. Cambridge: Cambridge University Press 1989.

[DM 78]   Dellacherie, C., Meyer, P.A.: Probabilities and potential. Paris: Hermann 1978.

[DM 82]   Dellacherie, C., Meyer, P.A.: Probabilities and potential B. Amsterdam - New York - Oxford: North-Holland 1982.

[DM 88]   Dellacherie, C., Meyer, P.A.: Probabilities and potential C. Amsterdam: Elsevier 1988.

[De 70]   Deny, J.: Méthodes Hilbertiennes et théorie du potentiel. Potential Theory, Centro Internazionale Matematico Estivo, Edizioni Cremonese: Roma 1970.

[DoM 91]   Dong, Z., Ma, Z.M.: An integral representation theorem for quasi-regular Dirichlet spaces. Preprint (1991). To appear in Chinese Science Bull.

[Dy 82]   Dynkin, E.B.: Green's and Dirichlet spaces associated with fine Markov processes. J. Funct. Anal. **47**, 381-418 (1982).

[Fe 71]   Feller, W.: An introduction to probability theory and its applications. Vol.II, Second edition. New York: Wiley 1971.

[FdL 78]   Feyel, D., de La Pradelle, A.: Dualité des quasi-résolvantes de Ray, Lect. Notes in Math. **713**, 67-88, Berlin: Springer 1979.

[FdL 79]   Feyel, D., de La Pradelle, A.: Processus associés à une classe d'espaces de Banach fonctionnels, Z. Wahrsch. verw. Geb. **43**, 339-351 (1978).

[Fi 89]   Fitzsimmons, P.: Markov processes and nonsymmetric Dirichlet forms without regularity. J. Funct. Anal. **85**, 287-306 (1989).

[Fi 90]   Fitzsimmons, P.: Private communication (1990).

[FiG 88]   Fitzsimmons, P., Getoor, R.: On the potential theory of symmetric Markov processes. Math. Annalen **281**, 495-512 (1988).

[F 71]   Fukushima, M.: Dirichlet spaces and strong Markov processes. Trans. Amer. Math. Soc. **162**, 185-224 (1971).

[F 73]   Fukushima, M.: On the generation of Markov processes by symmetric forms. Proceedings 2nd Japan - USSR symposium on probability theory. Lect. Notes in Math. **330**. Berlin: Springer 1973.

[F 76]   Fukushima, M.: Potential theory of symmetric Markov processes and its applications. In: Lecture Notes in Math. **550**, Berlin: Springer 1976.

[F 80]   Fukushima, M.: Dirichlet forms and Markov processes. Amsterdam-Oxford-New York: North Holland (1980).

[FO 88]   Fukushima, M., Oshima, Y.: On the skew product of symmetric diffusion processes. Preprint (1988). Forum Math. **1**, 103-142 (1989).

[G 75]   Getoor, R.: Markov processes: ray processes and right processes. Lect. Notes in Math. **440**. Berlin: Springer 1975.

[GG 83]   Getoor, R., Glover, J.: Markov processes with identical excessive measures. Math. Z. **184**, 287-300 (1983).

[GS 84]   Getoor, R., Sharpe, M.T.: Naturality, standardness and weak duality for Markov processes. Z. Wahrsch. verw. Geb. **64**, 1-62 (1984).

[GT 83]   Gilbarg, D., Trudinger, N.S.: Elliptic partial differential equations of second order. Berlin: Springer 1983.

[Go 85]   Goldstein, J.A.: Semigroups of linear operators and applications. Oxford Math. Monographs. New York: O. U. Press 1985.

[Gr 65]   Gross, L.: Abstract Wiener spaces. Proc. 5th Berkeley Symp. Math. Stat. Prob. **2**, 31-42 (1965).

[HH 68]   Hervé, M., Hervé, R.M.: Les fonctions surharmoniques associées à un opérateur elliptique du second ordre à coefficients discontinus. Ann. Inst. Fourier, Grenoble **19**, 305-359 (1968).

[HPS 88]   Hida, T., Potthoff, J., Streit, L.: Dirichlet forms and white noise analysis. Commun. Math. Phys. **116**, 235-245 (1988).

[H 48]   Hille, E.: Functional analysis and semigroups. Amer. Math. Soc. Colloquium Publication **31**, New York 1948.

[H 52]   Hille, E.: On the generation of semi-groups and the theory of conjugate functions. Kungl. Fys. Sälls. I Lund Förhand **21**, 1-13 (1952).

[Hi 74]   Hirsch, F.: Conditions nécessaires et suffisantes d'existence de résolvantes. Z. Wahrsch. verw. Geb. **29**, 73-85 (1974).

[HoJ 92]   Hoh, W., Jacob, N.: Some Dirichlet forms generated by pseudo differential operators. Bull. Sci. Math. **116** (1992), in press.

[J 91]   Jacob, N.: Further pseudo differential operators generating Feller semigroups and Dirichlet forms. Preprint (1991). Rev. Math. Iberoamericana, in press.

[J 92]   Jacob, N.: Feller semigroups, Dirichlet forms, and pseudo differential operators. Forum Math. **4** (1992), in press.

[K 66]   Kato, T.: Perturbation theory for linear operators. Berlin: Springer 1966.

[Kuo 75]   Kuo, H.: Gaussian measures in Banach spaces. Lect. Notes in Math. **463**, 1-224, Berlin-Heidelberg-New York: Springer 1975.

[Ku 82]   Kusuoka, S.: Dirichlet forms and diffusion processes on Banach spaces. J. Fac. Sci. Univ. Tokyo, Sec. **IA 29**, 387-400 (1982).

[L72]   Landkof, N.S.: Foundations of modern potential theory. Berlin: Springer 1972.

[Le 77]   LeJan, Y.: Balayage et formes de Dirichlet. Z. Warsch. verw. Geb. **37**, 297-319 (1977).

[Le 78]   LeJan, Y.: Mesures associées à une forme de Dirichlet. Applications. Bull. Soc. Math. France **106**, 61-112 (1978).

[Le 82]   LeJan, Y.: Dual Markovian semigroups and processes. In: "Functional Analysis in Markov processes", 47-75, ed. M. Fukushima. Lect. Notes Math. **923**. Berlin: Springer 1982.

[Le 83]   LeJan, Y.: Quasi-continuous functions and Hunt processes. J. Math. Soc. Japan **35**, 37-42 (1983).

[LR 90]   Lyons, T., Röckner, M.: A note on tightness of capacities associated with Dirichlet forms. Preprint Edinburgh (1990). To appear in Bull. London Math. Soc. .

[M 91]   Ma, Z.M.: Some new results concerning Dirichlet forms, Feynman-Kac semigroups and Schrödinger equations. In: "Probability theory in China". Contemporary Mathematics **118**, Am. Math. Soc. (1991).

[Mai 72]   Maisonneuve, B.: Topologies du type de Skorohod. In: Séminaire de Probabilités VI. Lecture Notes in Math. **258**, 113-117, Berlin: Springer 1972.

[Ma 78]   Malliavin, P.: Stochastic calculus of variation and hypoelliptic operators. Proc. of the International Symposium on Stochastic Differential Equations Kyoto 1976, Tokyo 1978.

[O 88]   Oshima, Y.: Lectures on Dirichlet forms. Preprint Erlangen (1988).

[Pa 78]   Paclet, P., Espaces de Dirichlet et capacités fonctionnelles sur triplets de Hilbert-Schmidt. Séminaire P. Krée, no. 5 (1978).

[P 67]   Parthasarathy, K.R.: Probability measures on metric spaces. New York - London: Academic Press 1967.

[PR 90]   Potthoff, J., Röckner, M.: On the contraction property of energy forms on infinite dimensional space. J. Funct. Anal. **92**, 155-165 (1990).

[Pr 74]   Preston, C.: Maximum principles and contractions in the potential theory of a functional Hilbert space. Preprint (1974). Unpublished.

[ReS 72]   Reed, M., Simon, B.: Methods of modern mathematical physics I. Functional Analysis. New York - San Francisco - London: Academic Press 1972.

[ReS 75]   Reed, M., Simon, B.: Methods of modern mathematical physics II. Fourier Analysis, self-adjointness. New York - San Francisco - London, Academic Press 1975.

[ReS 78]   Reed, M., Simon, B.: Methods of modern mathematical physics IV. Analysis of Operators. New York - San Francisco - London: Academic Press 1978.

[RiN 55]   Riesz, F., Nagy, B.: Functional Analysis. New York: Unger 1955.

[R 85]   Röckner, M.: Generalized Markov fields and Dirichlet forms. Acta Appl. Math. **3**, 285-311 (1985).

[R 89]   Röckner, M.: On the parabolic Martin boundary of the Ornstein-Uhlenbeck operator on Wiener space. Preprint (1989). To appear in Ann. Prob.

[R 90a]   Röckner, M.: Dirichlet forms on infinite dimensional state space and applications. Lectures held at Silivri Summer School. Preprint (1990). To appear.

[R 90b]   Röckner, M.: Potential theory on non-locally compact spaces via Dirichlet forms. Preprint (1990). To appear as "main lecture" in: Proceedings "International conference on potential theory, Nagoya 1990".

[RS 91a]   Röckner, M., Schmuland, B.: Tightness of general $C_{1,p}$-capacities on Banach space. Preprint (1991). To appear in J. Funct. Anal. .

[RS 91b]   Röckner, M., Schmuland, B.: In preparation.

[RW 85]   Röckner, M., Wielens, N.: Dirichlet forms – closability and change of speed measure In: "Infinite dimensional analysis and stochastic processes", Research Notes in Math. **124**, 119-144, Editor: S. Albeverio, Boston - London - Melbourne: Pitman 1985.

[RZ 90]   Röckner, M., Zhang, T.S.: Uniqueness of generalized Schrödinger operators and applications. Preprint (1990). J. Funct. Anal. **105**, 187-231 (1992).

[Ro 82]   Rozanov, Yu.A.: Markov random fields. New York - Heidelberg - Berlin: Springer 1982.

[S 90]   Schmuland, B.: An alternative compactification for classical Dirichlet forms on topological vector spaces. Stochastics **33** (1990), 75-90.

[Sch 73]   Schwartz, L.: Radon measures on arbitrary topological space and cylindrical measures. London: Oxford University Press 1973.

[Sh 88]   Sharpe, M.T.: General theory of Markov processes. New York: Academic Press 1988.

[Shi Ta 91]   Shigekawa, I., Taniguchi, T.: Dirichlet forms on separable metric spaces. Preprint (1991).

[Si 74]   Silverstein, M.L.: Symmetric Markov Processes. Lecture Notes in Math. **426**. Berlin - Heidelberg - New York: Springer 1974.

[Si 76]   Silverstein, M.L.: Boundary theory for symmetric Markov processes. Lecture Notes in Math. **516**. Berlin: Springer 1976.

[Si 78]   Silverstein, M.L.: Application of the sector condition to the classification of sub-Markovian semigroups. Trans. Amer. Math. Soc. **244**, 103-146 (1978).

[St 64]   Stampacchia, G.: Formes bilinéaires coercitives sur les ensembles convexes. C.R. Acad. Sc., Paris t. **258**, Série I, 4413-4416 (1964).

[St 65]   Stampacchia, G.: Le problème de Dirichlet pour les équations elliptiques du second ordre à coefficients discontinus. Ann. Inst. Fourier **15**, 189 - 259 (1965).

[T 91]   Takeda, M.: The maximum Markovian self-adjoint extensions of generalized Schrödinger operators. Preprint (1990). J. Math. Soc. Japan, **44**, 113-130 (1992).

[Ta 75]   Taylor, J.C.: A characterization of the kernel $\lim\limits_{\lambda \downarrow 0} V_\lambda$ for sub-markovian resolvants (V_λ). Ann. Prob. **3**, 355-357 (1975).

[W 84]   Watanabe, S.: Lectures on stochastic differential equations and Malliavin calculus. Berlin - Heidelberg - New York - Tokyo: Springer 1984.

[Wl 82]   Wloka, J.: Partielle Differentialgleichungen. Stuttgart: Teubner 1982.

[Y 48]   Yosida, K.: On the differentiability and representation of one-parameter semi-groups of linear operators. J. Math. Soc. Japan **1**, 15-21 (1948).

Index

vaguely continuous convolution semi-
      group, 65
vanishing at infinity, 175

weak sector condition, 15
well-μ-admissible, 54

Universitext

Aksoy, A., Khamsi, M.A: Methods in Fixed Point Theory
Aupetit, B.: A Primer on Spectral Theory
Berger, M.: Geometry I
Berger, M.: Geometry II
Bliedtner, J. / Hansen, W.: Potential Theory
Booss, B. / Bleecker, D.D.: Topology and Analysis
Chandrasekharan, K.: Classical Fourier Transforms
Charlap, L.S.: Bieberbach Groups and Flat Manifolds
Chern, S.: Complex Manifolds without Potential Theory
Chorin, A.J. / Marsden, J.E.: Mathematical Introduction to
Fluid Mechanics
Cohn, H.: A Classical Invitation to Algebraic Numbers and
Class Fields
Curtis, M.L.:Abstract Linear Algebra
Curtis, M.L.: Matrix Groups
Dalen, D. van: Logic and Structure
Devlin, K.J.: Fundamentals of Contemporary Set Theory
Edwards, R.E.: A Formal Background to Higher Mathematics
I a, and I b
Edwards, R.E.: A Formal Background to Higher Mathematics
II a, and II b
Emery, M.: Stochastic Calculus in Manifolds
Foulds, L.R: Graph Theory Applications
Fuks, D.B. / Rokhlin, V.A.: Beginner's Course in Topology
Frauenthal, J.C.: Mathematical Modeling in Epidemiology
Gallot, S. / Hulin, D. / Lafontaine, J.: Riemannian
Geometry
Gardiner, C.F.: A First Course in Group Theory
Garding, L. / Tambour, T.: Algebra for Computer Science
Godbillon, C: Dynamical Systems on Surfaces
Goldblatt, R.: Orthogonality and Spacetime Geometry
Hájek, P. / Havránek, T.: Mechanizing Hypothesis Formation
Hlawka, E. / Schoißengeier, J. / Taschner, R.: Geometric
and Analytic Number Theory
Humi, M. / Miller, W.: Second Course in Ordinary
Differential Equations for Scientists and Engineers
Hurwitz, A. / Kritikos, N.: Lectures on Number Theory
Iversen, B.: Cohomology of Sheaves
Kelly, P. / Matthews, G.: The Non-euclidean Hyperbolic
Plane
Kempf,G: Complex Abelian Varieties and Theta Functions
Kostrikin, A.I.: Introduction to Algebra
Krasnoselskii, M.A. / Pokrovskii, A.V.: Systems with
Hysteresis
Luecking, D.H. / Rubel, L.A.: Complex Analysis.
A Functional Analysis Approach
Mac Lane, S. /Moerdijk, I.: Sheaves in Geometry and Logic
McCarthy, P.J.: Introduction to Arithmetical Functions
Marcus, D.A.: Number Fields
Meyer, R.M.: Essential Mathematics for Applied Fields
Meyer-Nieberg, P.: Banach Lattices
Mines, R. / Richman, F./ Ruitenberg, W.: A Course in
Constructive Algebra
Moise, E.E.: Introductory Problem Courses in Analysis and
Topology

Montesinos, J.M.: Classical Tessellations and Three-Manifolds
Nikulin, V.V. / Shafarevich, I.R.: Geometries and Groups
Oden, J.T. / Reddy, J.N.: Variational Methods in Theoretical Mechanics
Oksendal, B.: Stochastic Differential Equations
Porter, J.R. / Woods, R.G.: Extensions and Absolutes of Hausdorff Spaces
Rees, E.G.: Notes on Geometry
Reisel, R.B.: Elementary Theory of Metric Spaces
Rey, W.J.J.: Introduction to Robust and Quasi-Robust Statistical Methods
Rickart, C.E.: Natural Function Algebras
Rotman, J.J.: Galois Theory
Rybakowski, K.P.: The Homotopy Index and Partial Differential Equations
Samelson, H.: Notes on Lie Algebras
Smith, K.T.: Power Series from a Computational Point of View
Smorynski, C.: Self-Reference and Modal Logic
Smorynski, C.: Logical Number Theory I: An Introduction
Stillwell, J.: Geometry of Surfaces
Stroock, D.W.: An Introduction to the Theory of Large Deviations
Sunada, T.: The Fundamental Group and the Laplacian (to appear)
Sunder, V.S.: An Invitation to von Neumann Algebras
Tondeur, P.: Foliations on Riemannian Manifolds
Verhulst, F.: Nonlinear Differential Equations and Dynamical Systems
Zaanen, A.C.: Continuity, Integration and Fourier Theory

MIX
Papier aus verantwortungsvollen Quellen
Paper from responsible sources
FSC® C105338

If you have any concerns about our products,
you can contact us on
ProductSafety@springernature.com

In case Publisher is established outside the EU,
the EU authorized representative is:
Springer Nature Customer Service Center GmbH
Europaplatz 3, 69115 Heidelberg, Germany

Printed by Libri Plureos GmbH
in Hamburg, Germany